T0191624

Fast Radial Basis Functions for Engineering Applications

Marco Evangelos Biancolini

Fast Radial Basis Functions for Engineering Applications

Marco Evangelos Biancolini
Department of Enterprise Engineering
"Mario Lucertini"
University of Rome "Tor Vergata"
Rome
Italy

ISBN 978-3-030-09127-9 ISBN 978-3-319-75011-8 (eBook)
https://doi.org/10.1007/978-3-319-75011-8

Softcover re-print of the Hardcover 1st edition 2017

Printed on acid-free paper

This Springer imprint is published by the registered company Springer International Publishing AG part of Springer Nature
The registered company address is: Gewerbestrasse 11, 6330 Cham, Switzerland

Preface

This book is the result of the industrial and academic research path that I started to follow since 2007. At that time, I had a cooperation in place with a Formula 1 top team. When the need for a software tool suitable for advanced and high-performance mesh morphing, namely such to enable the update of the nodal positions of large computational grids in order to accommodate shape variations, was expressed, I considered many options. Among them, the use of a structural solution using finite elements method (FEM) produced very good results. However, I was looking for a meshless approach capable to fast propagate the action defined on the surface mesh into the volume mesh regardless of the mesh elements' typology and the presence of interfaces between different partitions when processing meshes ready for parallel computing. So I did some attempts adopting the boundary elements method (BEM) which allows to define a deformation field as a point function. My first readings about RBF were related to the wide literature of meshless methods for the solution of structural problems. I discovered that RBF had already proven to be an effective tool for mesh morphing as well (Jakobsson and Amoignon 2007; De Boer et al. 2007). After this finding, I presented the concept to the Formula 1 team with a first prototype implemented using the Mathcad tool and exchanging information with the CFD solver ANSYS Fluent through an user-defined functions (UDF). The project was funded and my studies concerning RBF for fast mesh morphing began.

Soon I discovered that the nice behaviour we registered in the first applications adopting the direct solver was not affordable going up with the problem size. We found out that a good quality and long-distance interactions in 3D cases were possible employing the bi-harmonic RBF which gives a dense matrix to be solved. The maximum size of the RBF was about 10,000 points that took approximately a couple of hours to fit the RBF-sought coefficients. The need was to face RBF problems that comprised hundreds of thousands of source points at least, but with the target to reasonably manage even millions of source points. Two orders of magnitude more!

It was a very hard and challenging task. I invented, and still not disclosed, some specific algorithms such as the local correction method (LCM) that is embedded in the RBF Morph technology I have authored, and I also adapted other algorithms already existing in the literature, but adopted for other applications. The industrial solution for mesh morphing resulting from the project was then released as a commercial software, called RBF Morph™, which was, at that time, the first industrial implementation for mesh morphing based on Fast RBF (Biancolini 2012).

During the last ten years, I have been involved in many industrial and research projects where Fast RBF were successfully exploited and applied. I think that there is still a great undiscovered potential of Fast RBF for engineering applications. Considering that, I have decided to share through this book my experience of passionate "practitioner" of RBF to put other researchers and engineers in the condition to understand the great potential of the method and to embrace RBF for new challenges.

Monte Compatri, Italy
June 2017

Prof. Marco Evangelos Biancolini

References

Biancolini ME (2012) Mesh morphing and smoothing by means of Radial Basis Functions (RBF): a practical example using Fluent and RBF Morph. In: Handbook of research on computational science and engineering: theory and practice, pp 34, https://doi.org/10.4018/978-1-61350-116-0.ch015

de Boer A, Van der Schoot MS, Bijl H (2007) Mesh deformation based on radial basis function interpolation. Comput Struct 85(11–14):784–795. https://doi.org/10.1016/j.compstruc.2007.01.013

Jakobsson S, Amoignon O (2007) Mesh deformation using radial basis functions for gradient based aerodynamic shape optimization. Comput Fluids 36(6):1119–1136. https://doi.org/10.1016/j.compfluid.2006.11.002

Contents

Abstract

Radial basis functions (RBFs) are a mathematical tool mature enough for consistently handling engineering applications. As their theoretical foundation is very well established and the mathematical framework they are based on can be adapted in several ways, such a numerical means has proven to be effective in many technical fields.

As a matter of fact, a candidate engineering application can be faced taking the advantage of the peculiar features of RBF such as the availability of many radial functions with global and compact support as well as the interpolation and regression they allow in a multidimensional space. Such a flexibility makes RBF really attractive for users but, actually, their great potential is just partially exploited. This is due to the difficulty in doing a first step towards the effective usage of RBF not only because they are not commonly part of the cultural background of an engineer, but also, and above all, for the numerical complexity of the RBF problems that scale-up very quickly with the number of the considered RBF centres. Fast RBF algorithms are available to alleviate this latter boundary and, additionally, high-performance computing (HPC) systems and solutions can give a further aid of course. Nevertheless, a consolidate tradition in using RBF for engineering applications is still missing and the beginners may be confused by open literature that, in many cases, is presented with language and symbolisms that are comfortable for mathematicians, but that sound rather cryptic and discouraging to engineers.

The aim of this book is to put the interested readers in a better position with respect to RBF providing them with a clear vision of their potential, computational tools ready for practical use and precise guidelines to let them implement their own customised workflows in order to handle engineering applications employing RBF.

To this end, the book is divided into two main parts. The first one covers the foundation of RBF, the tools available for their quick implementation and the guidelines for facing new challenges, whilst the second part is a collection of practical applications of RBF in several engineering sectors. Such applications deal with lots of technical and scientific topics including, among others, response surface interpolation in n-dimensional spaces, mapping of magnetic and pressure loads, upscaling of flow fields, stress and strain analysis utilising experimental displacement fields, implicit surfaces generation, as well as mesh morphing for crack propagation, shape optimisation for external aerodynamics, ice and snow accretion using computational fluid dynamics (CFD) data and surface sculpting using adjoint sensitivity data. For each one of the presented applications, the adopted procedure is clearly and completely exposed using the approach identified in the Part I of the book.

Chapter 1
Introduction

Abstract In this chapter the "Fast Radial Basis Functions for Engineering Applications" book is introduced. The basic Radial Basis Functions concepts and the needs to have fast implementations suitable for engineering applications are explained. The vision of the author toward the topic is presented together with an overview of the book contents described on a chapter by chapter basis.

The topic covered in this book is explained in the title "Fast Radial Basis Functions for Engineering Applications". The mathematical framework of the radial basis functions (RBF) is introduced aiming at its understanding rather than at the rigorous mathematical demonstration. The reader interested in a deeper knowledge of RBF background theory could consider the book written by Buhmann (2003).

The concept "Fast Radial Basis Functions" can sound strange to the reader without an experience in advanced engineering and scientific computing. The Fast feature is a key point to enable the usage of RBF for large problems solution due to the poor performances that are observed when dealing with large dataset. Some concepts for the acceleration of RBF calculation are exposed in this book. The purpose is to guide the reader in such a complex world, in which performances are usually achieved sacrificing the great flexibility of RBF, providing information useful for the definition of a proper strategy for the Fast RBF implementation.

A rich collection of Engineering Applications that can be faced using RBF is then provided showing how the Fast feature is a key enabler as it allows to boost calculations feasible using standard method and, in many cases, it also makes feasible a problem that is not doable at all with standard method.

This book should not be intended as a review of all possible RBF applications in engineering, but rather as a source of inspiration for the identification of new applications that can be tackled using RBF and as a guide for their effective solution implementation. To this end, the applications described and demonstrated in this book are limited to those for which the author has a direct experience.

RBF were introduced as interpolators of scattered data in sixties. The RBF equation is represented and explained in Fig. 1.1. Usually the interpolation comprises

M. E. Biancolini, *Fast Radial Basis Functions for Engineering Applications*,
https://doi.org/10.1007/978-3-319-75011-8_1

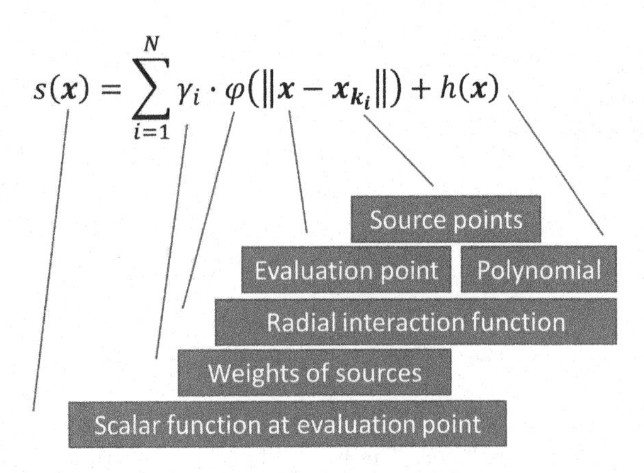

Fig. 1.1 Overview of a generic RBF with polynomial correction

a sum of weighted radial interactions and a polynomial correction. RBF are commonly used to interpolate a scalar function defined in a multi-dimensional space ($\mathbb{R}^n \rightarrow \mathbb{R}$). Interactions of a generic probe point $\boldsymbol{x}$ with all source points $\boldsymbol{x}_s$, also referred to as RBF centres in the book, are computed evaluating the inter-distance (radius $\mathbb{R}^n \rightarrow \mathbb{R}$), transforming the distance into a radial interaction using the radial interaction function ($\mathbb{R} \rightarrow \mathbb{R}$) and multiplying the result by a scalar weight that can be seen as the "intensity" of the source point. All the interactions are summed up and, in some cases, a polynomial term is added.

The concept of radial interaction is explained in Fig. 1.2. Source points are arranged in the plane and the coefficients of the RBF are computed so that the

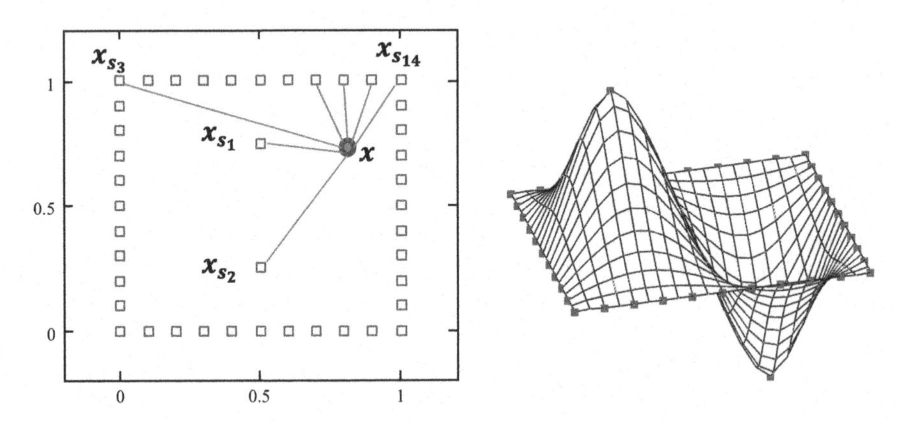

Fig. 1.2 RBF interactions: two-dimensional source points distribution (left) and three-dimensional plot constructed evaluating the RBF in an uniformly spaced grid defined inside the square (right)

function is zero at all the source points along the square edges and is ± 1 valued at two internal points. The interaction of the point $\boldsymbol{x}$ with all source points $\boldsymbol{x}_s$ can be repeated many times varying the position of the point $\boldsymbol{x}$ inside the square, and the resulting scalar value ($\mathbb{R}^2 \rightarrow \mathbb{R}$) can be represented as an height in the 3d plot.

In Chap. 2 the RBF mathematical concepts are exposed considering firstly the interpolation problem with the RBF function defined by known values at source points; a first hands-on example is provided showing how RBF work. Further topics of RBF theory are then introduced considering the differentiation of RBF, the fitting of an RBF with known values at locations different from the source points, the use of regression instead of exact interpolation and finally the management of noisy datasets.

Fast RBF foundation concepts are described in Chap. 3. Details are provided for an effective usage of the direct method using advanced linear algebra strategies and libraries. Methods for data compression which allow to represent the same information with a smaller and less expensive cloud are introduced. Space localization methods are successively considered firstly with compact supported RBF, whose interaction distance is limited in the RBF function itself, and then with partition of unity (POU) that consists of the superposition and blending of smaller clouds. Approximated evaluation of the far field of full supported RBF is described and an overview of the fast multipole method (FMM) is given. Information about iterative solvers and parallel computing are finally provided to complete the chapter.

Chapter 4 provides an overview of calculation tools that are usually required for the implementation of RBF based engineering workflows. The aim here is to provide the reader with a strategy to evaluate whether an application can be faced using RBF and how the various tools can be combined to make it effective.

The Engineering Applications Section begins with Chap. 5 where the implicit surface method is explained and demonstrated. The classical method that consists of the definition of a scalar valued function in the 3D space that is null at on-surface points and positive (negative) at off-surface points that are positioned outside (inside) the surface is firstly introduced. The projection onto the surface is achieved by computing the gradient of the implicit function. The second one is a novel method that consists of the creation of a projection field (vector valued RBF) defined on the surface and at off-surface points that is zero at on-surface points whereas it is equal to the vector moving to the surface at off-surface points. The effectiveness of both methods is demonstrated with practical examples.

Chapter 6 provides an introduction to the use of RBF for basic mesh morphing. A vector valued RBF, 2D or 3D, is used to propagate in the morphing domain the information known at a cloud of control points. As the method is meshless, the field can be defined at nodal positions of the mesh to be deformed or at other locations (on or off-surface). Usually the cloud that defines the source of morphing includes a set of points whose field is generated using simple rules for the generation of the displacements (rigid movement, rigid rotations, scaling). Auxiliary geometry can be adopted to generate more complex fields (constant offset vs. a surface, surface targets).

Examples of advanced mesh morphing for biomechanical applications are collected in Chap. 7 where the benefits of meshless deformation is demonstrated for situation in which an underlying CAD geometry is not available. Virtual surgery using CFD simulation of the hemodynamics is faced showing how a morphing interaction can provide the insight in the estimation of the evolution of a pathology or to plan the insertion of a cannula in a virtual environment where patient shapes and medical devices shapes coexist and can interact. The use of mesh morphing for the remodelling is then demonstrated on a femur and a tibia bones. FEA method is adopted for the structural assessment and mesh morphing can be used as a tool to transform a baseline (library) geometry onto the patient one. The approach is useful not just as a simplified mesh generation tool (adapting is faster than remeshing), but rather because it allows to enable the direct comparison between models preserving the one-to-one nodes correspondence.

Optimizations based on adjoint sensitivity data are presented in Chap. 8. In this case RBF are adopted to set up an advanced filtering tool suitable for removing the noise usually observed when shape sensitivities data are computed using CFD so as to enable the adjoint sculpting method where surfaces are updated according to the information provided by the flow solution to get the desired performances (as drag reduction or pressure loss control). Also in this case advanced mesh morphing is used to propagate, once properly filtered if needed, the shape data known at surfaces into the full volume mesh required for the calculation. The concept is demonstrated for FEM as well (in this case noise is not generated by the adjoint solution) showing how a bracket and a T-beam can be reshaped to control a target displacement. The adjoint preview approach, which consists of the computation of derivatives with respect of shape variations known in advance (as the ones described in Chap. 6) is also detailed. A collection of fluid shape optimizations, taking into account both internal and external flows, is provided at the end of the Chapter.

Evolutionary mesh morphing workflow are covered in Chap. 9. There are many situations where the local shape evolution is computed by the physics itself and unknown in advance. Ice and snow accretion examples are provided: mesh morphing is adopted to reshape wetted surface to account of the ice/snow thickness produced during the evolution and predicted by the multi-physics CFD model. A similar approach is then presented for mechanical applications where the evolution of an elliptical crack into a three-dimensional body is studied: the local crack speed computed using fracture mechanics can be easily converted in a local crack advance along the front. The Biological Growth Method (BGM) is finally demonstrated showing how mesh morphing can be used to support the reshape of structures miming the behaviour observed in nature: the local surface growth of trees trunk is promoted by the level of local stress so that the shape evolves toward reduced peak stress configurations.

Advanced workflow for Fluid Structure Interaction (FSI) modelling using CAE tools suitable for the simulation of fluids (CFD) and structures (CSM) are demonstrated in Chap. 10. In this context, RBF are adopted to interface the structure and the fluid. In the two-way approach loads computed using CFD (pressures and shear forces) are transferred to the structure using RBF interpolation for the mapping at surfaces (the mapping topic is further deepened in Chap. 13). Deformation computed using the CSM are then transferred to the CFD mesh using mesh morphing. The latter approach can be used to transfer on the CFD mesh modal shapes computed using eigenvalues extraction by FE analysis. The effectiveness of the two possible FSI approaches is demonstrated on practical applications considering aeronautical and motorsport fields. The reported FSI implementation can be used to tackle both steady and transient problems. The chapter is concluded showing how the method can handle vortex shedding excited oscillations and the vibration induced by the separation of a store from the wing of an aircraft.

In Chap. 11 a collection of RBF applications for the interpolation in multi-dimensional spaces is reported. RBF are in this case used as interpolator for the definition of a surrogate model in the parametric design space according to the optimization method based on the response surface that is suitable for objective functions that have an high evaluation cost. Design space is first sampled according to the design of experiment (DOE) approach. In the proposed application design points (DPs), which are typically computed or measured, identify the performance evaluated using CFD. The system response is eventually evaluated on the RBF interpolated model. The method is demonstrated with two aeronautical applications: the optimization of a NACA0012 aerofoil using 2D CFD and the optimization of a glider using 3D CFD. The chapter is concluded with a study of parametric sails that produces a surrogate model suitable to be coupled with velocity prediction programs (VPP) useful for mission performance prediction.

A collection of advanced post-processing methods based on RBF fields is supplied in Chap. 12. The idea is that field information known at discrete points in a continuum domain (displacement field of an elastic body subjected to load and deformed, local speed and pressure in a fluid) can be interpolated so that scattered information becomes continuous and, in some cases, differentiable. Examples based on the theory of elasticity are firstly provided: the post-processing of FEA solution available as a displacement field can be used to compute strains and stress with the benefit to make them mesh independent (and so ready to be used for the generation of various plot) and better resolved in the space despite the coarseness of the field. Experimental data can be processed according to the same principle so that RBF become an useful tool for strain and stress evaluation by image processing. The chapter then addresses another useful application of RBF mapping suitable for the compensation of metrological data: a complex environment that loads the structure and can adversely affect the quality of the measurement can be subtracted by the inverse mapping of the displacement field achieved by FEA so that measured points can be corrected and positioned in their undeformed configuration for an effective comparison with reference CAD geometry. The chapter is concluded showing how RBF can be used for the interpolation of experimentally acquired flow information related to hemodynamics and showing the potential of an upscaling post processing toward the reduction of the resolution of in vivo acquired flow field (and so the exposition time of the patient).

Chapter 13 details a generic mapping algorithm based on RBF. Mapping allows to interpolate information known at discrete locations across not-matching numerical grids of the same geometrical entities. After an overview of mapping procedures, examples of CFD computed pressure field mapping onto structural mesh adopted for FEA based CSM (see Chap. 11) is demonstrated starting with a simple reference geometry (a catenoid) and then with actual complex three-dimensional shapes of aircraft wings. The volume mapping is then addressed with a specific focus of the mapping of electro magnetic (EM) loads onto structural meshes. Examples in which magnetic loads are already available as force density or as magnetic field and density current (so that load density is computed on the FEA as a Lorentz Force). Practical examples involving the toroidal field (TF) Coil of DEMO superconductor magnet are shown with details of the stress response of the structure subjected to the EM loads.

Since many commercial software are cited in the text, the author wants to clarify that Mathcad® is a registered trademark of PTC Inc., MATLAB® is a registered trademark of MathWorks Inc., ANSYS® ACT™, ANSYS® DesignXplorer™, ANSYS® Fluent™, ANSYS® ICEM CFD™, ANSYS® Mechanical™ are ANSYS, Inc. and its subsidiaries' registered and unregistered trademarks in the U.S. and other countries, Abaqus® is a registered trademark of Abaqus, Inc. in the United States and other countries, Patran® is a registered trademark of MSC.Software Corporation, FEMAP® is a trademark of SIEMENS Product Lifecycle Management Software Inc., RBF Morph™ is a registered trademark of RBF Morph company, OPENFOAM® is a registered trademark of OpenCFD Ltd (ESI Group) and CFD++ is a registered trademark of Metacomp Technologies, Inc., TOSCA solver is the name of a legacy version of the Opera-3d Static Electromagnetics Module of OPERA Simulation software by Cobham plc.

Reference

Buhmann MD (2003) Radial basis functions: theory and implementation. Cambridge University Press, New York

Part I
Fast RBF for Engineering Applications

This part presents the theory of RBF, its numerical implementations and the tools available for their use.

Chapter 2
Radial Basis Functions

Abstract In this chapter the RBF mathematical concepts are exposed considering firstly the interpolation problem with the RBF function defined by known values at source points; a first hands-on example is provided showing how RBF work. Further topics of RBF theory are then introduced considering the differentiation of RBF, the fitting of an RBF with known values at locations different from the source points, the use of regression instead of exact interpolation and finally the management of noisy datasets. The chapter contents will not provide the details of mathematical theory but they will be mainly focused on the relevant results suitable for a wise and aware practical application of RBF.

2.1 RBF Theory Background

Radial basis were born as an interpolation method for scattered data (Hardy 1971, 1990) and consist of a very powerful tool because they are able to interpolate everywhere in the space a scalar function defined at discrete points giving the exact values at original points. The behaviour of the function between points depends on the kind of the adopted radial function. The RBF, here introduced as an interpolation method that recovers exactly scalar fields known at original points, can be extended to include the case in which the scalar information is known at locations not coincident with the source points of the RBF (see Sect. 2.5) and the case of regression (Sect. 2.6) in which the information known at many locations is approximated by a smaller set of source points.

The radial function can be fully or compactly supported. In some situations a polynomial corrector is added to guarantee solvability and uniqueness of the fit. As it will be shown in detail, a linear system of order equal to the number of the introduced source points, needs to be solved for sought coefficients calculation. Once the unknown coefficients are determined, the scalar function at an arbitrary

M. E. Biancolini, *Fast Radial Basis Functions for Engineering Applications*,
https://doi.org/10.1007/978-3-319-75011-8_2

location inside or outside the domain (interpolation/extrapolation) results to be expressed by the summation of the radial contribution of each source point (RBF centre) and of a polynomial term (if any).

The interpolation function, composed by the radial basis and the polynomial, is defined as follows:

$$s(\boldsymbol{x}) = \sum_{i=1}^{N} \gamma_i \varphi(\|\boldsymbol{x} - \boldsymbol{x}_{s_i}\|) + h(\boldsymbol{x}) \tag{2.1}$$

The $s(\cdot)$ scalar function is defined for an arbitrary sized variable $\boldsymbol{x}$ and represents a transformation $\mathbb{R}^n \to \mathbb{R}$. At a given point $\boldsymbol{x}$ the value of the RBF is obtained accumulating the interactions with all source points $\boldsymbol{x}_s$ gained computing the radial distance between $\boldsymbol{x}$ and each $\boldsymbol{x}_{s_i}$ processed by the radial function $\varphi(\cdot)$, consisting of a transformation $\mathbb{R} \to \mathbb{R}$, which is then multiplied by the weight γ_i. The polynomial term is then added. The degree of the polynomial has to be chosen depending on the kind of the adopted radial function. The radial basis fit exists if the weights γ and the coefficients of the polynomial can be found such that the desired function values g_s are obtained at source points:

$$s(\boldsymbol{x}_{s_i}) = g_{s_i}, \quad 1 \leq i \leq N \tag{2.2}$$

The number of equations is not enough as the number of unknowns is equal to the number of weights (N) plus the number of coefficients of the polynomial. The system is completed if the orthogonality conditions:

$$\sum_{i=1}^{N} \gamma_i p(\boldsymbol{x}_{s_i}) = 0 \tag{2.3}$$

are verified for all polynomials p with a degree less or equal than that of polynomial h. The minimal degree of polynomial h depends on the choice of the basis function. A good reference in which a table of polynomial augmentations is collected is reported in Boyd and Gildersleeve (2011). An unique interpolant exists if the basis function is a conditionally positive definite function. If the basis functions are conditionally positive definite of order m $\leq$ 2, a linear polynomial can be used. The subsequent exposition will assume valid the aforementioned hypothesis. A consequence of using a linear polynomial is that rigid body translations are exactly recovered. In a 3D space the linear polynomial has the form:

$$h(\boldsymbol{x}) = \beta_1 + \beta_2 x + \beta_3 y + \beta_4 z \tag{2.4}$$

and the orthogonality conditions (2.3) can be expressed as:

$$\sum_{i=1}^{N} \gamma_i = \sum_{i=1}^{N} \gamma_i x_{s_i} = \sum_{i=1}^{N} \gamma_i y_{s_i} = \sum_{i=1}^{N} \gamma_i z_{s_i} = 0 \tag{2.5}$$

The values for the weights of RBF vector γ and the coefficients vector $\boldsymbol{\beta}$ of the linear polynomial can be obtained by solving the system obtained imposing the conditions (2.2) and (2.5) that can be compactly expressed as:

$$\begin{pmatrix} \boldsymbol{M} & \boldsymbol{P}_s \\ \boldsymbol{P}_s^T & 0 \end{pmatrix} \begin{pmatrix} \gamma \\ \boldsymbol{\beta} \end{pmatrix} = \begin{pmatrix} \boldsymbol{g}_s \\ 0 \end{pmatrix} \tag{2.6}$$

where $\boldsymbol{g}_s$ contains the known values at the source points. $\boldsymbol{M}$ is the interpolation matrix defined calculating all the radial interactions between source points:

$$M_{ij} = \varphi\left(\left\|\boldsymbol{x}_{s_i} - \boldsymbol{x}_{s_j}\right\|\right), \quad 1 \le i \le N, 1 \le j \le N \tag{2.7}$$

and $\boldsymbol{P}_s$ is a constraint matrix that arises balancing the polynomial contribution and contains a column of “1” and the (x, y, z) positions of source points in the others three columns:

$$\boldsymbol{P}_s = \begin{pmatrix} 1 & x_{s_1} & y_{s_1} & z_{s_1} \\ 1 & x_{s_2} & y_{s_2} & z_{s_2} \\ \vdots & \vdots & \vdots & \vdots \\ 1 & x_{s_N} & y_{s_N} & z_{s_N} \end{pmatrix} \tag{2.8}$$

Radial basis interpolation works for scalar fields, but the fit can be repeated many times using the same interpolation and constraint matrixes. In this case, the $\boldsymbol{g}$ vector is replaced by a rectangular matrix and solved on a column wise fashion computing the coefficients γ and $\boldsymbol{\beta}$ related to each column (see Sect. 2.3 for further details).

If a deformation vector field has to be fitted in 3D (space morphing), each component of the displacement prescribed at the source points is interpolated as follows:

$$\begin{cases} s_x(\boldsymbol{x}) = \sum_{i=1}^{N} \gamma_i^x \varphi(\|\boldsymbol{x} - \boldsymbol{x}_{s_i}\|) + \beta_1^x + \beta_2^x x + \beta_3^x y + \beta_4^x z \\ s_y(\boldsymbol{x}) = \sum_{i=1}^{N} \gamma_i^y \varphi(\|\boldsymbol{x} - \boldsymbol{x}_{s_i}\|) + \beta_1^y + \beta_2^y x + \beta_3^y y + \beta_4^y z \\ s_z(\boldsymbol{x}) = \sum_{i=1}^{N} \gamma_i^z \varphi(\|\boldsymbol{x} - \boldsymbol{x}_{s_i}\|) + \beta_1^z + \beta_2^z x + \beta_3^z y + \beta_4^z z \end{cases} \tag{2.9}$$

If the evaluation is required at given set of evaluation points $\boldsymbol{x}_e$, a direct mapping matrix able to directly provide the interpolated scalar quantities at evaluation points as a linear transformation of $\boldsymbol{g}_s$ at source points can be constructed. The interpolation matrix, which relates evaluation points $\boldsymbol{x}_e$ and source points $\boldsymbol{x}_s$, is first defined:

$$A_{ij} = \varphi\left(\boldsymbol{x}_{e_i} - \boldsymbol{x}_{s_j}\right), \quad 1 \leq i \leq N_e, 1 \leq j \leq N_s \tag{2.10}$$

In the case of pure RBF, namely without the polynomial term, the scalar at evaluation points $\boldsymbol{g}_e$ can be written as a linear function of the scalar at source points $\boldsymbol{g}_s$:

$$\boldsymbol{g}_e = \boldsymbol{A}\boldsymbol{M}^{-1}\boldsymbol{g}_s = \boldsymbol{H}\boldsymbol{g}_s \tag{2.11}$$

It is worth to notice that the $\boldsymbol{H}$ matrix can be precomputed in advance and allows to give a direct transformation from the source points to the evaluation points. Nevertheless, this approach is not commonly used in engineering applications because of the high cost required for the computation and the storage of the inverse matrix.

2.2 Radial Functions

The RBF fit guarantees the passage of the interpolated function through all the points of the original dataset with the prescribed value. The behaviour of the function between points (interpolation) or outside the dataset (extrapolation) depends on the radial function used.

The complexity of the fit is strongly related to the radial function. Many options are available and some of them are summarized in Table 2.1. We can roughly classify the radial functions in two families: the global supported ones and the compact supported ones.

Table 2.1 Radial basis functions with global and compact support

RBF with global support	$\varphi(r)$
Spline type (R_n)	r^n, n odd
Thin plate spline (TPS_n)	$r^n \log(r)$, n even
Multiquadric (MQ)	$\sqrt{1+r^2}$
Inverse multiquadric (IMQ)	$\frac{1}{\sqrt{1+r^2}}$
Inverse quadratic (IQ)	$\frac{1}{1+r^2}$
Gaussian (GS)	e^{-r^2}
RBF with compact support	$\varphi(r) = f(\xi), \xi \leq 1, \xi = \frac{r}{R_{sup}}$
Wendland C^0 (C0)	$(1-\xi)^2$
Wendland C^2 (C2)	$(1-\xi)^4(4\xi+1)$
Wendland C^4 (C4)	$(1-\xi)^6\left(\frac{35}{3}\xi^2+6\xi+1\right)$

When the global support is used, all the points of the cloud interact themselves. High accuracy is achieved using global support but the numerical problem becomes very challenging because a dense fully populated matrix has to be solved in the fit stage. Fast methods can be used to accelerate both the fit process and the evaluation of global supported RBF but only for specific cases (see Sect. 3.6 for further details).

Compact supported RBF are used when long distance interaction is not desired. The interaction radius can be defined. The Wendland functions (Wendland 1995) with different continuity class are collected in the Table 2.1. The function is active only inside the sphere, identified by the interaction radius, whilst it is zero outside. It is worth to notice that the compact support affects not only the fitting and interpolation behaviour, but it can also simplify the fitting process as the matrix of the system becomes sparse and, as explained in Sect. 3.4, can be managed with numerical algorithms specifically conceived for this class of problems.

2.3 A First Example of RBF at Work

In this section a first simple example in which RBF theory is put at work using Mathcad[1] is described. Detailed tools suitable for facing this problem are provided in Sect. 4.1. In this example the RBF fit is demonstrated for a 2D topometric sculpting application. Flat surfaces are reshaped in curved ones adding a third coordinate (altitude) that is controlled using RBF centres both on the boundary of the flat domain and inside it.

2.3.1 Definition of the Source Points

The first shape examined is a rectangle of sides a and b. The points spacing along the boundary is controlled by the variable n_{edge} constraining at zero the altitude at such points, while a single point located at the centre of the rectangle is used to control the altitude. Mathcad constants[2] are used for the definition of the sides:

$$\begin{aligned} a &:= 1.0 \\ b &:= 1.0 \end{aligned} \tag{2.12}$$

[1]Mathcad software allows to use live equations and to introduce short piece of code suitable for setting up and demonstrating algorithms. Functions and constants can be defined in any place of the worksheet and are evaluated in the definition order. Vector and matrixes can be zero based or one based. In this book the one base convention is adopted. Plotting features are available and used in the book.

[2]Mathcad variables, constants and functions can be assigned using the := operator.

The points on the boundary are generated using the following piece of code[3]:

$$
P_{square} := \left| \begin{array}{l} n_{edge} \leftarrow 5 \\ \begin{pmatrix} dx \\ dy \end{pmatrix} \leftarrow \begin{pmatrix} \dfrac{a}{n_{edge}} \\ \dfrac{b}{n_{edge}} \end{pmatrix} \\ \text{for } i \in 1 .. n_{edge} \\ \quad \left| \begin{array}{l} P_i \leftarrow \begin{pmatrix} dx \cdot i \\ 0.0 \end{pmatrix} \\ P_{i+n_{edge}} \leftarrow \begin{bmatrix} dx \cdot (i-1) \\ b \end{bmatrix} \\ P_{i+n_{edge} \cdot 2} \leftarrow \begin{bmatrix} 0.0 \\ dy \cdot (i-1) \end{bmatrix} \\ P_{i+n_{edge} \cdot 3} \leftarrow \begin{pmatrix} a \\ dy \cdot i \end{pmatrix} \end{array} \right. \\ P \end{array} \right. \tag{2.13}
$$

The spacing n_{edge} is first assigned as a local variable of the programming block and set to 5; the increments *dx* and *dy* in *x* and *y* directions are computed (notice that all global variable defined outside the programming block are available inside); the **for** block is then used with the range variable i looping from 1 to n_{edge}.[4] The subscripts with a proper numbering offset are used to generate simultaneously the points of the vector P on the four edges starting from the second point of the lower horizontal edge, the first point of the upper horizontal edge, the second point of the left vertical edge and the first point of the right vertical edge.[5] It is worth to notice

[3]In Mathcad a program block works as other assignments (in this case a constant) but allows to have more freedom to internally specify how a constant or a function is structured. Local variable assignments are possible using the operator $\leftarrow$, the variable n_{edge} is an example (it is local and it is not exposed outside the block). Global variables of the worksheet (a and b) can be used inside a block. Mathcad programming block convention is to output the last variable (or the result of last function).

[4]The range variable i is inserted using the .. operator which allows to define a list using the first value and the last one (if the increment is 1); it accepts 3 arguments (the second value of the list after a comma) when an increment different than 1 has to be prescribed. "1 .. 5" results in "1 2 3 4 5", "1, 3 .. 7" results in "1 3 5 7", "1, 1.1 .. 1.5" results in "1 1.1 1.2 1.3 1.4 1.5".

[5]Mathcad accepts nested vectors. In this example a cloud of point is stored in the vector P; each element of P is a two components vector. Direct assignment of vectors and matrixes components can be performed with the subscript operator.

that this algorithm does not generate duplicates at corners. According to Mathcad syntax, the last entry of the programming block is passed as final result. In this case the internal vector P is assigned to P_{square}. In this example the vector nesting is used: the vector P_{square} is a single column of length 20 (the number of points) in which each entry is a 2D vector containing the position (x, y) of the points. The array of the internal control points (one point in this first example) is directly assigned on a point-wise basis:

$$\mathrm{P_{control}} := \left[\begin{pmatrix} \frac{a}{2} \\ \frac{b}{2} \end{pmatrix} \right] \tag{2.14}$$

Generated data can be displayed on Cartesian plot using the following range variables:

$$\begin{aligned} &\mathrm{iplot1} := 1..\mathrm{rows}(\mathrm{P_{square}}) \\ &\mathrm{iplot2} := 1..\mathrm{rows}(\mathrm{P_{control}}) \\ &\mathrm{rows}(\mathrm{P_{square}}) = 20 \\ &\mathrm{rows}(\mathrm{P_{control}}) = 1 \end{aligned} \tag{2.15}$$

The range variables iplot1 and iplot2 range from the first index to the last one of the two arrays of points; the special function **rows(·)** returns the number of the rows of a matrix. The plot is shown in Fig. 2.1 and exploits the ability of Mathcad of showing multiple data on the same Cartesian graph. Notice that plotted points are visited according to range variables iplot1 and iplot2 used as array subscripts; components x and y are extracted using nested information subscripts 1 and 2.

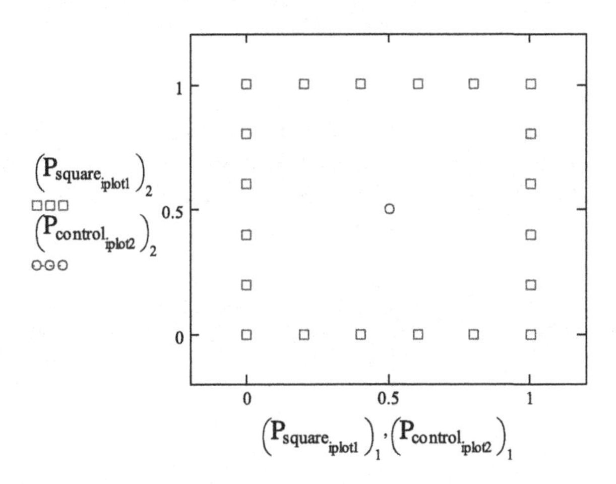

Fig. 2.1 Cartesian plot showing the points arrangement. Points on the boundary are plotted as red squares, internal control points are plotted as blue circles

The RBF problem can be posed defining the target values at source points that, in this case, are zero at the boundary and different from zero (0.5 in this example) at the control point:

$$x_s := \mathrm{stack}\left(P_{square}, P_{control}\right)$$

$$g_s := \left| \begin{array}{l} \text{for } i \in 1..\,\mathrm{rows}\left(P_{square}\right) \\ \quad g_i \leftarrow 0 \\ g_{1+\mathrm{rows}\left(P_{square}\right)} \leftarrow 0.5 \\ g \end{array} \right. \tag{2.16}$$

The function **stack(·)** gives as output a matrix obtained stacking by rows the input matrixes which must contain the same number of columns.

2.3.2 *Fitting the RBF*

The radial function is first defined and, to this end, the bi-harmonic spline in 2D, identified by the relationship $\varphi(r) = r^2\log(r)$ is used. The interpolation matrix, defined according to Eq. (2.7), is computed as:

$$\begin{array}{l} M := \text{for} \quad i \in 1\,..\,\mathrm{rows}(x_s) \\ \qquad \text{for} \quad j \in 1\,..\,\mathrm{rows}(x_s) \\ \qquad\quad M_{i,j} \leftarrow \phi\left(\left|x_{s_i} - x_{s_j}\right|\right) \end{array} \tag{2.17}$$

where the $|.|$ operator is used for the computation of the modulus of a vector. Considering that, data are stored as points, the subtraction operation between two points gives as a result the vector connecting the two points and, so, the radial distance is retrieved and processed by the radial function. On the other hand, the constraint matrix, defined according to Eq. (2.8), is computed as:

$$P := \text{for } i \in 1..\,\mathrm{rows}\left(x_s\right) \left| \begin{array}{l} P_{i,1} \leftarrow 1 \\ \text{for } j \in 1..\,\mathrm{rows}\left(x_{s_1}\right) \\ \quad P_{i,j+1} \leftarrow \left(x_{s_i}\right)_j \end{array} \right. \tag{2.18}$$

Notice that the size of the space (2 in this example) is retrieved measuring the size of the first point. An auxiliary function to define an arbitrary sized matrix with zero is first defined as:

$$\text{zero\_mat}(m,n) := \left| \begin{array}{l} \text{for } i \in 1..m \\ \quad \text{for } j \in 1..n \\ \quad\quad \text{mat}_{i,j} \leftarrow 0 \\ \text{mat} \end{array} \right. \tag{2.19}$$

The matrixes are finally assembled to form and solve the system of Eq. (2.6):

$$\begin{aligned} MP &:= \text{augment}\left(\text{stack}\left(M, P^T\right), \text{stack}(P, \text{zero\_mat}(\text{rows}(x_{s_1}) + 1, \text{rows}(x_{s_1}) + 1))\right) \\ g0 &:= \text{stack}(g_s, \text{zero\_mat}(\text{rows}(x_{s_1}) + 1, \text{cols}(g_k))) \\ \gamma\beta &:= \text{lsolve}(MP, g0) \\ \gamma &:= \text{submatrix}(\gamma\beta, 1, \text{rows}(x_s), 1, 1) \\ \beta &:= \text{submatrix}(\gamma\beta, \text{rows}(x_s) + 1, \text{rows}(\gamma\beta), 1, 1) \end{aligned} \tag{2.20}$$

The function **augment**(·) gives as output a matrix obtained pairing by columns the input matrixes (that must contain the same number of rows). The function **lsolve**(·) solves a linear system using the direct method; as referred in the Mathcad Guide this function is based on the BLAS/LAPACK libraries provided by Intel. The function **submatrix**(·) gives as output a sub-matrix and it is here used to split the coefficients γ of the RBF and the weights β of the polynomial.

2.3.3 *Evaluating the RBF*

Once the fitted coefficients are computed, the RBF can be evaluated as a point function using the Eq. (2.1) whose calculation is assigned in the worksheet as:

$$s_Z(x) := \sum_{i=1}^{\text{rows}(x_s)} (\gamma_i \cdot \phi(|x - x_{s_i}|)) + \beta \cdot \text{stack}(1, x) \tag{2.21}$$

It is worth to notice that this expression works for an arbitrary dimension of the space (2D in this example) because the module operator is dimension independent and the dot product spans the weights β of the polynomial on all the components (the 1 is stacked to the x vector to account for the constant term contained in the first row of β).

The correctness of the interpolation can be checked by computing the RBF at some of the input points:

$$\begin{aligned} s_Z(x_{s_1}) &= 0 \\ s_Z(x_{s_2}) &= 0 \\ s_Z(x_{s_{21}}) &= 0.5 \end{aligned} \tag{2.22}$$

2.3.4 Exploring the Interpolated Function

The RBF interpolation can be explored in other locations using plotting. For instance, a range variable can be used to move the coordinate along horizontal segments at different constant values of y; three different distances from the bottom edge (zero, one quarter and one half of the side) are shown in the plot of Fig. 2.2. It is interesting to notice that in the half side curve the imposed values (0.0, 0.5, 0.0) are retrieved at start, middle, end points whilst the function is gently interpolated in remaining points.

Moreover, it can be also noticed that the function on the bottom edge is not exactly zero as expected. This issue can be further explored in the plot of Fig. 2.3 where the ability to enforce the zero at the edge is highlighted; the error is indeed quite low and depends on the density of points used to set the zero value of the function at that edge. It is worth to highlight that the symmetric layout has produced a perfectly symmetric interpolation and the error is the same on both edges, whilst it vanishes, as expected, at the 6 RBF points locations.

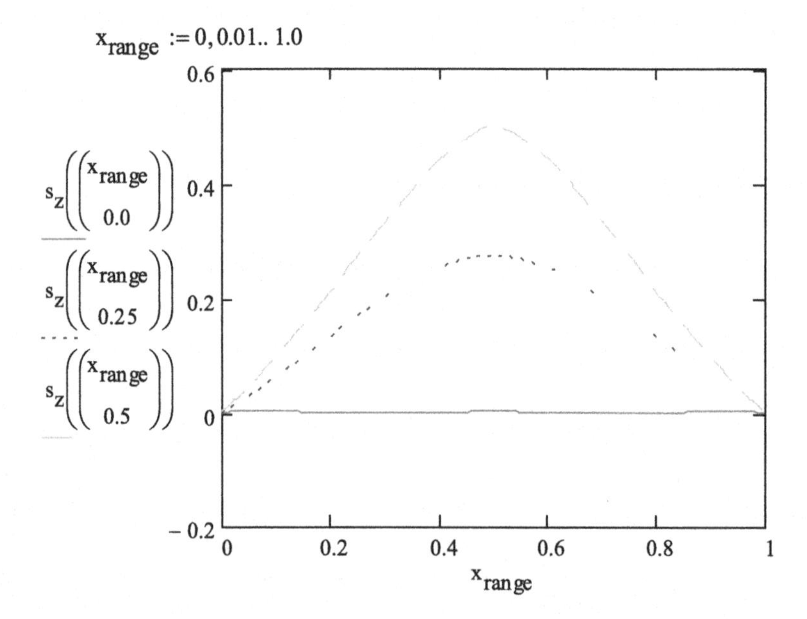

Fig. 2.2 Cartesian plot representing the interpolated function behaviour on 3 horizontal segments located at the bottom at one quarter and at one half of the height of the square domain

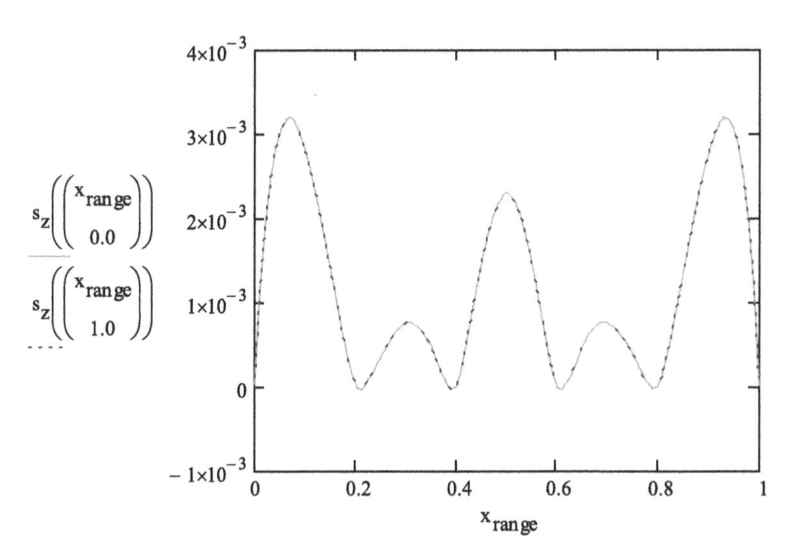

Fig. 2.3 Cartesian plot representing the interpolated function behaviour on 2 horizontal segments located at the top and the bottom of the square domain

The global behaviour of the function on the square domain is represented in Fig. 2.4. The surface plot is generated on a regular grid according to Eq. (2.23). The number of interval and the spacing in x and y directions are assigned as constants.

$$nx := 20. \quad ny := 20. \quad dx := \frac{1}{nx - 1} \quad dy := \frac{1}{ny - 1}$$

$$xyz_{plot} := \left| \begin{array}{l} id \leftarrow 1 \\ \text{for } i \in 1..nx \\ \quad \text{for } j \in 1..ny \\ \quad\quad \left| \begin{array}{l} Pxy_1 \leftarrow (i-1)\cdot dx \\ Pxy_2 \leftarrow (j-1)\cdot dy \\ xplot_{id} \leftarrow Pxy_1 \\ yplot_{id} \leftarrow Pxy_2 \\ zplot_{id} \leftarrow s_z(Pxy) \\ id \leftarrow id + 1 \end{array} \right. \\ \text{augment}(xplot, yplot, zplot) \end{array} \right. \tag{2.23}$$

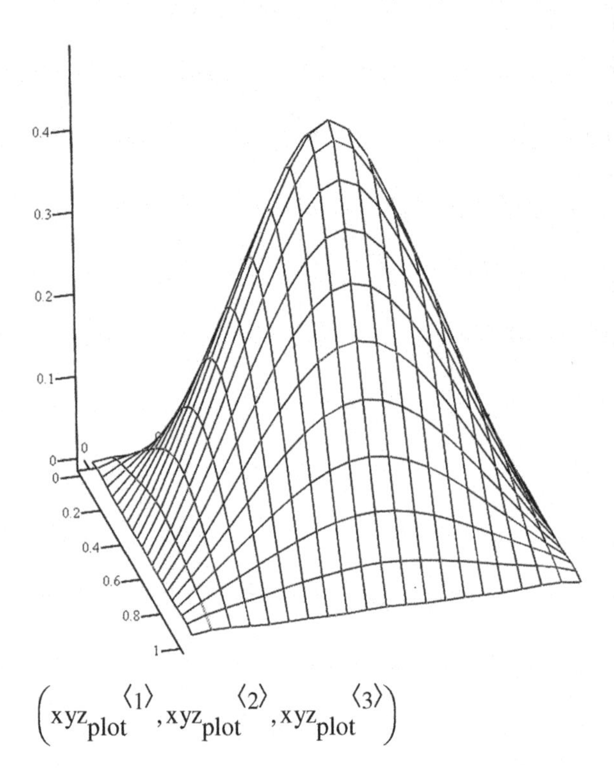

Fig. 2.4 Surface plot representing the interpolated function behaviour on the square domain

In the block the variable **id** is first set to 1, two nested cycles looping the variables **i** and **j** over the ranges [1,nx], [1,ny] are then defined; at each loop the coordinate **Pxy** of the point are defined on the regular grid and assigned, together with the RBF evaluated value, to local vectors **xplot, yplot**, **zplot**. The function **augment**(·) is used to build a three columns matrix containing xyz information. The surface plot expects as input three columns delimited by parentheses. Column extraction is achieved using the operator $^{<\cdot>}$ that allows to extract data from matrixes on a column by column basis.

2.3.5 Exploring Different Arrangements of Control Points

The just described experiment can be enriched changing the position of the control point or adding extra control points as shown in Fig. 2.5.

The effect of radial function using the linear ($\varphi(r) = r$) and the cubic ($\varphi(r) = r^3$) ones is shown in Fig. 2.6.

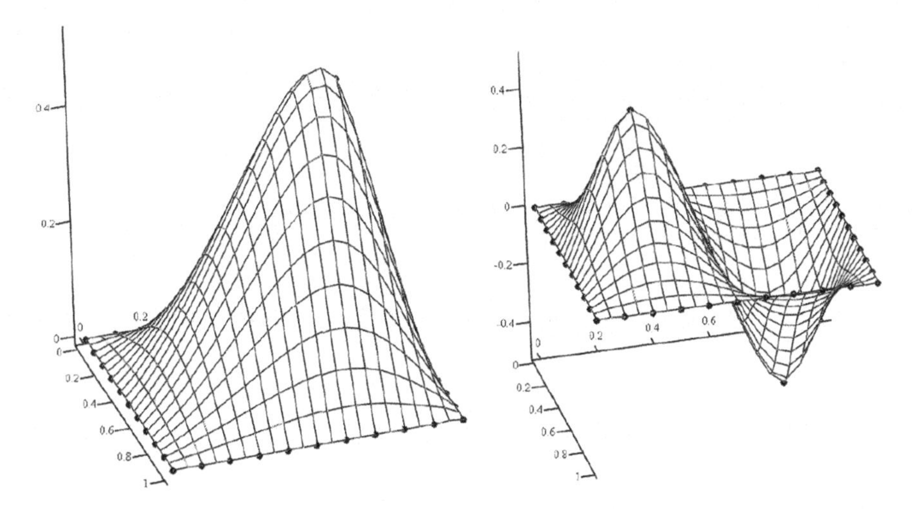

Fig. 2.5 The numerical experiment is herein enriched by moving the internal control point (left) and then adding a second control point with opposite altitude. RBF points are represented in the same plot; notice that with adopted radial function the interpolated function can be higher than the controlled value

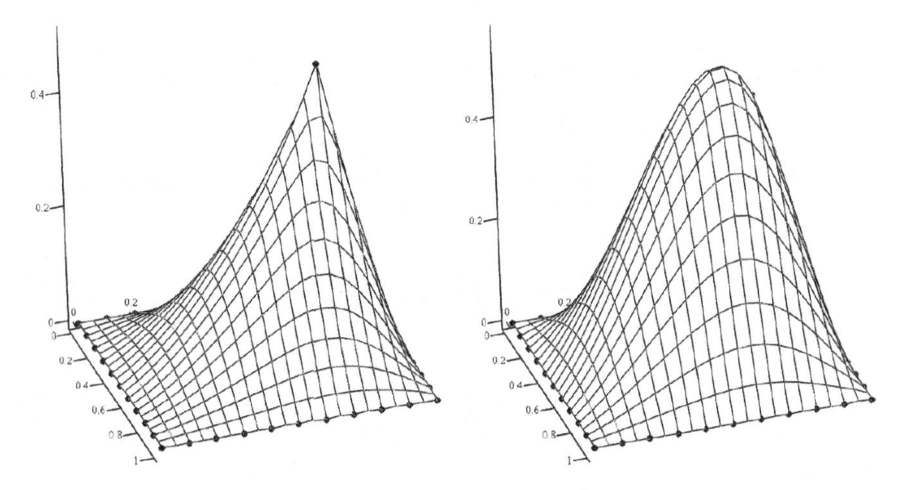

Fig. 2.6 The effect of the radial function is shown for the same arrangement of Fig. 2.5 (left). A linear radial function (left) and a cubic one (right) are tested

2.4 Differentiation of RBF

The RBF expansion of the interpolated function can be differentiated bearing in mind that the order of differentiation depends on the selected RBF kernel. The concept is demonstrated using a three-dimensional case in which a linear polynomial is added to the RBF, but it can easily extended to other cases. The expansion is available as a closed form:

$$s(\boldsymbol{x}) = \sum_{i=1}^{N} \gamma_i \varphi(\|\boldsymbol{x} - \boldsymbol{x}_{s_i}\|) + \beta_1 + \beta_2 x + \beta_3 y + \beta_4 z \tag{2.24}$$

that can be detailed as follows:

$$s(\boldsymbol{x}) = \sum_{i=1}^{N} \gamma_i \varphi\left(\sqrt{\left(x - x_{s_{ix}}\right)^2 + \left(y - x_{s_{iy}}\right)^2 + \left(z - x_{s_{iz}}\right)^2}\right) + \beta_1 + \beta_2 x + \beta_3 y + \beta_4 z \tag{2.25}$$

2.4.1 First Derivative

The first partial derivative is obtained considering the derivation chain rule for functions of functions (i.e. the derivative of the radial function $\frac{\mathrm{d}\varphi(\mathrm{r})}{\mathrm{dr}}$ multiplied by the derivative of the square root $\frac{\mathrm{d}\sqrt{f}}{\mathrm{d}f} = \frac{1}{2\sqrt{f}}$ multiplied by the derivative of the squared term $\frac{d\left(x - x_{s_{ix}}\right)^2}{dx} = 2\left(x - x_{s_{ix}}\right)$):

$$\frac{\partial s(\boldsymbol{x})}{\partial x} = \sum_{i=1}^{N} \gamma_i \frac{\mathrm{d}\varphi(\mathrm{r})}{\mathrm{dr}} \frac{1}{2\sqrt{\left(x - x_{s_{ix}}\right)^2 + \left(y - x_{s_{iy}}\right)^2 + \left(z - x_{s_{iz}}\right)^2}} 2\left(x - x_{s_{ix}}\right) + \beta_2 \tag{2.26}$$

that in compact form is:

$$\frac{\partial s(\boldsymbol{x})}{\partial x} = \sum_{i=1}^{N} \gamma_i \frac{\partial \varphi(\|\boldsymbol{x} - \boldsymbol{x}_{\boldsymbol{k}_i}\|)}{\partial \mathrm{r}} \frac{\left(x - \boldsymbol{x}_{\boldsymbol{k}_{ix}}\right)}{\boldsymbol{x} - \boldsymbol{x}_{\boldsymbol{k}_i}} + \beta_2 \tag{2.27}$$

whilst the full gradient is:

$$\nabla s(\boldsymbol{x}) = \begin{pmatrix} \frac{\partial s(\boldsymbol{x})}{\partial x} \\ \frac{\partial s(\boldsymbol{x})}{\partial y} \\ \frac{\partial s(\boldsymbol{x})}{\partial z} \end{pmatrix} = \sum_{i=1}^{N} \gamma_i \frac{\frac{\partial \varphi\left(\|\boldsymbol{x} - \boldsymbol{x}_{s_i}\|\right)}{\partial \mathrm{r}}}{\|\boldsymbol{x} - \boldsymbol{x}_{s_i}\|} \begin{pmatrix} x - x_{s_{ix}} \\ y - x_{s_{iy}} \\ z - x_{s_{iz}} \end{pmatrix} + \begin{pmatrix} \beta_2 \\ \beta_3 \\ \beta_4 \end{pmatrix} \tag{2.28}$$

also valid in the generic n-dimensional case:

$$\nabla s(\boldsymbol{x}) = \begin{pmatrix} \frac{\partial s(\boldsymbol{x})}{\partial x_1} \\ \frac{\partial s(\boldsymbol{x})}{\partial x_2} \\ \vdots \\ \frac{\partial s(\boldsymbol{x})}{\partial x_N} \end{pmatrix} = \sum_{i=1}^{N} \gamma_i \frac{\frac{\partial \varphi(\|\boldsymbol{x}-\boldsymbol{x}_{s_i}\|)}{\partial \mathrm{r}}}{\|\boldsymbol{x}-\boldsymbol{x}_{s_i}\|} \begin{pmatrix} x_1 - x_{s_{i_1}} \\ x_2 - x_{s_{i_2}} \\ \vdots \\ x_N - x_{s_{i_N}} \end{pmatrix} + \begin{pmatrix} \beta_2 \\ \beta_3 \\ \vdots \\ \beta_{N+1} \end{pmatrix} \tag{2.29}$$

2.4.2 *Example of First Derivatives Exploitation*

The closed form representation can be used for many applications as demonstrated in Chap. 12 where the derivatives are used for the computation of strain from the RBF interpolated displacement field. In this section the example of Sect. 2.3 is further extended demonstrating how the global maximum shown in Fig. 2.6 (right) can be located searching for the point in which the gradient is null. The expression of the gradient can be computed according to Eq. (2.29) that in the Mathcad worksheet is formulated as follows:

$$\mathrm{grad}(x) := \sum_{i=1}^{\mathrm{rows}(x_s)} \left[\frac{\gamma_i \cdot \phi p(|x - x_{s_i}|)}{2 \cdot |x - x_{s_i}|} \cdot \begin{bmatrix} 2 \cdot [x_1 - (x_{s_i})_1] \\ 2 \cdot [x_2 - (x_{s_i})_2] \end{bmatrix} \right] + \begin{pmatrix} \beta_2 \\ \beta_3 \end{pmatrix} \tag{2.30}$$

Local maxima/minima can be computed using the gradient method. The algorithm starts with a guess point (in the x-y plane for this example) and it continuously updates the position searching a new point in the direction of the gradient. A simple implementation is given by:

```
LocateMax := | x_1 ← (0.75, 0.25)
             | dx ← 0.1
             | id ← 1
             | tol ← 0.001
             | while |grad(x_id)| > tol
             |     | x_id+1 ← x_id + grad(x_id)·dx
             |     | id ← id + 1
             | x
```
(2.31)

In this simple code fragment, the point is updated using a step that is obtained multiplying the gradient by a given *dx*. As the gradient is not normalized, the

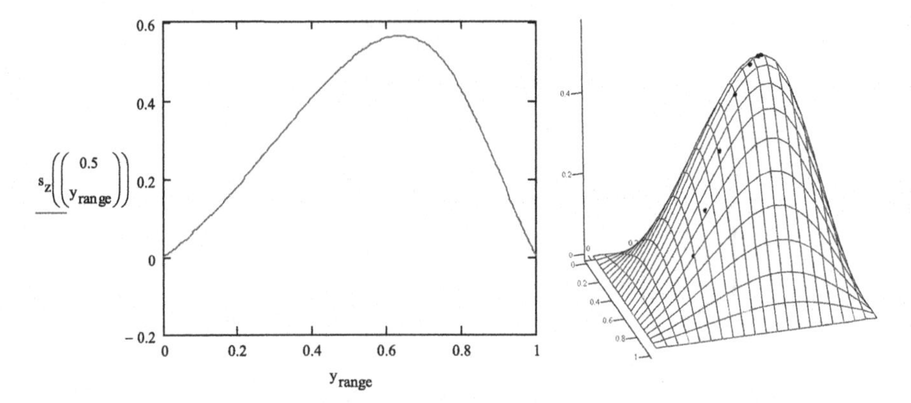

Fig. 2.7 The maximum location can be easily located using a Cartesian plot representing the behaviour of the function along the symmetry segment (left); the research path of the gradient method is highlighted on the surface plot (right) where the convergence can be noticed (search point density increases at maximum site)

increment is automatically reduced when the solution approaches the maximum. The algorithm is iterated (adopting the Mathcad **while** block) until the modulus of the gradient is higher than the given tolerance *tol*. The solution path is highlighted in Fig. 2.7 (right). In this special case, due to the symmetry, the maximum can be located seeking the root of the second component of the gradient as expressed in:

$$\begin{aligned} &\mathrm{LocateMax}_{\mathrm{rows(LocateMax)}} = \begin{pmatrix} 0.5 \\ 0.637 \end{pmatrix} \\ &\mathrm{root}\left[\mathrm{grad}\left(\begin{pmatrix} 0.5 \\ \mathrm{y} \end{pmatrix}\right)_2, \mathrm{y}, 0.0, 1.0\right] = 0.637 \end{aligned} \tag{2.32}$$

The **root(·)** function provided by Mathcad allows to numerically locate the zero of a function; it takes as arguments the function (in this case the second component of the gradient), the variable to be varied and its range of variation. Using the given tolerance values, the maximum is properly located.

2.4.3 Higher Order Derivatives

Higher order derivatives can be computed using a similar approach. In the one-dimensional case we have:

$$\mathrm{s}(t) = \sum_{i=1}^{N} \left[\gamma_i \varphi\left(\sqrt{(t - \boldsymbol{t}_{\boldsymbol{k}_i})^2}\right)\right] + \beta_1 + \beta_2 t \tag{2.33}$$

that can be differentiated as

$$\frac{\mathrm{ds}(t)}{\mathrm{dt}} = \sum_{i=1}^{N} \left[\gamma_{\mathrm{i}} \frac{\mathrm{d}\varphi\left(\sqrt{(t - t_{k_i})^2}\right)}{\mathrm{dr}} \frac{(t - t_{k_i})}{\sqrt{(t - t_{k_i})^2}} \right] + \beta_2 \tag{2.34}$$

and can be differentiated again in:

$$\frac{\mathrm{d}^2\mathrm{s}(t)}{\mathrm{dt}^2} = \sum_{i=1}^{N} \left[\gamma_{\mathrm{i}} \frac{\mathrm{d}^2\varphi\left(\sqrt{(t - t_{k_i})^2}\right)}{\mathrm{dr}^2} + \frac{\mathrm{d}\varphi\left(\sqrt{(t - t_{k_i})^2}\right)}{\mathrm{dr}} \frac{1}{(t - t_{k_i})^2} \right] \tag{2.35}$$

2.5 Fitting an RBF at Arbitrary Locations of the Known Data

The common usage is to self-fit the RBF, namely imposing the known function at source locations. The method can be easily generalised by setting the known function at locations that differ from the source points. In this case, the Eq. (2.2) is replaced by:

$$s(\boldsymbol{x}_{f_i}) = g_{f_i}, \quad 1 \leq i \leq N \tag{2.36}$$

where $\boldsymbol{x}_f$ are the locations at which the g_f is known. In this case the system (2.6) becomes:

$$\begin{pmatrix} \boldsymbol{M} & \boldsymbol{P}_f \\ \boldsymbol{P}_s^T & 0 \end{pmatrix} \begin{pmatrix} \boldsymbol{\gamma} \\ \boldsymbol{\beta} \end{pmatrix} = \begin{pmatrix} \boldsymbol{g}_f \\ 0 \end{pmatrix} \tag{2.37}$$

where the $\boldsymbol{P}_f$ matrix is defined as:

$$\boldsymbol{P}_f = \begin{pmatrix} 1 & x_{f_1} & y_{f_1} & z_{f_1} \\ 1 & x_{f_2} & y_{f_2} & z_{f_2} \\ \vdots & \vdots & \vdots & \vdots \\ 1 & x_{f_N} & y_{f_N} & z_{f_N} \end{pmatrix} \tag{2.38}$$

A detailed example of this approach is given in Sect. 3.6.2 where a multipole expansion is approximated using a small set of RBF points.

2.5.1 A Simple Example

The example of Sect. 2.3 is here adapted to demonstrate the type of fitting just introduced. The arrangement of source and fit locations is defined using a slightly modified version of Eq. (2.13) where source points are distributed on a square with 1.2 side centred in the origin. An extra control point located in the origin is added; fit points are arranged in the same fashion but using a square with side 1.0 as represented in Fig. 2.8 (left).

The behaviour of the function interpolated at source points with prescribed fit points is represented in Fig. 2.8 (right) where the surface plot is extended to the square of source points' locations. In this plot, fit points are highlighted and it is clear that the fit conditions are properly met.

The system (2.37) is solved in this case adapting the commands (2.20) as follows:

$$\begin{aligned}
&\mathrm{MP} := \mathrm{augment}\left(\mathrm{stack}\left(\mathrm{M}, \mathrm{P}_s^T\right), \mathrm{stack}(\mathrm{P_f}, \mathrm{zero_mat}(\mathrm{rows}(\mathrm{x}_{s_1}) + 1, \mathrm{rows}(\mathrm{x}_{s_1}) + 1))\right)\\
&\mathrm{g0} := \mathrm{stack}(\mathrm{g_f}, \mathrm{zero_mat}(\mathrm{rows}(\mathrm{x}_{s_1}) + 1, \mathrm{cols}(\mathrm{g_f})))\\
&\gamma\beta := \mathrm{lsolve}(\mathrm{MP}, \mathrm{g0})\\
&\gamma := \mathrm{submatrix}(\gamma\beta, 1, \mathrm{rows}(\mathrm{x_s}), 1, 1)\\
&\beta := \mathrm{submatrix}(\gamma\beta, \mathrm{rows}(\mathrm{x_s}) + 1, \mathrm{rows}(\gamma\beta), 1, 1)
\end{aligned} \tag{2.39}$$

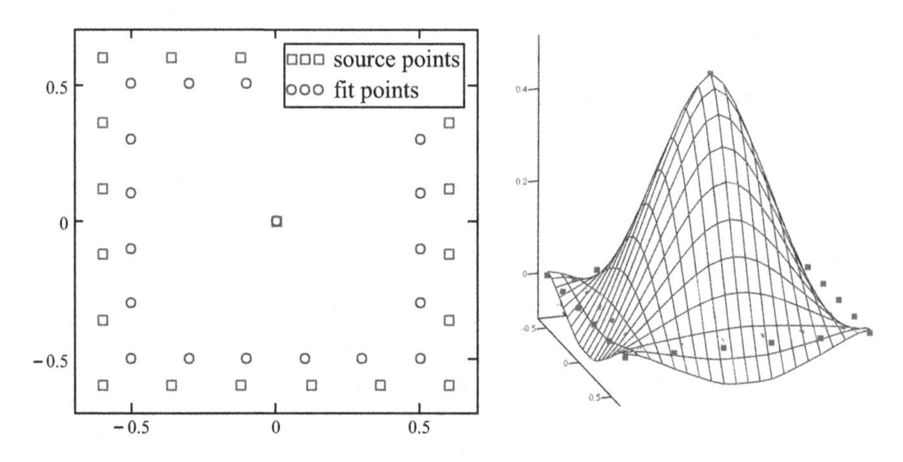

Fig. 2.8 The source points are distributed at locations different from the fit points (left). A square of side 1.2 is used for source positions and the function is imposed equal to zero on a square of side 1.0. Central point is used both as a source and as a fit point. The behaviour of the function is represented as a surface plot in which the locations and the values at fit points and source points are highlighted (right)

2.6 Interpolation Versus Regression

The theory detailed in Sect. 2.1 concerning RBF interpolation, will be the one mainly adopted in this book. However, there are situations in which the regression can be the best choice. The main difference between RBF interpolation and RBF regression is that in the case of interpolation the number of centres at which the RBF coefficients are fitted and the number of centres at which the scalar function is known are coincident. In the regression problem the number of source points is lower than the one at which the function is known and, so, an exact fit cannot be guaranteed. In this latter scenario, an error is accepted and the fitting procedure, identified by the solution of system (2.6), is replaced by a minimization procedure that searches for coefficients that give the best approximation at all the points at which the g function is known.

It is worth to notice that the regression can be preferred mainly for two reasons that are its ability to filter noise and the performance saving due to problem size reduction. Details of noise filtering are given in Sect. 2.7. As far as the numerical methods are concerned, a common strategy for the reduction of the problem complexity consists of retaining just a small subset of the original data centres as deepened in Sect. 2.3. The input value at retained centres is commonly used as the input for the approximated solution; such an approach is based on the same numerical tools valid for the complete fit. A better global accuracy can be achieved replacing the fitting method with a regression method because it allows to get the minimum error on all the original dataset using the retained centres as sources.

The mathematical problem has yet the form of Eq. (2.37) but the matrix of the system becomes rectangular because the number of fit points is larger than the number of source points. It cannot be solved directly, as an exact solution does not exist, but it can faced using a minimization approach in which the unknown weights γ and coefficients $\boldsymbol{\beta}$ are varied until the least squares error is minimized.

The regression problem is represented by the expression:

$$\min_{\mathrm{s}} \frac{1}{\mathrm{N}} \sum_{j=1}^{N} \left[\mathrm{s}\left(\mathbf{x}_{\mathrm{f}_j}\right) - \mathrm{g}_{\mathrm{f}_j} \right]^2 \tag{2.40}$$

The solution block (2.39) can be extended using the **Given** and **Minerr(·)** functions of Mathcad that allows to solve an arbitrary system of equations even if over constrained by minimizing the error; many algorithms are available in Mathcad, and the least squares one is used for linear systems. The following Mathcad code is used for the solution of the system:

$$\begin{aligned}
&\gamma\beta := \text{zero_mat}(\text{cols}(\text{MP}), 1)\\
&\text{Given}\\
&\text{MP} \cdot \gamma\beta - \text{g0} = 0\\
&\gamma\beta_{\text{sol}} := \text{Minerr}(\gamma\beta)\\
&\gamma := \text{submatrix}(\gamma\beta_{\text{sol}}, 1, \text{rows}(\text{x}_\text{s}), 1, 1)\\
&\beta := \text{submatrix}(\gamma\beta_{\text{sol}}, \text{rows}(\text{x}_\text{s}) + 1, \text{rows}(\gamma\beta_{\text{sol}}), 1, 1)
\end{aligned} \tag{2.41}$$

An initial value, zero in this case, has first to be provided for the vector $\boldsymbol{\gamma\beta}$ that is the unknown items of the system. The keyword **Given** is used to start the definition of the system of equations; the system of equations is then defined as a function of the unknown $\boldsymbol{\gamma\beta}$ and of the known data. The **Minerr** function ends the solution block and allows to compute the solution $\boldsymbol{\gamma\beta_{sol}}$ that is successively employed to get the weights of the RBF and the coefficients of the polynomial.

The least squares problem can be faced solving directly the normal equations (Press et al. 1992):

$$\begin{pmatrix} \boldsymbol{M} & \boldsymbol{P_f} \\ \boldsymbol{P_s^T} & 0 \end{pmatrix}^{\text{T}} \begin{pmatrix} \boldsymbol{M} & \boldsymbol{P_f} \\ \boldsymbol{P_s^T} & 0 \end{pmatrix} \begin{pmatrix} \boldsymbol{\gamma} \\ \boldsymbol{\beta} \end{pmatrix} = \begin{pmatrix} \boldsymbol{M} & \boldsymbol{P_f} \\ \boldsymbol{P_s^T} & 0 \end{pmatrix}^{\text{T}} \begin{pmatrix} \boldsymbol{g_f} \\ 0 \end{pmatrix} \tag{2.42}$$

Using Mathcad software the solution block (2.41) becomes:

$$\begin{aligned}
&\gamma\beta_{\text{sol}} := \text{lsolve}\left(\text{MP}^{\text{T}} \cdot \text{MP}, \text{MP}^{\text{T}} \cdot \text{g0}\right)\\
&\quad\gamma := \text{submatrix}(\gamma\beta_{\text{sol}}, 1, \text{rows}(\text{x}_\text{s}), 1, 1)\\
&\quad\beta := \text{submatrix}(\gamma\beta_{\text{sol}}, \text{rows}(\text{x}_\text{s}) + 1, \text{rows}(\gamma\beta_{\text{sol}}), 1, 1)
\end{aligned} \tag{2.43}$$

Identical results are obtained with Mathcad using the solution of the normal equations and the default settings of **Minerr(·)** solver.

2.6.1 A Simple Example of RBF Regression

The example of Sect. 2.3 is here adapted to demonstrate RBF regression. The arrangement of source and fit locations is defined using a slightly modified version of Eq. (2.13); source points are distributed on a square with 1.0 side centred in the origin, an extra control point located in the origin is added; fit points are arranged in the same fashion but with a finer spacing as represented in Fig. 2.9 (left). The behaviour of the function interpolated at source points with prescribed fit points is represented in the same Fig. 2.9 (right) in which the surface plot is extended to the square of source points locations; fit points are highlighted and it is clear that the fit conditions are properly met.

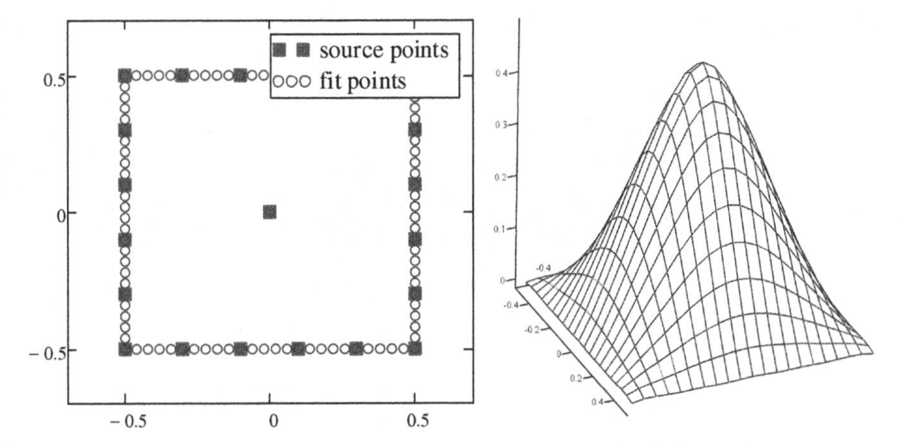

Fig. 2.9 A coarse distribution of source points at square sides is used to impose a zero on a fine distribution of fit points. Central point is used both as a source and as a fit point (left). The behaviour of the interpolated function looks almost identical to the one of Fig. 2.4

It is interesting to explore the behaviour of the function along the bottom and top edge of the square. Figure 2.10 represents the value of the function on both segments, which is identical for symmetry reasons. The error path is very similar to the one of Fig. 2.3 as the number of points used to control the RBF is the same. However, in this case the error in representing a perfect zero is uniformly distributed along the segment (positive and negative error), whilst in the case of fitting at source location the error is always positive.

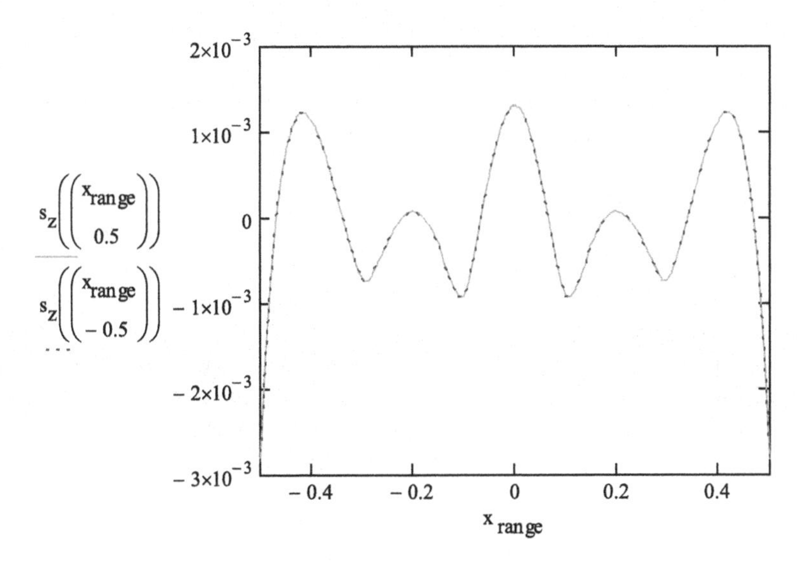

Fig. 2.10 Cartesian plot representing the interpolated function behaviour on 2 horizontal segments located at the top and the bottom of the square domain

2.7 Approximation of Noisy Data

RBF can be adopted as a tool to process, and accordingly represent, experimental or numerical data that are affected by noise. The source of the noise could be of various nature and a filtering is often desired.

Noise filtering could be required when RBF are used for image processing if raw data are not perfectly smooth. A similar scenario occurs during the reconstruction of surfaces based on laser scan data that might show local irregularities (crispy surface) that need to be filtered. Noise might also appear during numerical analyses as well. An example is the processing of shape sensitivity information at surfaces coming from adjoint CFD solutions (Sect. 8.6.1.2) that need to be filtered and used for surface sculpting (see Sect. 8.3.3). Another example is the case of optimal surface generated during topological optimisation workflow which is based on the extraction of an isodensity function.

The noise can be filtered by either modifying the RBF fitting strategies of the interpolation method, foreseeing the processing of all the input points as sources, or reducing the number of sources and adopting an RBF regression that accommodates all the noisy data.

2.7.1 *Noisy Data Fitting with a Relaxation Function*

The first approach for noisy data filtering, proposed by Carr et al. (2001), is based on a modified version of the system (2.6):

$$\begin{pmatrix} \boldsymbol{M} - 8N_s\pi\rho\boldsymbol{I} & \boldsymbol{P}_s \\ \boldsymbol{P}_s^T & 0 \end{pmatrix} \begin{pmatrix} \boldsymbol{\gamma} \\ \boldsymbol{\beta} \end{pmatrix} = \begin{pmatrix} \boldsymbol{g}_s \\ 0 \end{pmatrix} \tag{2.44}$$

This allows to relax the exact fit hypothesis when solving the system. The parameter ρ balances smoothness against fidelity to the data.

To demonstrate this concept, a parabolic function is disturbed using a random noise of a given amplitude. The synthetic dataset is comprised of 2500 points that are generated on a regular 50×50 grid using the parabolic function $z(x, y) = -1.0(x^2 + y^2) + 1.0$ to which a random noise of amplitude 0.05 is added. A cubic radial function $\varphi(r) = r^3$ is adopted and Fig. 2.11 shows the effect of the filter parameter ρ on this synthetic noisy function.

The crispy behaviour at grid points is exactly recovered when filtering is off (Fig. 2.11 top-left); the effect of the filter parameter can be observed changing its intensity (Fig. 2.11 top-right, bottom-left and Fig. 2.11 bottom-right). It is worth to notice how the filter reduces the spatial wave amplitude retained in the fit.

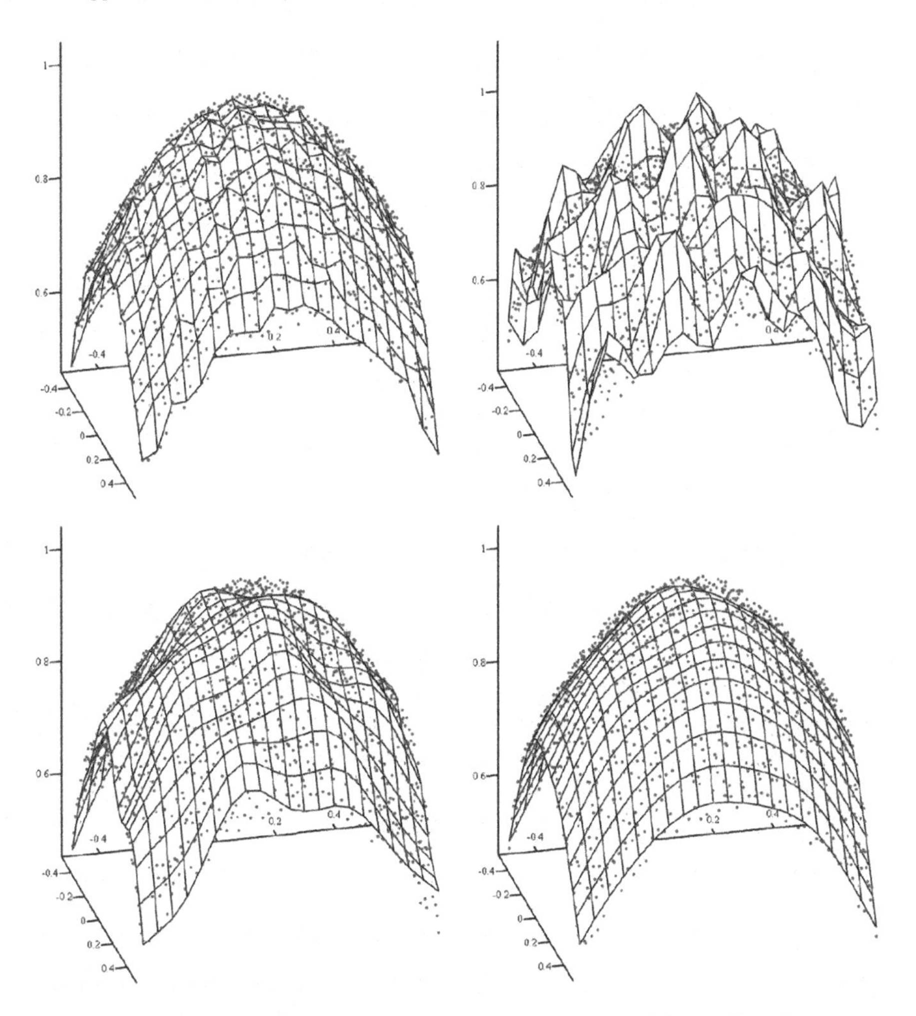

Fig. 2.11 The effect of filtering is demonstrated. Original data (top left) are filtered using $\rho = \lambda \times 10^{-6}$ with $\lambda = 0.1$ (top right), $\lambda = 0.5$ (bottom left) and $\lambda = 1.0$ (bottom right)

2.7.2 Noisy Data Fitting with Regression

The fitting of noisy data addressed in Sect. 2.7.1 foresees the use of all the original dataset. A meaningful alternative to such approach consists of the use of just a part of the original locations as RBF sources keeping the entire dataset as a target for the RBF regression. The filtering is achieved by controlling the spacing of retained points because the RBF becomes not anymore capable to represent high frequency spatial oscillations; the coarser the spacing of retained points, the lower the maximum spatial oscillation that can be reproduced. The same dataset of Sect. 2.7.1 is here investigated but a coarser subset comprising 25 RBF centres is used as sources

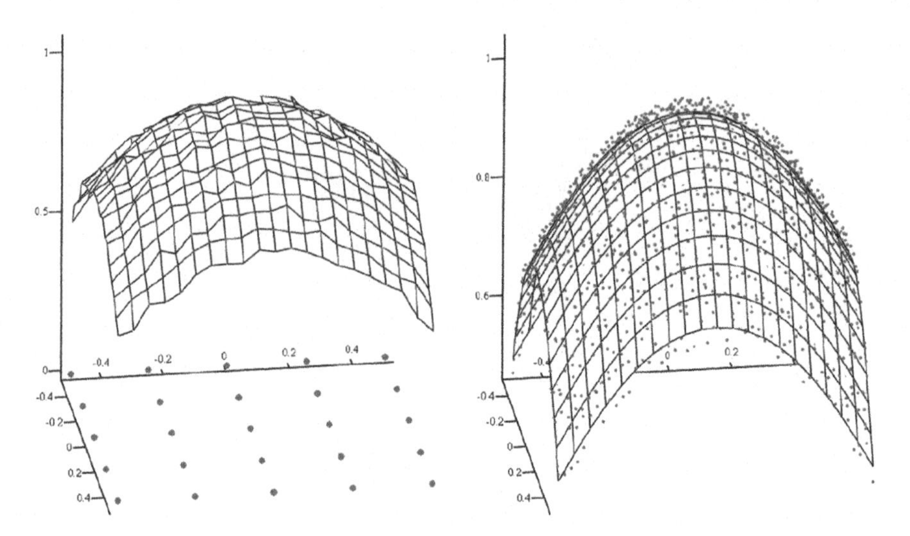

Fig. 2.12 The noisy dataset is known on a regular grid 50 × 50 and is represented by a surface plot (left); the same plot shows the grid 5 × 5 used as locations for RBF sources. The smooth surface plot of the evaluated RBF (right) is plotted together with noisy points interpolated

and is represented in the plot of Fig. 2.12 (left) where the surface plot of the original dataset is plotted as well. Noisy data and RBF regression evaluated on the fine grid are represented in Fig. 2.12 (right) where the ability to reconstruct the data filtering out the noise is clearly demonstrated.

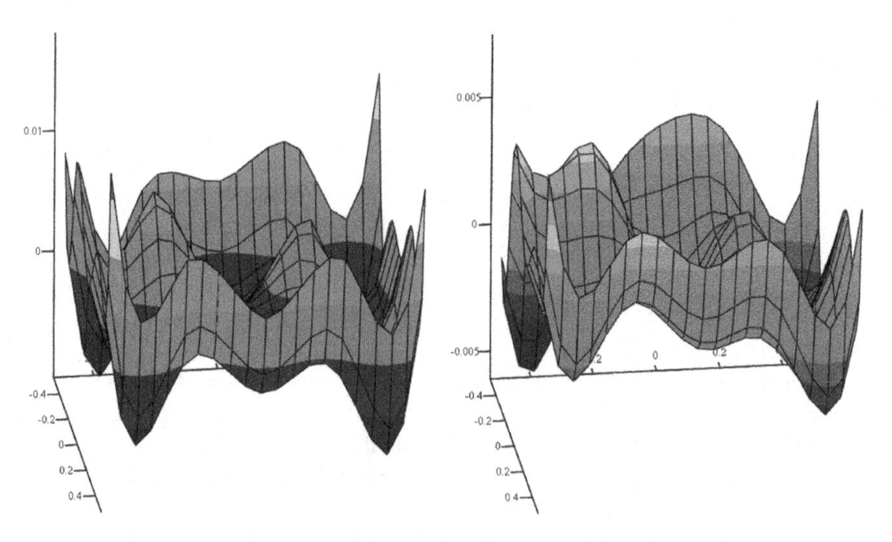

Fig. 2.13 The ability of filter out the artificial noise is inspected computing the error function that is the difference between obtained RBF regression and the original parabolic function. Registered error ranges [min, max] are [−0.0068, 0.0173] for the spline RBF and [−0.0034, 0.0078] for the cubic RBF

A quantitative analysis of the ability of removing the artificial noise can be conducted by calculating the difference between filtered RBF regression and original function known in this case. Two radial functions are investigated for this purpose: the 2D bi-harmonic spline $\varphi(r) = r^2\log(r)$ and the cubic function $\varphi(r) = r^3$. The error over the interpolated domain is represented in Fig. 2.13 as a coloured surface map. In both cases a small error is registered but the performances of the cubic function are significantly higher.

References

Boyd PB, Gildersleeve KW (2011) Numerical experiments on the condition number of the interpolation matrices for radial basis functions. Appl Numer Math 61:443–459. https://doi.org/10.1016/j.apnum.2010.11.009

Carr JC, Beatson R, Cherri J, Mitchell T, Fright W, McCallum B (2001) Reconstruction and representation of 3D objects with radial basis functions. In: Proceedings of the 28th annual conference on computer graphics and interactive techniques, Los Angeles, CA, pp 67–76

Hardy RL (1971) Multiquadric equations of topography and other irregular surfaces. J Geophys Res 76(8):1905–1915

Hardy RL (1990) Theory and applications of the multiquadric-biharmonic method, 20 years of discovery, 1968–1988. Comp Math Applic 19(8/9):163–208

Press WH, Flannery BP, Teukolsky SA (1992) Numerical recipes in C. The art of scientific computing, 2nd edn. ISBN 0-521-43108-5

Wendland H (1995) Piecewise polynomial, positive definite and compactly supported radial basis functions of minimal degree. Adv Comput Math 4(1):389–396. ISSN 1019-7168

Chapter 3
Fast RBF

Abstract Fast RBF foundation concepts are described in this chapter. Details are provided for an effective usage of the direct method using advanced linear algebra strategies and libraries. Methods for data compression which allow to represent the same information with a smaller and less expensive cloud are introduced. Space localization methods are successively considered firstly with compact supported RBF, whose interaction distance is limited in the RBF function itself, and then with partition of unity (POU) that consists of the superposition and blending of smaller clouds. Approximated evaluation of the far field of full supported RBF is described and an overview of the fast multipole method (FMM) is given. Information about iterative solvers and parallel computing are finally provided to complete the chapter.

At a first glance RBF implementation seems to be an easy task, but what usually happens to practitioners is that this first positive experience is followed by a substantial disappointment when the limit of the simple approach is discovered. The complexity of the fit grows with a power of 3 with respect to the number of RBF centres, whilst the cost of an evaluation of the RBF at a single point and storage space grows with a power of 2. This means that the practical limit of a problem that can be managed is about 10,000 points, value that is high enough to experiment with engineering problems, but not satisfactory for the production of effective industrial solutions. This is why RBF have a bad reputation and are often perceived as an interesting tool suitable for theoretical investigations but useless for practical applications.

The purpose of this book is to overtake this barrier. Fast RBF implementations capable to deal with large problems (up to millions of RBF centres) are available. An overview of fast methods will be provided, not with the aim to put the reader in a position to write her/his own fast RBF solver but rather to gain a clear understanding of what is feasible. An high level of complexity is required to get desired fast performances but such a working context is quite common in Engineering Sciences. How many users of FEM solver are able to implement a structural solver

M. E. Biancolini, *Fast Radial Basis Functions for Engineering Applications*,
https://doi.org/10.1007/978-3-319-75011-8_3

with performances comparable to the ones of industrial solutions? How many CFD users can implement a flow solver?

As for other solvers technologies there are many levels of approach. Relating to RBF problems, industrial solutions are available as well but they are not as widespread and popular as in other CAE fields. This means that when a new application that can benefit of RBF is identified, it can be first investigated using simple tools based on direct method and then accelerated using fast methods. The acceleration could be complex and costly, and usually the full flexibility that can be accessed using direct methods cannot be accessed. Practical examples that demonstrate this process are part of this book.

3.1 Fast RBF Foundation

Unique quality performances of RBF interpolation are due to the meshless nature of the method, that makes them attractive for scattered dataset, and to their capability to account for far field interactions between centres when global support is adopted, that allows to precisely capture complex behaviours. Such quality performances come with an high numerical cost. Dense systems have to be solved and a single evaluation can become very expensive.

A compromise between quality of the interpolation and numerical performances needs to be set to enable fast RBF for the solution of industrial problems. There are many options and, in all the cases, some kind of localization is required; this means that the full exact interaction hypothesis has to be relaxed accepting some simplifications whilst monitoring the quality of the achieved RBF fit. Some of the possible strategies are deepened in the next sections.

3.2 Direct Methods with Fast Libraries for Linear Algebra

The maximum flexibility for the solution of an RBF system is provided by direct methods. The linear system, that comprises the interpolation matrix and, optionally, the constraint matrix, can be solved using standard algorithms suitable for linear systems. Such a flexibility comes with an high cost because the complexity scales as a cubic power with respect to the number of points of the cloud. Nevertheless, the direct solver could be the best choice in the case of small dataset and resources can be anyway saved reusing the same partitioning if more than one scalar field needs to be fitted as the interpolation and constraint matrices stays the same.

Two very well established sources of routines for linear algebra are the LAPACK (2017a) and the Numerical Recipes (Press et al. 1992).

LAPACK is a standard software library for numerical linear algebra. It provides routines for solving systems of linear equations and linear least squares, eigenvalue problems and singular value decomposition. It also includes routines to implement the associated matrix factorizations such as LU, QR, Cholesky and Schur decomposition. LAPACK was originally written in FORTRAN 77, but moved to Fortran 90 in version 3.2. The routines handle both real and complex matrices in both single and double precision. LAPACK is designed to effectively exploit the caches on modern cache-based architectures and has also been extended to run on distributed-memory systems in later packages such as ScaLAPACK (2017).

The LAPACK routine DSPTRF and DSPTRS (LAPACK 2017b), which are based on the Bunch-Kaufman diagonal pivoting method, proven to be robust and reliable for the solution of linear problems required for the computation of the coefficients of RBF.[1]

LU decomposition implemented in ***ludcmp*** and ***lubksb*** routines of the numerical recipes (Press et al. 1992) has been successfully tested for RBF as well.[2]

It is worth to add that, in both cases, the factorization is computed once and can be reused for multiple vectors of coefficients. This is a common situations for instance in three-dimensional mesh morphing applications presented in Chap. 6 of this book where the three components of the vector field are managed as three different linear systems using each component as a scalar problem; the linear system is the same, only the coefficients to be used for the fitting are different.

3.3 Reduction of the Cloud Size (Greedy Methods)

A well accepted strategy for the reduction of the complexity of a given RBF cloud consists of the use of a (possible small) subset of the original dataset provided that the error is kept below a given threshold. It is worth to notice that this method can be used adopting two mainly different approaches. The first one pertains the computation of an exact fit of the RBF using retained sources (RBF interpolation), whereas the second one has the target to minimise the overall error and, therefore, the exact fit is not guaranteed at original sites (see Sect. 2.5).

[1]"DSPTRF computes the factorization of a real symmetric matrix A stored in packed format using the Bunch-Kaufman diagonal pivoting method ($A = U \cdot D \cdot U^T$ or $A = L \cdot D \cdot L^T$) where U (or L) is a product of permutation and unit upper (lower) triangular matrices, and D is symmetric and block diagonal with 1-by-1 and 2-by-2 diagonal blocks. DSPTRS solves a system of linear equations A*X = B with a real symmetric matrix A stored in packed format using the factorization $A = U \cdot D \cdot U^T$ or $A = L \cdot D \cdot L^T$ computed by DSPTRF."

[2]"Suppose we are able to write the matrix A as a product of two matrices LU where L is lower triangular and U is upper triangular, we can solve the linear set $A \cdot x = (L \cdot U) \cdot x = L \cdot (U \cdot x) = b$ by first solving $L \cdot y = b$ and then solving $U \cdot x = y$. The advantage is that the solution of a triangular set of equations is quite trivial. $L \cdot y = b$ can be solved by forward-substitution and $U \cdot x = y$ by back-substitution".

The concept is quite straightforward. For a given RBF cloud, an initial subset of a minimum acceptable size is provided using a point decimation criterion that is usually based on the distance between points (see Sect. 4.3). This selection can be problem wise so that some centres may have a priority with respect to others. The first point of the set is retained, then the algorithm visits the next point and if it is separated from the first retained point by a given distance, it is added to retained points otherwise it is discarded. All the other points are visited and only the ones separated from all the already retained ones are kept. Obviously the results is affected by the original order of the points and a pre-processing operation in which the original data are sorted, can drive the method. For instance if the cloud of data comes from points laying on a given surface, a special sorting in which the first points are the ones on vertexes of the boundary, followed by the ones on the edges and then the ones internal to the surface, could be used. This way to process data will assure that the final subset will be respectful of original one and that geometry boundaries will be in any case represented.

Usually the greedy methods are iterative. A first guess small dataset is defined and an RBF fit computed; RBF interpolation is then used to compute the approximation on discarded points. If the error is higher than the prescribed threshold, new points are added using error information as a way to pick new points. The method can be iterated until the required tolerance is met.

Special solution strategies exist that allow to reuse previous system solution so that the extra computation required for adding new points is lower than a full re-computation of the overall system. Botsch and Kobbelt (2005) presented a study where a least squares regression based on an incremental QR solver allows to refine the cloud of retained points until the desired tolerance is met. In such a study a cubic full supported RBF is adopted as a tool to model tessellated surfaces by mesh deformation. A speed up in the range 3x up to 84x is observed with respect to a full LU solution based on LAPACK libraries. The best speed up is registered for the "Bunny" test case where the same effects (i.e. desired tolerance) is obtained adopting just 280 RBF centres out of the original 4913 ones (the full LU took 212 s).

Rendall and Allen (2009) demonstrated how greedy methods can be optimised for the management of mesh deformation required for fluid structure interaction (FSI); compact supported functions are adopted and a substantial speed up with respect to the full evaluation are registered (up to 12x). The time required for a cloud of 3881 points, that takes 77 s using the full method on a 2.2 GHz Opteron CPU, is reduced to about 6 s retaining 200 points.

The reduction of the dataset size can be beneficial not just as a tool for the reduction of the problem size; there are some situations in which many RBF centres are clustered but the function to be fitted is regular. In this case the sub-sampling helps in the regularization of the dataset which facilitates the RBF fitting process and automatically removes duplicates.

3.3.1 A Simple Mesh Morphing Problem

In this example the commercial software RBF Morph (Biancolini 2012) is used to demonstrate the effectiveness of greedy methods. The mesh for the test represents a 1 m size cube immersed in a box-shaped wind tunnel (10 m length, 5 m width, 3 m height); the cube is centred with respect to the width and the front face is located at 3 m from the inlet so that 6 m are left for the development of the wake. The volume mesh used for modelling the fluid includes 56,824 tetrahedrons connected by 11,696 nodes, 4165 of which are located at the wetted surfaces (inlet, outlet, tunnel, ground and cube).

The mesh morphing action consists of an uniform scaling of the sides of the cube performed with respect to the centre of its bottom face. All the nodes at surfaces are managed as follows: the cube is directly controlled using the desired scaling action, keeping fixed the inlet, the outlet and the tunnel while the ground is controlled by a deforming action computed by an auxiliary RBF acting on its edges (Fig. 3.1).

The initial morphing problem comprises all the 4165 points that belong to controlled surfaces. The cloud is first sub-sampled using a spacing of 0.25 m and reduced at 1120 points. It is then automatically refined using six extra iterations adding extra points at each iteration so that the final cloud includes 1279, namely about 30% of original cloud. A comparison between the original point distribution and the optimized one is shown in Fig. 3.2.

As it can be clearly observed from Fig. 3.3, the error is substantially reduced, from about 20 mm to about 2 mm, just adding 159 extra points.

As already explained, when the direct method is adopted the cost of the RBF scales as N^3. This means that the average cost of each iteration is less than 3% with respect of the fitting of the large cloud. Since a total of seven RBF fits (including the first fit) of the reduced clouds is performed, the total cost in terms of computing time should sum to be about the 18.5% of the time needed to accomplish the full fit.

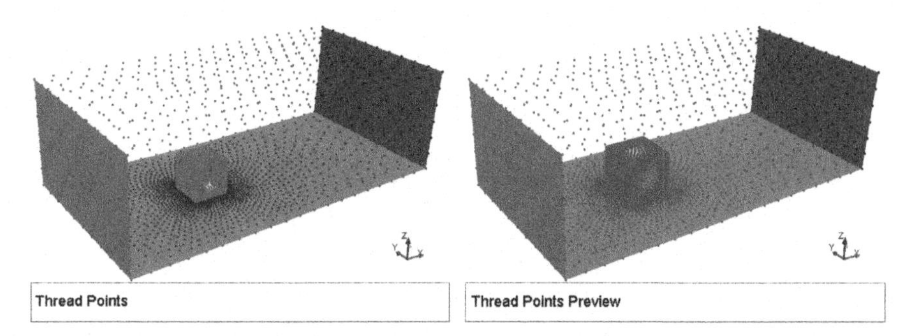

Fig. 3.1 The mesh used for the CFD analysis of a cube immersed in a wind tunnel is partially shown. Tunnel faces are turned off to see inside. Inlet, outlet, ground and cube faces are shown. Nodes extracted from the mesh are used as RBF centres using three different sets, each one with a specific input rule: cube points are scaled, tunnel points are fixed and ground points are deformed so that the scaling of cube base can be accommodated. Source points arrangement coloured by red is represented on the left, their preview in the deformed position is represented on the right

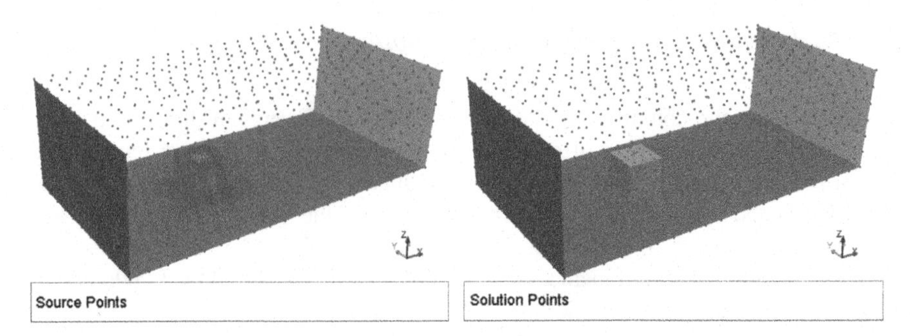

Fig. 3.2 The original cloud of source points (left) is compared with the optimised reduced one (right) used for the solution of the RBF

To check such as a rough estimate, some wall clock measurements are collected. All the tests are conducted using a standard laptop equipped with an Intel i7 processor Q720@1.6 GHz and 16 Gb RAM. In this case since an high order RBF is used (Wendland C2), RBF Morph uses LU partitioning for solution achievement; the time required for fitting the complete cloud is 559 s, whilst the one used for the refined greedy method requires 90 s (16.1%). It is interesting to observe that if the default RBF is selected (in this case RBF Morph uses a fast iterative solver of the family presented in Sect. 4.5) the time for fitting the full cloud is reduced to just 8 s. The effect of RBF mesh morphing is demonstrated in Fig. 3.4 by previewing the morphing on the boundary surfaces and on two internal cuts of the volume mesh.

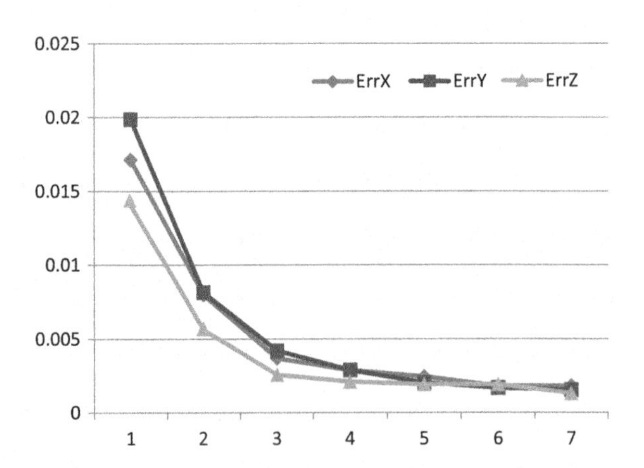

Fig. 3.3 This chart shows the convergence of the total error (meters) in the 3 directions in the refinement iterations. Considering the maximum displacement in the three directions the relative error results to be less than 2%

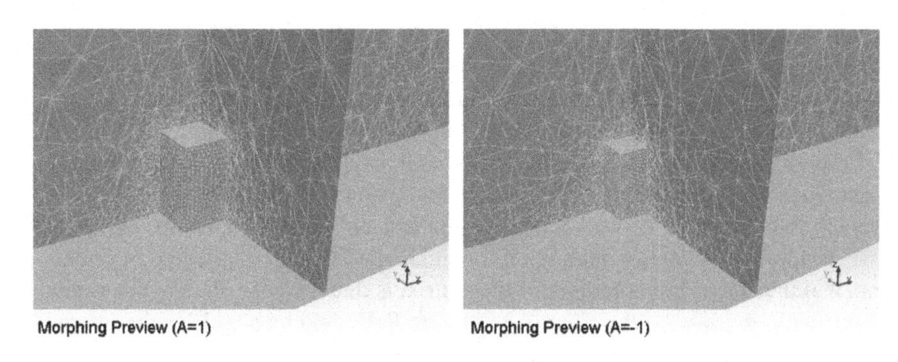

Fig. 3.4 The mesh morphing action is represented considering both a positive sign of the amplification (+20% scaling on the left) and both a negative sign (−20% scaling on the right)

3.4 Use of Compact Supported RBF

A well recognized solution to reduce the complexity of RBF problem consists in the use of compact supported RBF. A well-established set of compact supported RBF are the Wendland ones (see Table 2.1). The support limits the influence of the RBF and the radial function is defined so that it is zero valued on the support sphere and null outside.

Compact supported RBF are usually chosen for their local effect. A common application is to use them in mesh morphing problems in which a field is generated on a surface, and it must vanish on the far field. Using RBF the cloud of control points also defines an envelope domain which is affected by the RBF field and the far field is left unaffected. When global supported RBF are used, the far field behaviour requires a direct control by placing RBF centres with null movement that are not required when using CS RBF.

The morphing problem example of Sect. 3.3.1 has been used to explore this effect comparing the mesh quality achieved. The set-up with compact supported RBF envisions to adopt only the 476 nodes at cube surface as input points; the radius is varied in the range 1 m (the side of the cube) up to 2 m (maximum distance allowed to not touch the wind tunnel surfaces). Table 3.1 shows that the best quality is achieved adopting the maximum allowed radius and the C2 function.

Table 3.1 effect of compact supported RBF on the quality of volume mesh deformation. The maximum skewness before morphing is 0.746, the skewness achieved after positive scaling (20%) of cube size is 0.858 using full supported linear RBF. The skewness is explored for various values of the support radius and for C0, C2 and C4 RBF

CS radius (m)	C0	C2	C4
2.0	0.895	0.871	0.897
1.5	0.936	0.927	0.940
1.0	0.978	0.978	0.991

In order to compare the quality of CS RBF with the full supported one the morphing is repeated using the linear RBF ($\varphi(r) = r$); in this case the nodes on the wind tunnel walls are added as a fixed boundary getting a slight better mesh quality (0.858 versus 0.871) while the RBF problem size is doubled (1009 points).

From a numerical efficiency perspective, it happens that the local behaviour leads to sparse system instead of a fully populated one. The sparsity depends on the average inter distance between cloud points and on the chosen radius. When the RBF is applied on dataset defined on a regular spaced grid, the degree of interaction, and thus the degree of filling of the matrix, is directly controlled by the radius. An example is given in the Chap. 12.7 where RBF are employed for the interpolation of hemodynamics flow patterns. Sparse system can be faced using specific iterative solvers reducing substantially the complexity of the problem. An example of numerical performances optimisation using a sparse solver has been presented by Kojekine et al. (2002) for the implementation of real time deformation using CS RBF.

3.5 Partition of Unity Methods (POU)

The localization of RBF interaction can be also achieved using partition of unity (POU) methods (Babuška and Melenk 1997; Wendland 2002). The localization in this case is imposed without acting on the kernel of the RBF (and so leaving to the engineer all possible options), but instead by a decomposition of the full RBF problem in many smaller ones. Smaller problems are extracted from the original cloud using a set of overlapping sub domains. This means that many of the points are repeated in more than one subset. Each local problem is fitted using a specific RBF; as the local RBF come from the global one, in the overlapping areas the approximation is similar. The global continuity when crossing overlapping areas is guaranteed by blending functions that weights the contribution of simultaneously acting RBF using a distance criterion.

There are many options to define a POU. For instance, the original domain could be wrapped into a box that can be subdivided in a regular spaced array of sub boxes; if the cloud is unevenly distributed and there are areas in which RBF centres are more clustered, a variation can be given by a decomposition using Octal Trees (Tobor et al. 2006) or Binary Trees (Bentley 1975). A distribution of overlapping spheres can be used as well. In this case sphere centres can be located at points of the original cloud; the original cloud size is first reduced using the algorithm described in Sect. 4.3, and then a sphere with a proper radius (1.5–2.5 times the spacing) is built for each of the retained points.

Regardless the POU approach chosen, the advantage in numerical complexity is great because the original problem is replaced by many smaller ones. If a generic RBF is adopted and a direct solver exploited, the cost of RBF fitting grows with a cubic law. Regardless the extra work due to the fact that because of overlapping there are repeated points, it's cheaper to fit many small problems rather than a single large one.

A quick and rough estimation can be given considering a cloud of point that is almost uniformly distributed in a rectangular box. Let us subdivide the original box in a regular grid of small boxes (a total of N_{boxes}) with the same proportion of the original one but sides enlarged of a factor λ_{box} so that each box in the grid overlaps with all its 6 neighbours. The average content of points in each box can be estimated as

$$N_{pbox} = \frac{N_p}{N_{boxes}} \lambda_{box} \tag{3.1}$$

where N_p is the total number of points to be fitted.

Given that, the numerical cost for fitting the RBF with the direct method is:

$$cost_{direct}(\mathrm{N}) = \lambda_{direct} N^3 \tag{3.2}$$

We can easily understand how:

$$\lambda_{direct} N_p^3 \gg \lambda_{direct} N_{boxes} \left(\frac{N_p}{N_{boxes}} \lambda_{box} \right)^3 \tag{3.3}$$

if the side of the small boxes is selected to limit N_{pbox} to a certain given target value we have:

$$\lambda_{direct} N_p^3 \gg \lambda_{direct} N_p \frac{\lambda_{box}}{N_{pbox}} N_{pbox}^3 \tag{3.3bis}$$

This means that, regardless the size of the cloud, POU performances scales in a linear fashion with respect to the number of points.

The key working principle of POU consists in putting together the local RBF solved in the overlapping domains through blending functions so that the continuity of the complete field can be recovered. The smoothness of the global solution at x can be guaranteed combining the local RBF of all the i domains that contain x. It is worth to observe that the POU subdivision has to be selected so that the while evaluation space required is covered by the overlapping domains; this means that the overlapping amount should be addressed with in mind the requirements of some extrapolation outside the original cloud locations and the absence of internal holes when the overlapping is achieved by spherical or ellipsoidal shapes. POU evaluation results as a combination of local RBF:

$$f_{POU}(\boldsymbol{x}) = \sum_i w_i(\boldsymbol{x}) f_{RBF_i}(\boldsymbol{x}) \tag{3.4}$$

The $w_i(x)$ is defined to get a smooth transition across overlapping domains and to have, in any case, the proper intensity. The latter aspect is simply achieved adopting a normalization procedure:

$$w_i(x) = \frac{W_i(x)}{\sum_j W_j(x)} \tag{3.5}$$

which guarantees that the condition $\sum w_i = 1$ is satisfied (at the same location the contributions of overlapping domains are close each other as they approximate the same function).

The weighting functions W_i can be defined as the composition of a distance function d_i and a decay function V_i. The distance function has to satisfy the condition $d_i(x) = 1$ at the boundaries of a subdomain. The decay function is defined through the distance function, and its degree can defined by the user. Examples of decay functions with growing degree of continuity (C^0, C^1, C^2) are:

$$\begin{aligned} V^0(d) &= 1 - d \\ V^1(d) &= 2d^3 - 3d^2 + 1 \\ V^2(d) &= -6d^5 + 15d^4 - 10d^3 + 1 \end{aligned} \tag{3.6}$$

Given that, in the overlapping regions the solution is an averaged weighted function of the local solutions of the overlapping subdomains.

The shape of a subdomain can be arbitrarily chosen. For a spherical subdomain, the distance function assumes the form:

$$d(x) = \frac{r(x)}{R} \tag{3.7}$$

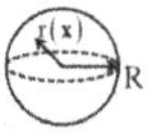

where $r(x)$ is the distance from the centre and R is the radius of the sphere.

For boxes (parallelepipeds) the distance function assumes the form:

$$d(x) = 1 - \prod_{r \in (x,y,z)} \frac{4(x_r - S_r)(T_r - x_r)}{(T_r - S_r)^2} \tag{3.8}$$

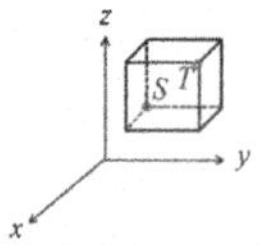

where the points S and T are the corners of the box. The weight function at a certain point results to be:

$$W(x) = V(d(x)) \tag{3.9}$$

Numerical implementations usually perform the POU evaluation at a certain point accumulating in separate storage variables the numerator and the denominator of:

$$f_{POU}(x) = \sum_i w_i(x) f_{RBF_i}(x) = \sum_i \frac{W_i(x) f_{RBF_i}(x)}{\sum_j W_j(x)} = \frac{\sum_i W_i(x) f_{RBF_i}(x)}{\sum_i W_i(x)} \tag{3.10}$$

Engineering applications usually require the management of cloud of points that are non-uniformly distributed. In this case POU methods can be adapted to better fit the problem whilst giving high performances. Tobor et al. (2006) presented a method based on POU for the reconstruction of implicit surfaces (see Sect. 5.2 to learn more about implicit surface with RBF). The original cloud is hierarchically subdivided using a balanced binary tree with a partial overlapping zone between each pair of three dimensional boxes. The tree structure is here used to limit the RBF evaluation at leaf level, the value at a given point is then computed.

3.5.1 Example of POU for Surface Projection

Performances of the POU solver are demonstrated in this section using RBF Morph Fluent Add On. The RBF Morph software has a surface projection tool that is based on the implicit surface method described in Sect. 5.3. Projection algorithm requires the solution of an RBF problem comprises on-surface points (could be the vertices or the centroids of a triangulated surface mesh) and a pair of off-surface points for each surface point used. The related RBF problem size can grow very fast as the detail and the complexity of the surface is increased. The POU approach is herein adopted as an effective way to boost performances and a distribution of overlapping spheres is adopted for this purpose. The original cloud of on-surface points is processed with a distance based sub sampling method similar to the one described in Sect. 4.3. A collection of overlapping spheres centred at retained points and with a radius that is 1.5 times of the one used for data decimation is adopted to work the projection across the whole surface. In the presented example the projection onto a new position of the surface of a motorbike driver is addressed; new surface consists of a small rotation of the driver to change the hunching angle. The target surface size is changed taking the full surface, an half and a quarter as exposed in Fig. 3.5, in fact surface spacing is not uniform and a clip of the full one is used so that the

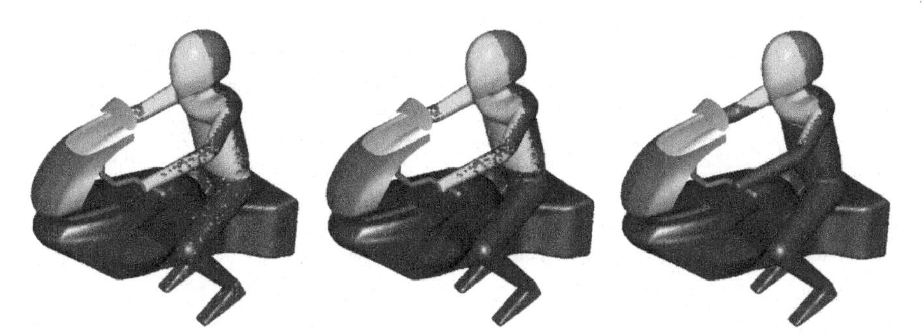

Fig. 3.5 Example of surface projection; the new position of the driver is given as an STL file (in yellow). The full surface, an half and a quarter surface are on the left, centre and right

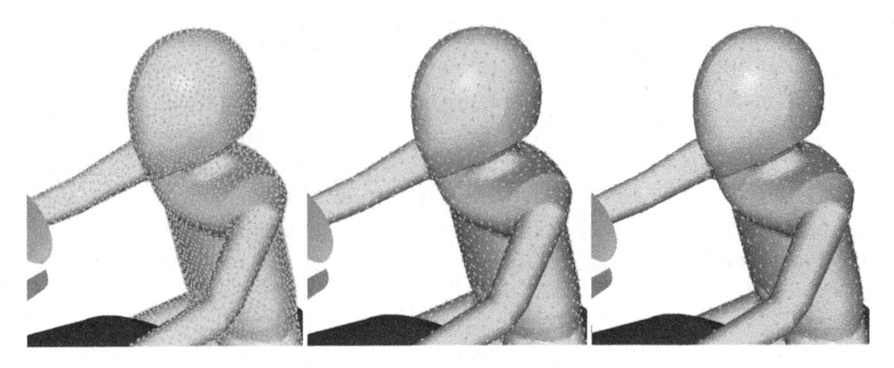

Fig. 3.6 Details of POU sphere centres distribution for three values of the spacing (0.01, 0.02 and 0.03 from left to right). The POU spheres contain a variable number of points; min/max number of points contained are for the three spacing 9/97, 48/347, 93/820

size change of the RBF to be fitted is representative of the actual geometrical complexity.

Three values of the spacing for POU definition are considered: 0.01, 0.02 and 0.03 (units are meters) as depicted in Fig. 3.6. The spacing should be as low as possible to have a fast solution but large enough to guarantee enough points in each sphere (so that surface curvature is captured) and to assure that the projection field influence domain is far enough from the surface.

The numerical test conducted on a laptop i7-4750HQ 2.00 GHz demonstrates the effectiveness of the POU approach. Results are collected in Table 3.2 and show how with POU the computational cost grows in a linear fashion with the number of centres and that the reduction of the number of points in each sphere allows to reduce the computation time. The use of the direct solver, embedded in this version of RBF Morph, is accelerated by a fast iterative solver of the family described in Sect. 3.7 and leads to a cost almost proportional to the square of the number of RBF centres making almost unaffordable the fit. Use of POU allows to reduce hours in minutes.

Table 3.2 Complexity test using RBF Morph 1.7 running on a laptop i7-4750HQ 2.00 GHz

Surface target	#RBF centres	0.01 (s)	0.02 (s)	0.03 (s)	Full (s)
Quarter	17.679	2	7	21	163
Half	32.307	5	14	37	634
Full	60.678	15	28	80	2655

The number of RBF centres is provided in the second column the time in seconds required to fit the RBF system with POU and with full method is reported in columns 3, 4, 5 and 6

3.6 Fast Multipole Methods (FMM)

FMM can be effectively used to accelerate the evaluation of various RBF (Beatson and Greengard 1997). The method has been introduced by Greengard and Rokhlin (1987) and has been ranked one of the top ten algorithms of the 20th century (Cipra 2010). It has been developed first for astrophysics studies as a tool to manage n-bodies interaction; a fast implementation on GPU is available for this specific problem (Yokota and Barba 2011). It can be also implemented for RBF and is suitable for multi-interactions between RBF centres in which the support is global and, thus, all the centres give a contribution in a generic probe point.

The method is based on the concepts of the far and near field. The evaluation of the RBF on a single point has a cost proportional to N, where N is the number of RBF centres of the cloud; each inter distance interaction requires the evaluation of the Euclidean distance and its processing through the radial function. Even if all the points contribute to the summation:

$$s(\boldsymbol{x}) = \sum_{i=1}^{N} \gamma_i \varphi(\|\boldsymbol{x} - \boldsymbol{x}_{s_i}\|) \quad (3.11)$$

An approximate evaluation can be given using the full evaluation of just a few points close to the probe (near field) whilst the effect of the remaining ones (far field) can be substituted using approximate functions:

$$s(\boldsymbol{x}) = \sum_{i=1}^{N_{near}} \gamma_i \varphi(\|\boldsymbol{x} - \boldsymbol{x}_{s_i}\|) + s_{far}(\boldsymbol{x}) \quad (3.12)$$

Replacing the full evaluation with an approximated far field introduces an error (that needs to be evaluated and controlled) and allows to drastically reduce the cost of the evaluation of the RBF. To better understand how the $s_{far}(\boldsymbol{x})$ function is defined, the multipole expansion concept is introduced. It consists of the substitution of a cluster of points with a simpler and less costly function (that can be obtained using several strategies); the truncation error with respect to the exact evaluation decreases with the distance between the probe and the centre of gravity of the cluster itself. The expansion accuracy can be controlled acting on the truncation error. The far field effect is achieved replacing all the points outside the near field with their approximated multipole expansions (hierarchical method). The complete FMM method consists in the definition of an inner field, valid in the domain where the near field is computed, which is built summing up in advance all the contribution of far field expansions.

To fully exploit the multipole expansion, a proper space framework is required; if the RBF centres are distributed with a not too much varying space density, a regular array of boxes can be used. Otherwise, a common choice is to use binary or octal trees; the domain is iteratively subdivided until the number of points is higher than a certain threshold. Expansions are first evaluated for all the leaves using a direct evaluation; then the expansion of other nodes are computed using direct evaluation or the combination of the expansion already computed for the children (exploiting the concept of translation of the expansion).

Hierarchical data structure is used for the definition of the near and far field at a given point. Near field is obtained exploring the neighbours of the leaf that contain the evaluation points, whilst the far field effect is obtained using the expansions at a depth as lower as possible in the tree. In this case It is worth that the concept of "well separated" domain is applied for octal trees, while an expansion accuracy level (i.e. degree of polynomial included in the expansion) is used for binary balanced tree; in the first case the maximum poly order is defined in advance on the basis of geometrical data because the far field is built using only well separated domains (the octal decomposition is regularly spaced); in the second case the binary decomposition is balanced (it means that the size of sub-boxes is adapted to match the number of points) and so a dynamic control of the degree of polynomial has to be managed on the basis of the effective distance of the domain.

Regardless the implementation, it is important to observe that a full self RBF evaluation, typically adopted in iterative algorithms for RBF data fitting, has a cost of N^2 using direct evaluation; this cost is reduced to $N \log N$ if a hierarchical multipole approach is adopted. A further reduction to N can be achieved if the full multipole method is implemented. In the full approach the far field is not evaluated each time, but just once at set-up for each leaf; this is achieved combining all the far field expansions in a single one which is valid inside the leaf. It is worth to notice that the performances that can be achieved switching between direct evaluation, hierarchical multipole and full multipole depends on the problem size and on the computational resources. RBF summation can be easily accelerated if parallel calculation is available. A recent study of Yokota and Barba (2011) demonstrated that using a single GPU the direct method results to be the fastest up to 40.000 RBF centres, and that beyond this threshold the hierarchical method is fastest up to 1 million of RBF centres. Given that, in such a scenario the full FMM is not useful at all. The same benchmark conducted on CPU (small parallelism) demonstrates that the full FMM is faster than the direct method only beyond the 3000 RBF centres, and that the full FMM is faster than the hierarchical method over the full explored range.

The FMM framework exists for different class of RBF. The full theoretical foundation for setting up a 3D FMM solver for poly-harmonic splines is given by Beatson et al. (2007) and is herein reported in Sect. 4.6.1 for the case of the bi-harmonic kernel adopting spherical functions. This is of specific interest as the poly-harmonic splines in 3D guarantee an high smoothness that make them attractive for surface reconstruction (Carr et al. 2001, 2003) and for mesh morphing (Sieger et al. 2014). An expansion based on Taylor expansion for multi-quadrics is presented by Krasny and Wang (2011). A kernel independent FMM method is presented by Lexing (2006); the key concept is that the effect of an arbitrary distribution of points can be obtained by a regular arrangement of sources (for instance on the border of a box) providing that the equivalent distribution match the original one at given remote location (for instance a box larger than the one used for the definition of the equivalent cloud). The concept is demonstrated in Sect. 4.6.2 adopting spheres as equivalent and matching surfaces.

3.6.1 Multipole Expansion of the Bi-Harmonic Kernel

The basic idea of the multipole expansion is that the far effect of a cluster of points can be approximated by means of an analytical expression that is independent by the number of points, although the coefficients are calculated accounting for all the contributions. This concept is quite intuitive and can be easily understood considering the bi-harmonic kernel $\varphi(r) = r$ for which an expansion based on spherical harmonics exists. In a very far point of evaluation distant R from the centre of gravity of the cluster the contributions of each points can be approximated considering r = R and so multiplying R for the sum of the weights. This crude approximation is valid only if the evaluation point is very far from the cluster. For closer locations a better approximation can be expressed adopting analytic expansion, which is presented in this section, or by means of the field produced by a fictitious set of source points (indeed a reduced set) able to replace the original cluster using the same kernel (Sect. 4.6.2).

The multipole expansion for the bi-harmonic kernel is well known (Beatson et al. 2007). Only the essential formulas are herein reported.

The basic concept is to approximate the effect of the cluster composed by N points located in the $\boldsymbol{x}_{si}$ positions (with centre of gravity located in $\mathbf{c}$) and γ_i intensities:

$$s(\boldsymbol{x}) = \sum_{i=1}^{N} \gamma_i \|\boldsymbol{x} - \boldsymbol{x}_{si}\| \tag{3.13}$$

with the following expansion:

$$s_{far}(\boldsymbol{x}) = \|\boldsymbol{x} - \mathbf{c}\|^2 \left\{ \sum_{n=0}^{p} \sum_{m=-n}^{n} N_{nm} O_n^m(\boldsymbol{x} - \boldsymbol{c}) \right\} + \left\{ \sum_{n=0}^{p-2} \sum_{m=-n}^{n} M_{nm} O_n^m(\boldsymbol{x} - \boldsymbol{c}) \right\} \tag{3.14}$$

where p is the degree of the expansion, the matrixes M_{nm} and N_{nm} are calculated by the intensities according to the following summations:

$$N_{nm} = -\frac{(-1)^n}{2n-1} \sum_{i=1}^{N} \gamma_i I_n^{-m}(\boldsymbol{x}_{s_i} - \boldsymbol{c}) \tag{3.15}$$

$$M_{nm} = -\frac{(-1)^n}{2n+3}\sum_{i=1}^{N}\gamma_i\|\boldsymbol{x}_{s_i} - \boldsymbol{c}\|^2 I_n^{-m}(\boldsymbol{x}_{s_i} - \boldsymbol{c}) \tag{3.16}$$

where $O_n^m(\boldsymbol{x})$ and $I_n^m(\boldsymbol{x})$ are respectively the Outer and Inner expansions:

$$O_n^m(\boldsymbol{x}) = E_n^m r^{-n-1} P_n^m(\cos\theta)e^{im\varphi} \tag{3.17}$$

$$I_n^m(\boldsymbol{x}) = F_n^m r^n P_n^m(\cos\theta)e^{im\varphi} \tag{3.18}$$

that are point functions expressed in spherical coordinates that depend on the constants E_n^m and F_n^m:

$$E_n^m = (-i)^m(n-m)! \tag{3.19}$$

$$F_n^m = \frac{(-1)^n i^m}{(n+m)!} \tag{3.20}$$

and $P_n^m(u)$ is the associated Legendre function:

$$P_n^m(u) = \frac{(n+m)!}{2^n n!(n-n)!}\left(1-u^2\right)^{-\frac{m}{2}}\frac{d^{n-m}}{du^{n-m}}\left(u^2-1\right)^n, -1 \le \mathrm{u} \le 1 \tag{3.21}$$

Such functions can be evaluated in a closed form or numerically adopting a recursive formula; the same precision and computational effort has been observed. An expansion can be easily computed using Mathcad symbolic processing and is here reported for convenience. The data can be retrieved using the following formula:

$$\mathrm{P_{Leg}(n,m,u)} := \begin{cases} \left(\mathrm{Poly_{Leg}(u)_{n+1}}\right)_{\mathrm{m+n+1}} & \text{if } \mathrm{n} \ge 0 \wedge -\mathrm{n} \le \mathrm{m} \le \mathrm{n} \\ 0 & \text{otherwise} \end{cases} \quad \mathrm{Poly_{Leg}(u)} := \begin{pmatrix} \mathrm{P0(u)} \\ \mathrm{P1(u)} \\ \mathrm{P2(u)} \\ \mathrm{P3(u)} \\ \mathrm{P4(u)} \\ \mathrm{P5(u)} \end{pmatrix} \tag{3.22}$$

where the terms up to the third degree are computed according to the following polynomials:

$$\mathrm{P0(u)} := (1)\ \mathrm{Pl(u)} := \begin{pmatrix} \frac{\sqrt{1-u^2}}{2} \\ u \\ -\sqrt{1-u^2} \end{pmatrix} \mathrm{P2(u)} := \begin{pmatrix} \frac{1}{8}-\frac{u^2}{8} \\ \frac{u\cdot\sqrt{1-u^2}}{2} \\ \frac{3\cdot u^2}{2}-\frac{1}{2} \\ -3\cdot u\cdot\sqrt{1-u^2} \\ 3-3\cdot u^2 \end{pmatrix}$$

$$\mathrm{P3(u)} := \begin{pmatrix} \frac{(1-u^2)^{\frac{3}{2}}}{48} \\ -720\cdot u\cdot\left(\frac{u^2}{5760}-\frac{1}{5760}\right) \\ \frac{(360\cdot u^2-72)\cdot\sqrt{1-u^2}}{576} \\ u^2+\frac{3\cdot u\cdot(u^2-1)}{2} \\ \frac{24\cdot u^2\cdot(u^2-1)+6\cdot(u^2-1)^2}{4\cdot\sqrt{1-u^2}} \\ -15\cdot u.(u^2-1) \\ -15\cdot(1-u^2)^{\frac{3}{2}} \end{pmatrix} \tag{3.23}$$

The most expensive task is the evaluation of the matrixes M_{nm} and N_{nm} and a trade-off has been observed: for each value of p there is a minimum number of points below which the direct evaluation becomes cheaper than the expansion. A specific tuning needs to be performed as the break even point depends on the degree of optimisation of the implementation; typical figures are 70 centres for third degree and 150 for fifth degree.

Multipole expansion is a very powerful tool because it allows to replace the complete calculation with its approximated expansion. However, the truncation error depends on the degree of the expansion and on the distance of evaluation. A calculation algorithm that relies on this kind of expansion has to be controlled for truncation error to be usable.

The error can be estimated to be lower than:

$$\mathrm{err(r,R,p,sum)} := \mathrm{sum}\cdot\frac{2\cdot r}{2\cdot p+1}\cdot\frac{1}{1-\frac{R}{r}}\cdot\left(\frac{R}{r}\right)^{p+1} \tag{3.24}$$

where sum is the sum of the weights of all points in the cluster, p is the degree of the expansion, r is the distance of the evaluation point from the centre of the cluster, R is the maximum distance calculated for all the points of the cluster with respect to the cluster centre itself. Taking R as the maximum leads to a safe (but expensive) error bounding. A typical plot of the error estimation is represented in Fig. 3.7 where R = 1, sum = 1 and p change from 4 to 8.

Error bounding can be used conveniently to choose the degree of expansions for a prescribed error. Consider for instance the following equation:

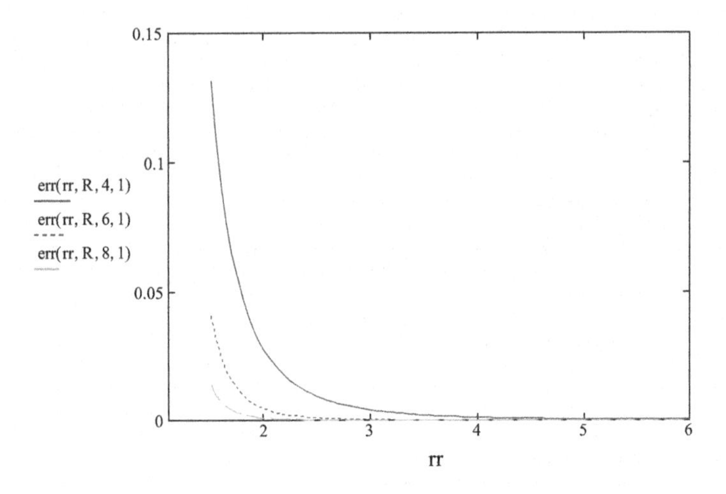

Fig. 3.7 Expansion error as a function of the distance for increased value of the degree (4, 6, 8)

$$\text{dist\_min}(\text{p}, \text{sum}) := \text{root}(\text{err}(\text{r}, \text{R}, \text{p}, \text{sum}) - 0.01, \text{r}, \text{R} \cdot 1.1, 20 \cdot \text{R}) \qquad (3.25)$$

Here we are looking for the distance that produces a 0.01 error for a given R (R=1 in the example) changing r between $R \cdot 1.1$ and $R \cdot 20$, given sum. The plot of Fig. 3.8 shows how such a minimum distance depends upon the expansion degree for three values of the intensities sum.

It is clear that the absolute error is proportional to the sum of the contributions and that the greater is p the better is the accuracy of the approximation at points

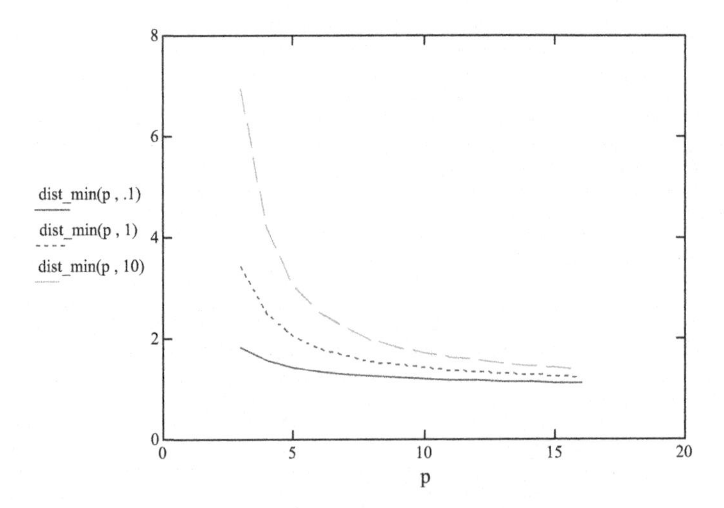

Fig. 3.8 Minimum distance for a prescribed error (0.01) as a function of the expansion degree for various values of the sum of intensities (0.1, 1, 10)

closer to the cluster itself. The presented plot can be used as the control mechanism for selecting and accepting an expansion optimising on a point by point basis the degree (and the cost) of the expansion to be used. A heuristic selection for R is usually adopted by replacing the safe R_{max} value with $\lambda \cdot R_{mean}$ where R_{mean} is the mean distance from the centre and λ is in the range 1.15–1.2.

3.6.2 *Approximated Expansion Using an RBF Function*

The concept of a kernel independent FMM method presented by Lexing (2006) opens to a great flexibility in the choice of the full supported RBF function that can be used. The remote effect is in this case gained by replacing the original cloud with a regular (and coarser) arrangement of sources defining the coefficient so that the two clouds have the same far field.

An original adaption of the approach, where boxes are replaced by spheres, is here presented. The function for generating a uniform distribution of points on the surface of sphere of Sect. 4.6.3 is adopted. Two auxiliary RBF points distribution are defined: source points at a radius suitable to wrap the cluster for which the far field approximation is required, and matching points at a higher radius where the far field approximation validity begins. The coefficients of the RBF sources are computed so that the far field approximation match the original RBF at the outer sphere, furthermore an extra equation is added to balance the summation of the weights of both RBF (the original and the far field one). No polynomial correction is added. The system matrix is defined according to Eq. (3.26) where the interpolation matrix entries are defined according to Sect. 2.5 where $\mathbf{P_{eq}}$ are the points on the inner sphere where the source points of the far field RBF are located and $\mathbf{P_{check}}$ are the points on the outer sphere where the matching between the original RBF and the far field one is imposed. Last equation (with coefficient 1) serves to apply the constraint of far summation of the coefficients.

$$
M_{eq} := \left|\begin{array}{l}
\text{for } i \in 1..\text{rows}(P_{eq}) - 1 \\
\quad \text{for } j \in 1..\text{rows}(P_{eq}) \\
\quad\quad MEQ_{i,j} \leftarrow \phi\left(\left|P_{check_i} - P_{eq_j}\right|\right) \\
\text{for } j \in 1..\text{rows}(P_{eq}) \\
\quad MEQ_{\text{rows}(P_{eq}),j} \leftarrow 1
\end{array}\right. \tag{3.26}
$$

The system known vector of (3.27) is built by imposing the far field RBF and the original one to have the same value at match points (here we assume that the weights γ_1 of the original RBF are known in advance), the summation of RBF original weights is added in the last row.

$$u_B := \left| \begin{array}{l} \text{for } i \in 1..\text{rows}(P_{check}) \\ \quad UB_i \leftarrow \sum_{j=1}^{\text{rows}(\gamma_1)} \left[(\gamma_1)_j \cdot \phi\left(\left| \text{Points}_j - P_{check_i} \right| \right) \right] \\ UB_{\text{rows}(P_{eq})} \leftarrow \sum_{i=1}^{\text{rows}(\gamma_1)} (\gamma_1)_i \end{array} \right. \tag{3.27}$$

The linear system is solved as usual (3.28)

$$\varphi_{Bu} := \text{lsolve}(M_{eq}, u_B) \tag{3.28}$$

and both RBF (3.29), the original $\mathbf{sol_{RB}}$ and the far field $\mathbf{sol_{far}}$, can then be evaluated everywhere in the space.

$$\begin{aligned} sol_{RB}(x) &:= \left[\sum_{i=1}^{\text{rows(Points)}} \left[(\gamma_1)_i \cdot \phi(|x - \text{Points}_i|) \right] \right] \\ sol_{far}(x) &:= \left[\sum_{i=1}^{\text{rows}(P_{eq})} \left(\varphi_{Bu_i} \phi\left(\left| x - P_{eq_i} \right| \right) \right) \right] \end{aligned} \tag{3.29}$$

The concept is demonstrated with a simple benchmark. A distribution of random points is defined in the unit side cube, the cube is wrapped with an inner sphere at the radius that is 50% larger than the cube semi-diagonal, an outer sphere, with the same number of points, is defined at a radius that is two-times of the inner sphere and is used to match the far field. The geometrical arrangement of RBF centres is demonstrated in Fig. 3.9 where a fine and a coarse distribution of auxiliary points is represented for the auxiliary inner and outer sphere together with the points of the original cluster defined in the unit side cube.

The coarse spacing is enough to get a good far field approximation and a finer spacing is not required and has been observed to not improve substantially the far field behaviour, with the exception of the region close to outer sphere. The accuracy of the far field function has been here evaluated comparing the original RBF and the far field approximation on the point of a far sphere with a fine spacing (1320 points). The behaviour of the function on the evaluation sphere is demonstrated in Fig. 3.10 for a sphere at a radius 4 times the inner sphere one. The function behaviour is properly captured and, as can be noticed inspecting Fig. 3.11, the relative error is lower than 0.5%. The peak of the absolute value of the relative error has then be extracted changing the radius of the evaluation sphere. A plot in log scale of the maximum relative error is represented in Fig. 3.12 that shows how the

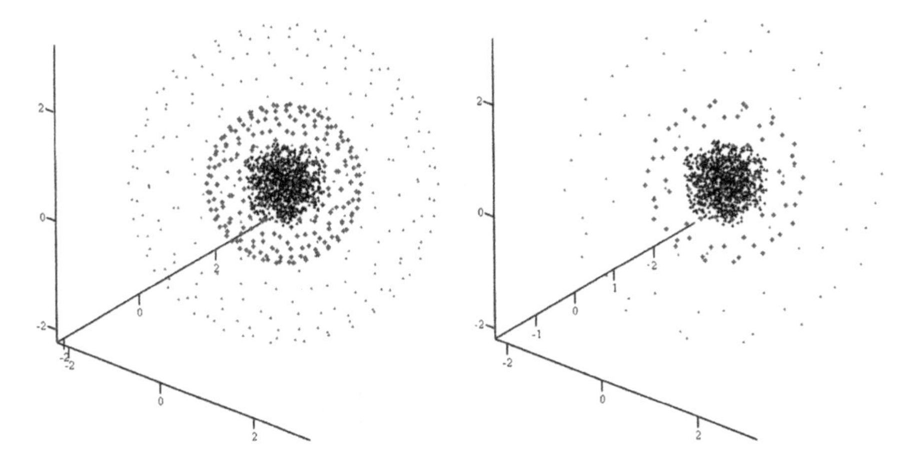

Fig. 3.9 Arrangement of source points for different resolutions of the auxiliary spheres (234 points on the left 70 points on the right). The original RBF is defined as a random function (valued in [0,1]) on a random distribution of 1000 points in the unit side cube. Far field RBF is defined on red points. Equivalence between the original and the far field RBF is imposed at blue points

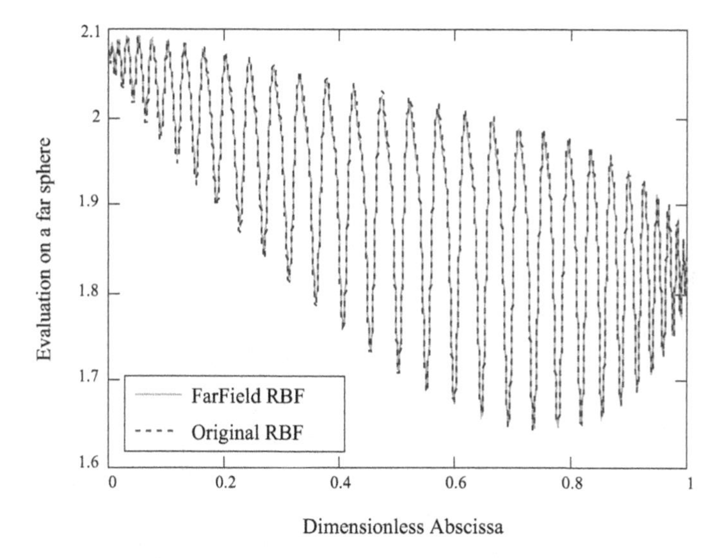

Fig. 3.10 The original RBF and the far field approximation are compared onto a sphere at a dimensionless radius 4 with respect to the radius of the inner sphere. The dimensionless abscissa represents a trajectory onto the surface of the sphere that is visited parallel by parallel

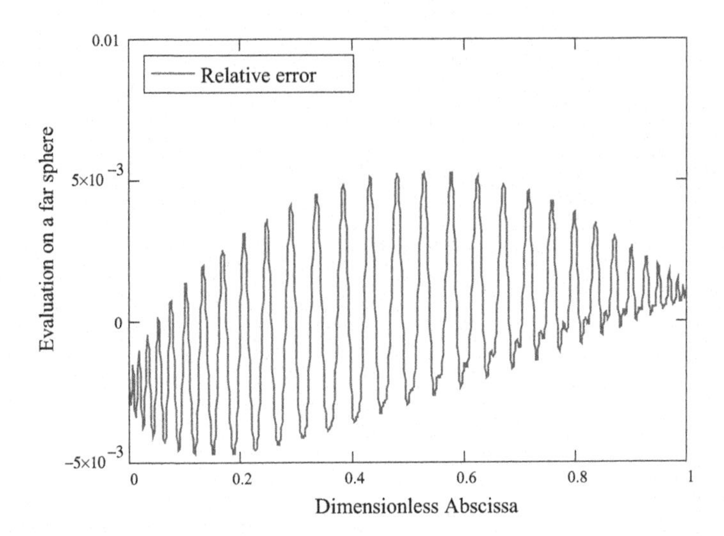

Fig. 3.11 The relative error of the far field approximation, computed with respect to the original RBF as evaluated onto a sphere at a dimensionless radius 4 with respect to the radius of the inner sphere. The dimensionless abscissa represents a trajectory onto the surface of the sphere that is visited parallel by parallel

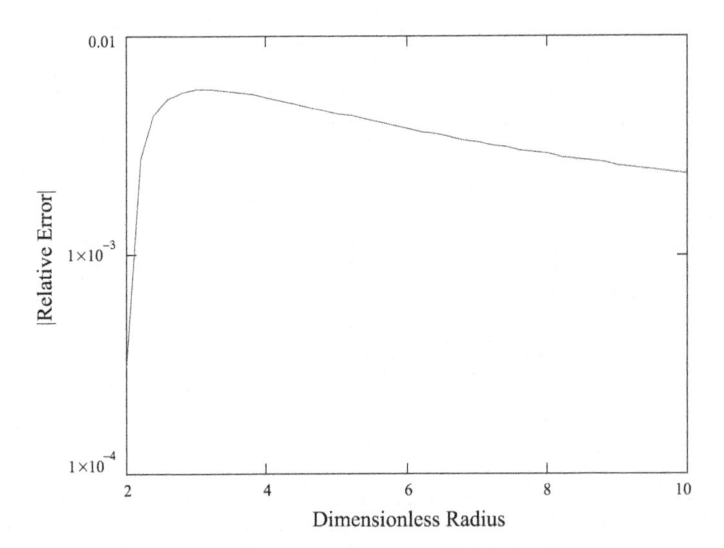

Fig. 3.12 The maximum of the absolute value of the relative error of the far field approximation, computed with respect to the original RBF as evaluated onto a sphere at a dimensionless radius between 2 and 10

minimum error (0.03%), as expected, is at the outer sphere where the matching is imposed, the maximum error (0.6%) is at a dimensionless radius in the range 3-3.2, then the error decreases at about 0.2% at 10.

3.6.3 *Scaling Test Example Using Hierarchical FMM*

An FMM evaluator based on the theoretical framework of Sect. 3.6.1 has been implemented in FORTRAN 90 using Octree and is currently part of the RBF Morpher Tool of the EU project RBF4AERO (European Commision 2013). The full RBF fit requires multiple self-evaluations of the cloud because the solution is performed by an iterative solver (see Sect. 3.7). A numerical benchmark consisting of the replication (with space offset) of the points distribution of the problem represented in Sect. 3.3.1 is adopted: the original cloud, that contains 2063 points (inlet, outlet and cube) is copied along z (adding a clearance to avoid duplicates) in a range from 26 up to 251 times, z values are then compressed so that the overall volume of the cloud is preserved.

The complexity test reported in Table 3.3 shows clearly how the FMM method allows to substantially reduce the cost of the evaluation (FMM iteration) that grows as $N^{1.2}$; because of the set-up stage, the overall complexity of the solution (Total time) results to grow as $N^{1.47}$, a very good result if compared with the overall time (Full) reported in Table 3.2 which results to grow as $N^{2.25}$. Even if the CPU are not the same (laptop adopted for Table 3.2 test is faster than the one used for Table 3.3), it is worth to notice how FMM allows to manage a 300,000 points problem in a time comparable to solve a 60,000 points one with the direct method.

Table 3.3 Complexity test using FMM solver

#RBF centres	Total time (s)	Time set-up (s)	FMM iterations (s)	#Iteration	FMM iteration (s)
53.638	201	171	30	6	5.0
105.213	495	423	72	5	14.4
156.788	887	758	129	6	21.5
208.363	1416	1086	330	11	30.0
259.938	1830	1401	429	11	39.0
311.513	2482	1843	639	13	49.2

The "Test Cube" benchmark is run for a number of centres in the range 50 k-300 k on a dual core processor. The total time is divided in the set-up time (serial operation) and iteration time (2 cores). The number of iterations required to reach the prescribed tolerance and the time for a single FMM iteration are provided

3.7 Iterative Solvers

The direct fit of the RBF using a linear solver gives the maximum flexibility but it is not the most effective approach for performances. As stated in the introduction of this chapter, there are mainly two reasons that limit the use of the direct approach. The first one is related to memory usage; if the radial function is global supported the system matrix is dense and a lot of memory is required to store it as the size grows as N^2 (8 Mb are required for storing an order 1000 square matrix in double precision, 0.8 Gb for 10,000 and 80 Gb for 100,000). This means that it is not possible at all to assemble the complete system. The second reason is that the direct method complexity grows as N^3 and becomes too slow for large dataset. The problem can be substantially alleviated when an iterative solver is adopted in which the cost of a single iteration (residual computation) consists of a self-evaluation of the RBF at all the centres (whose cost is N^2). If the algorithm is capable to achieve an acceptable error in just a few iterations (typically 20–50 depending on the requested tolerance) the complexity is reduced by an order of magnitude (parabolic versus cubic), and the memory limit problem is solved as well because the required interpolation matrix entries are not stored but computed on the fly. An optional partial pre computation of part of the matrix is usually used exploiting the available memory and saving the time required for re computing the missing entries. If parallelism is available (multi core CPUs and/or GPUs) performance are further increased as the cost of a single RBF evaluation is distributed.

The strategy that has to be adopted for a successful iterative solution of the RBF fit depends on the RBF kernel. Unfortunately, out of the box iterative solvers for generic linear systems perform bad with RBF and require so many iterations to converge that the cost becomes similar to the one of the direct solution. Special pre-processing is required to implement a fast iterative method suitable for the computation of the unknown coefficients in a reasonable number of iterations. Iterative solutions of multiquadric and polyharmonic RBF can be enabled with preconditioners that use cardinal functions based on a subset of the interpolating points. This concept was firstly introduced by Beatson and Powell (1993). The idea of selecting a set of data near each point, a mix of near and far points, and via modified approximate cardinal functions that have a specified decay behaviour was presented by Beatson et al. (1999). An effective algorithm capable to deal with multi quadric in generic n-dimensional spaces was published by Faul et al. (2005), which uses the cardinal function idea, and in which for each point a carefully selected set of q points was used to construct the preconditioner. The method has been cleared explained by Buhman (2003) which baptised it as BFGP "…which we shall call after its inventors the BFGP (Beatson–Faul–Goodsell–Powell) method…".

The algorithm has been further studied by Gumerov and Duraiswami (2007) who demonstrate how to reduce the complexity of the method for the 3D case; the preconditioner set-up cost and the evaluation cost (that scales as N^2 in the original work) are reduced using fast search methods during the preconditioner set-up and

FMM during iteration getting an implementation that scale as $Nlog(N)$. Complexity test with up to 1,000,000 RBF centres are reported observing that the complexity of the method grows in the range $N^{1.15-1.4}$.

Examples of the performances of the BFGP are already reported in Table 3.2 and Table 3.3 of previous sections. A further numerical example will be given in Sect. 3.8.

3.8 Parallel Solvers

Parallel calculation introduces an high level of complexity in the implementation. However, there are applications that can benefit of parallelism with very low efforts and RBF evaluation can be for sure included in this set. For large dataset it is possible to benefit of parallelism at single point evaluation level (each processor takes care of the RBF evaluation of an individual point) or at summation level (the overall summation is split in parts, each part is summed by a processor and the result is then reduced). The acceleration achieved can be really very high (close to 100% efficiency) because the main cost is due to the evaluation of the inter distance (square root computation). It is worth to notice that such a cost can be avoided implementing specific algorithms in which space partitioning is used to rearrange the problem as the combination of sub problems (POU methods exposed in Sect. 4.5) or using an approximated expansion of the far field (FMM methods exposed in Sect. 4.6).

The simpler implementation consists in the use of shared memory parallelism using the OpenMP protocol so that all the cores can contribute to the calculation (it is quite common to have 20 cores at disposal for the calculation). A further acceleration can be achieved adopting vector computation (SSE and AVX) available in modern processor; exploiting special registers (128bit up to 512bit), it is possible to compute simultaneously many square roots using a single core (4x to 16x in single precision, 2x to 8x in double precision). A further acceleration can be achieved if the GPU is used for the summation (as an example 2880 CUDA cores are available in a Tesla K40 GPU).

3.8.1 Example of Acceleration Achieved with Parallel Solving Technology

A benchmark has been conducted using the RBF solver of the RBF Morph ACT Extension software that works in the FEA software ANSYS Mechanical. The benchmark concerns a simple mesh morphing problem (Fig. 3.13). A sphere with unit radius is meshed in the volume using parabolic tetra elements; the spacing at the surface can be changed so that the count of nodes onto the surface and inside the

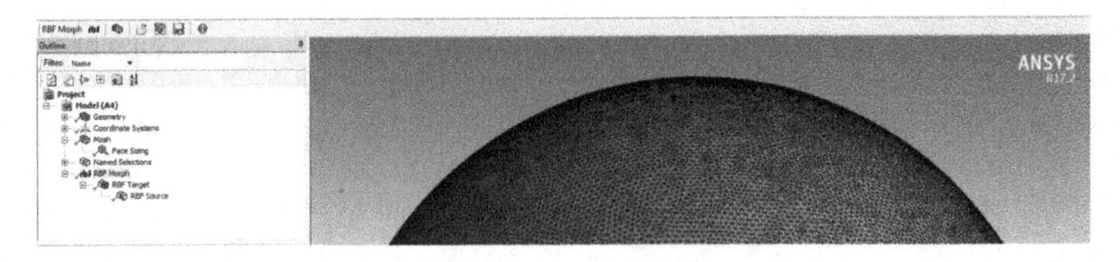

Fig. 3.13 Sphere benchmark. A mesh spacing is prescribed onto the surface of a sphere of 1 m radius. The RBF Morph component is inserted in the tree; all the nodes within the ball are prescribed as morphing target, all the nodes on the surface of the sphere are controlled using a constant offset. The spacing is varied to change the complexity of the benchmark

volume can be controlled. The nodes on the surface are used as RBF centres, an offset with respect to the spherical surface (refer to Sect. 6.3.4 for the specific shape modification description) is prescribed so that each point is subjected to a vector with all the three components active. The fitted RBF is then evaluated to update the position of all the nodes of the volume mesh.

The results of the benchmark are summarized in Table 3.4. Both RBF solvers are based on BFGP are parallel (shared memory) and features a fast search neighbours algorithms for the construction of the preconditioner. The 2017 version has SSE enabled that provides a 4x acceleration in the internal RBF evaluations: distances are computed using 128 bit registry (4 in a single operation because single precision is adopted). The parallelisation of 2017 version is not limited to RBF evaluation but extended to other steps of the calculation as well. Performances are substantially improved running on the same laptop thanks to proper parallelisation of the algorithm.

The growing laws are $1.107 \times 10^{-9}\ N^{2.12}$ for the 2017 library and 3.43×10^{-9} $N^{2.2}$ for the 2015 one. In both cases we are very close to the quadratic expected behaviour (closer for the 2017); the different value of the constant is mainly due to better parallelisation effect enabled using SSE. Data of Table 3.2 (complete solver scales as $4.4 \times 10^{-8}\ N^{2.25}$) are obtained with the same laptop but with a serial library that comes with a different level of optimisation. If compared with 2015 of this example, we notice a similar power law and a ratio between the constant close to 13x; parallelism explains a 4x factor, the further acceleration is due to a better optimisation of the algorithm.

Table 3.4 Complexity test using iterative parallel solver (laptop i7-4750HQ 2.00 GHz)

Spacing (m)	#RBF centres	#Targets	2017 solver (s)	2015 solver (s)	Speed-up
0025	85.434	205.866	38	263	6.92
0.02	132.598	323.464	76	635	8.36
0015	234.522	574.960	270	2270	8.41

Three different values of the spacing (coarse, medium and fine) are explored. Resulting nodes count increases both on the surface and in the volume. The wall clock time is for the full process consisting in RBF coefficient fitting and in the update of all the volume mesh

References

Babuška I, Melenk JM (1997) The partition of unity method. Int J Numer Meth Eng 40(4):727–758, https://doi.org/10.1002/(SICI)1097-0207(19970228)40:4:727::AID-NME86:3.0.CO;2-N

Beatson RK, Greengard L (1997) A short course on fast multipole methods. In: Ainsworth M, Levesley J, Light WA. Marletta M (eds) Wavelets, multilevel methods and elliptic PDEs, Oxford University Press, pp 1–37

Beatson RK, Powell MJD (1993) An iterative method for thin plate spline interpolation that employs approximations to Lagrange functions. Numerical Analysis, Griffiths D.F & Watson G.A. (eds), London: Longmans, pp 17–39

Beatson RK, Cherrie JB, Mouat CT (1999) Fast fitting of radial basis functions: methods based on preconditioned GMRES iteration, Adv Comput Math 11:253–270

Beatson RK, Powell MJD, Tan AM (2007) Fast evaluation of polyharmonic splines in three dimensions. IMA J Numer Anal 27:427–450. https://doi.org/10.1093/imanum/drl027

Bentley JL (1975) Multidimensional binary search trees used for associative searching. Commun ACM 18(9):509–517

Biancolini ME (2012) Mesh morphing and smoothing by means of Radial Basis Functions (RBF): a practical example using Fluent and RBF morph. In: Handbook of research on computational science and engineering: theory and practice, pp 34, https://doi.org/10.4018/978-1-61350-116-0.ch015

Botsch M, Kobbelt L (2005) Real-time shape editing using radial basis functions. In: Marks J, Alexa M (eds) Computer graphics, 24(3), Eurographics

Buhmann MD (2003) Radial basis functions: theory and implmentation. Cambridge University Press, New York

Carr JC, Beatson R, Cherri J, Mitchell T, Fright W, McCallum B (2001) Reconstruction and representation of 3D objects with radial basis functions. In: Proceedings of the 28th annual conference on Computer graphics and interactive techniques, Los Angeles, CA, p 67–76

Carr JC, Beatson RK, McCallum BC, Fright WR, McLennan TJ, Mitchell TJ (2003) Smooth surface reconstruction from noisy range data. In: First international conference on computer graphics and interactive techniques, p 119

Cipra BA (2010) The best of the 20th century: Editors name top 10 algorithms. SIAM News Soc Ind Appl Math. 33 (4):2. Retrieved 23 Dec

European Commision (2013) RBF4AERO Project. http://cordis.europa.eu/project/rcn/109141_en.html

Faul AC, Goodsell G, Powell MJD (2005) A Krylov subspace algorithm for multiquadric interpolation in many dimensions. IMA J Numer Anal 25:1–24

Greengard L, Rokhlin V (1987) A fast algorithm for particle simulations. J Comput Phys 73:325–348

Gumerov NA, Duraiswami R (2007) Fast radial basis function interpolation via preconditioned Krylov Iteration. SIAM J Sci Comput 29(5):1876–1899. https://doi.org/10.1137/060662083

Kojekine N, Savchenko V, Seni M, Hagiwara I (2002) Real-time 3D Deformations by means of compactly supported radial basis functions, EUROGRAPHICS 2002/ Navazo Al I and Slusallek P (Guest Editors)

Krasny R, Wang L (2011) Fast evaluation of multiquadric RBF fums by a Cartesian Treecode. SIAM Soc Ind Appl Math 33:2341–2355

LAPACK (2017a) Linear Algebra PACKage. http://www.netlib.org/lapack/

LAPACK (2017b) http://www.netlib.org/lapack/explore-html/d1/dcd/dsptrf_8f.html

Lexing Y (2006) A kernel independent fast multipole algorithm for radial basis functions. J Comput Phys 21:451–457

Press WH, Flannery BP, Teukolsky SA (1992) Numerical recipes in C. In: The art of scientific computing, 2nd edn. ISBN 0-521-43108-5

Rendall TCS, Allen CB (2009) Efficient mesh motion using radial basis functions with data reduction algorithms. J Comput Phys 228(17):6231–6249

ScaLAPACK (2017) Scalable linear algebra package. http://www.netlib.org/scalapack/
Sieger D, Menzel S, Botsch M (2014) RBF morphing techniques for simulation-based design optimization. Eng Comput 30(2):161–174
Tobor I, Reuter P, Schlick C (2006) Reconstructing multi-scale variational partition of unity implicit surfaces with attributes. Graph Models 68(1):25–41. https://doi.org/10.1016/j.gmod.2005.09.003
Wendland H (2002) Fast evaluation of radial basis functions: methods based on partition of unity. In: Chui CK, Schumaker LL, Stöckler J (eds) Approximation theory X: wavelets, splines, and applications, Vanderbilt University Press, Nashville, pp 473–483
Yokota R, Barba LA (2011) Treecode and fast multipole method for n-body simulation with cuda. GPU Comput Gems Emerald Ed, p 113

Chapter 4
RBF Tools

Abstract This chapter provides an overview of calculation tools that are usually required for the implementation of RBF based engineering workflows. The aim here is to provide the reader with a strategy to evaluate whether an application can be faced using RBF and how the various tools can be combined to make it effective.

In this chapter an overview of the available RBF tools is presented. Details of the basic algorithms usually required to set-up a workflow based on RBF are here presented with worksheet implementations based on the Mathcad software. Mathcad code snippet can be easily adapted to other high level mathematics environments as Matlab (Fasshauer 2007), Maple or Excel. Furthermore such snippets can be effectively used as a guide for more efficient code developing by means of advanced languages as Python, FORTRAN or C.

4.1 RBF Fitting and Evaluating Functions

Usually working with RBF requires multiple usage of an RBF fit. For example, multiple local RBF fits are blended using the POU method (see Sect. 3.5), and a repeated use of RBF fits is used in Greedy method when the number of points is iteratively increased (see Sect. 3.3).

A demonstration of the Mathcad implementation of a toolkit function widely used in this book is given in Eq. (4.1). The function **RBF_fit**$(\cdot)$ takes as input a vector containing the points of the cloud $\boldsymbol{x}_s$ and a matrix $\boldsymbol{g}_s$ which contains by columns the values to be fitted. It is worth to notice that many scalar functions are usually interpolated using the same cloud; an example is mesh morphing in which the three components of a displacement field are interpolated on the RBF centres as formulated in Eq. (2.9). The output are the coefficients of the RBF γ and β related to each column. The implementation is supplied as program block and the steps are similar to the steps defined through Eqs. (2.17), (2.18) and (2.20) of Sect. 2.3.

M. E. Biancolini, *Fast Radial Basis Functions for Engineering Applications*,
https://doi.org/10.1007/978-3-319-75011-8_4

The number of RBF centres N_C, the size of the interpolation space N_d and the number of scalar function to be fitted N_{RBF} are first extracted examining the size of input vector; then the interpolation and constraint matrixes are built and the augmented system MP is finally built and solved using the **lsolve(·)** function. The output is then compacted in a single vector which contains the matrix of γ coefficients at the first row and the one of β coefficients at the second one, whilst each column of the matrix is related to the column of input $\boldsymbol{g}_s$.

$$
\mathrm{RBF\_fit}(x_s, g_s) := \left|
\begin{array}{l}
N_c \leftarrow \mathrm{rows}(x_s) \\
N_d \leftarrow \mathrm{rows}\left(x_{s_1}\right) \\
N_{RBF} \leftarrow \mathrm{cols}(g_s) \\
\text{for } i \in 1..N_c \\
\quad \text{for } j \in 1..N_c \\
\qquad M_{i,j} \leftarrow \phi\left(\left|x_{s_i} - x_{s_j}\right|\right) \\
\text{for } i \in 1..N_c \\
\quad \left|
\begin{array}{l}
P_{i,1} \leftarrow 1 \\
\text{for } j \in 1..N_d \\
\quad P_{i,j+1} \leftarrow \left(x_{s_i}\right)_j
\end{array}\right. \\
MP \leftarrow \mathrm{augment}\left(\mathrm{stack}\left(M, P^T\right), \mathrm{stack}\left(P, \mathrm{zero\_mat}\left(N_d + 1, N_d + 1\right)\right)\right) \\
\gamma\beta \leftarrow \mathrm{lsolve}\left(MP, \mathrm{stack}\left(g_k, \mathrm{zero\_mat}\left(N_d + 1, N_{RBF}\right)\right)\right) \\
\gamma \leftarrow \mathrm{submatrix}\left(\gamma\beta, 1, N_c, 1, N_{RBF}\right) \\
\beta \leftarrow \mathrm{submatrix}\left(\gamma\beta, N_c + 1, N_c + N_d + 1, 1, N_{RBF}\right) \\
out \leftarrow \begin{pmatrix} \gamma \\ \beta \end{pmatrix}
\end{array}\right.
\tag{4.1}
$$

The coefficients calculated using the **RBF_fit(·)** function are used by the **RBF_eval(·)** function:

$$
\mathrm{RBF\_eval}(x, ic, x_s, \gamma, \beta) := \sum_{i=1}^{\mathrm{rows}(x_s)} \left[\left(\gamma^{\langle ic\rangle}\right)_i \cdot \phi\left(\left|x - x_{s_i}\right|\right)\right] + \beta^{\langle ic\rangle} \cdot \mathrm{stack}(1, x)
\tag{4.2}
$$

In this form the **RBF_eval(·)** function takes as input the evaluation point x, the index ic that points the RBF to be used (in case of multiple RBF fitted on the same cloud), the vector containing the position of the RBF centres and the matrixes containing the interpolated RBF γ and β coefficients.

Toolkit functions usage is hereinafter demonstrated adopting the example shown in Sect. 2.3. The case of two control points shown in Fig. 2.5 (right) is re-executed considering two different inputs for controlling the values at the two control points obtaining a new version of Eq. (2.14):

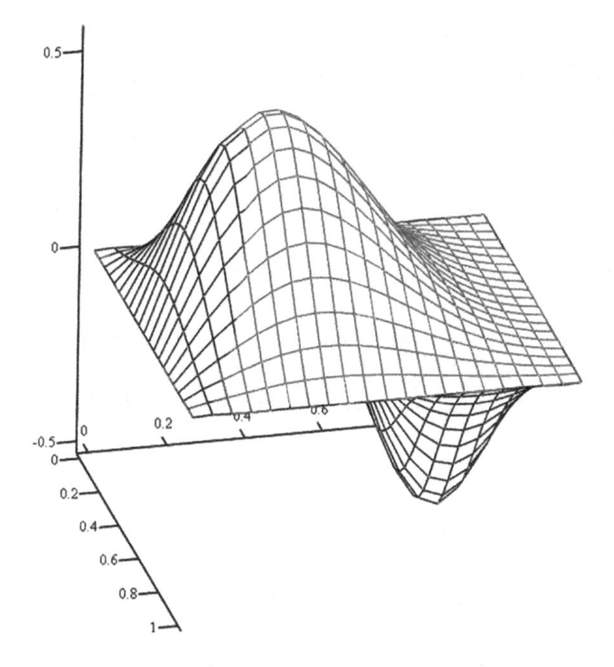

Fig. 4.1 The usage of toolkit functions is demonstrated for the same arrangement of Fig. 2.5 (right). The surface with new values at internal control points is plotted together with the original one

$$P_{control} := \begin{bmatrix} \begin{pmatrix} \frac{1}{2} \\ \frac{1}{4} \end{pmatrix} \\ \begin{pmatrix} \frac{1}{2} \\ \frac{3}{4} \end{pmatrix} \end{bmatrix} \qquad g_{control} := \begin{pmatrix} 0.5 & 0.5 \\ -0.5 & 0.1 \end{pmatrix} \tag{4.3}$$

Once the RBF problem is completed with the points on the square according to Eq. (2.16) the new coefficients are computed using the **RBF_fit(·)** function:

$$\begin{pmatrix} \gamma_{2points} \\ \beta_{2points} \end{pmatrix} := \mathrm{RBF\_fit}(x_s, g_s) \tag{4.4}$$

The two interpolated functions are compared in Fig. 4.1; the maps adopted for plotting are generated adapting the Eq. (2.23) where the z component is now computed using the **RBF_eval(·)** function as follows:

$$\begin{aligned} xyz_{plot}(ic) := \; & id \leftarrow 1 \\ & \text{for } i \in 1..nx \\ & \quad \text{for } j \in 1..ny \\ & \qquad xplot_{id} \leftarrow (i-1) \cdot dx \\ & \qquad yplot_{id} \leftarrow (j-1) \cdot dy \\ & \qquad zplot_{id} \leftarrow \mathrm{RBF\_eval}\left[\begin{pmatrix} xplot_{id} \\ yplot_{id} \end{pmatrix}, ic, x_k, \gamma_{2points}, \beta_{2points}\right] \\ & \qquad id \leftarrow id + 1 \\ & \text{augment}(xplot, yplot, zplot) \end{aligned} \tag{4.5}$$

Notice that in this case the map generation function takes as input the variable *ic* to select the proper fitted column.

The use of Eq. (4.2) is not optimal if multiple columns are to be evaluated simultaneously because the high cost of inter-distance evaluation is repeated for each column. In the case the two functions are interpolated and are to be evaluated simultaneously, the evaluation time can be halved adopting the following equation:

$$\mathrm{RBF\_eval\_2}(\mathrm{x}, \mathrm{x_s}, \gamma, \beta) := \sum_{i=1}^{\mathrm{rows}(\mathrm{x_s})} \left[\begin{pmatrix} \gamma_{i,1} \\ \gamma_{i,2} \end{pmatrix} \cdot \phi(|\mathrm{x} - \mathrm{x_{s_i}}|) \right] + \begin{pmatrix} \beta^{\langle 1 \rangle} \cdot \mathrm{stack}(1, \mathrm{x}) \\ \beta^{\langle 2 \rangle} \cdot \mathrm{stack}(1, \mathrm{x}) \end{pmatrix} \tag{4.6}$$

In this case the summation is conducted in vector form.

4.2 Aligned Points Check

The RBF fit operation can fail when the dataset exhibits special arrangements. Common scenarios are when all the points belong to a plane in a 3D problem or a line in a 2D problem. The solution fails if a polynomial term is associated with the RBF because the system matrix becomes singular. Singularity can be avoided adopting an RBF without the polynomial part or adding an auxiliary point (usually with zero input) that is not on the same line (2D) or plane (3D).

4.3 Reducing the Size of the RBF Cloud (Sub-sampling)

There are many situations in which the original dataset is very large and unevenly distributed. Some points are clustered all together whilst other are scattered. There are algorithms to extract a partial dataset from the original one and are usually based on the mutual inter distance between points.

In this book a simple sub-sampling algorithm is introduced. It can be used to regularize the spacing of the cloud before computing an RBF fit or it can be used in iterative methods, known as Greedy methods (Rendall and Allen 2010), where we search for a smaller cloud capable to represent the same information of the original one while reducing the computational complexity.

It is worth to observe that if the sub-sampling is used, duplicates are automatically removed.

The sub-sampling algorithm can be time consuming especially if a very large cloud needs to be processed; fast method based on space partitioning can be employed to alleviate this. Nevertheless, the algorithm herein introduced exhibits very good performances if the final cloud is small, even if we start with a huge dataset.

The algorithm is quite straightforward and takes as input the original dataset and a sub-sampling radius, whilst the output is a subset containing points of the original one separated at least by the sub-sampling radius. Obviously the order in which the points are processed substantially affects the final result and, so, some sorting strategies could be used.

The algorithm can be described as follows:

1. pick the first point and keep it in the retained set;
2. visit the subsequent point, if it's separated by a sub-sampling radius from all the retained points keep it otherwise skip it and repeat the step until all the points are processed.

The computational cost can be very high if the sub-sampling radius is very small and almost all the points are retained; this is because each point is compared with all the previous ones. If the retained dataset size is small if compared with the original one, each point is compared with a dataset that is, at worst, large as the final one.

The sub-sampling routine is an important building block of the Toolkit and its Mathcad implementation is here reported (4.7).

The input data consist in the original cloud of points (the vector $\mathbf{x_{file}}$ where points are stored), the known input (the vector $\mathbf{g_{file}}$ where known scalar field are stored by columns) and the subsampling distance $\boldsymbol{\delta}$. The retained points are stored in the vectors $\mathbf{x_{new}}$ and $\mathbf{g_{new}}$ (the latter stored in transposed form). The first point is retained and then other points are visited and compared with all the retained ones deciding whether adding them using the **addi** variable. The variable is set first to "1" (i.e. by default consider the new point as a candidate to be retained), then the point is compared with all the retained points and if its distance is lower than the prescribed distance $\boldsymbol{\delta}$ the point is excluded setting **addi** to "0". Once all the retained points are compared with the candidate the decision is made inspecting the value of the **addi** variable; if it is still equal to "1" the point is added and the **idnew** variable is increased. The resulting vectors $\mathbf{x_{new}}$ and $\mathbf{g_{new}}$ are the output of the function.

$$
xg(\delta, x_{file}, g_{file}) := \left|
\begin{array}{l}
x_{new_1} \leftarrow x_{file_1} \\
g_{new}^{\langle 1 \rangle} \leftarrow \left(g_{file}^{T}\right)^{\langle 1 \rangle} \\
idnew \leftarrow 2 \\
\text{for } i \in 2..\,rows(x_{file}) \\
\quad \left|
\begin{array}{l}
addi \leftarrow 1 \\
\text{for } j \in 1..\,idnew - 1 \\
\quad addi \leftarrow 0 \text{ if } \left| x_{file_i} - x_{new_j} \right| < \delta \\
\text{if } addi = 1 \\
\quad \left|
\begin{array}{l}
x_{new_{idnew}} \leftarrow x_{file_i} \\
g_{new}^{\langle idnew \rangle} \leftarrow \left(g_{file}^{T}\right)^{\langle i \rangle} \\
idnew \leftarrow idnew + 1
\end{array}
\right.
\end{array}
\right. \\
out \leftarrow \begin{pmatrix} x_{new} \\ g_{new}^{T} \end{pmatrix}
\end{array}
\right. \tag{4.7}
$$

The method is often used for 3D cases but it can be used also in n-d cases. A typical 2D scenario is demonstrated in Figs. 4.2 and 4.3: a cloud of point coming from a mesh with a local refinement at the central square (unit size) has to be regularized with a constant sizing. The case of the coarse spacing is represented in Fig. 4.2 where the effect of the sorting of the cloud can be appreciated. A better recovering of corners and edges is achieved sorting the cloud so that the first points are the ones on the edges and, among them, the ones on the corners are the firsts.

The same approach is reported in Fig. 4.3 for a fine spacing. It is worth to notice that the number of retained points depends on the points numbering order.

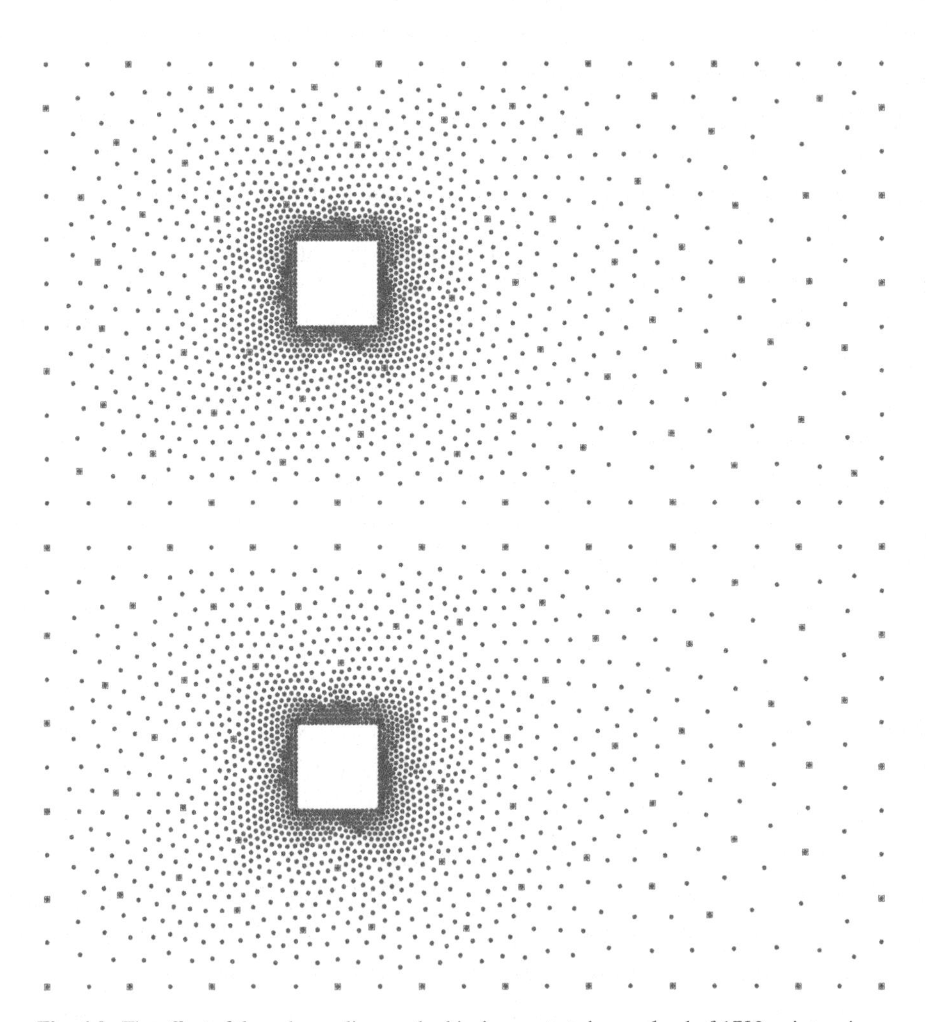

Fig. 4.2 The effect of the subsampling method is demonstrated on a cloud of 1722 points using a distance $\delta = 0.75$. In the original node numbering edges and corners are not preserved (top 72 points retained). The same algorithm applied on sorted points allows to preserve geometrical features (bottom 76 points retained)

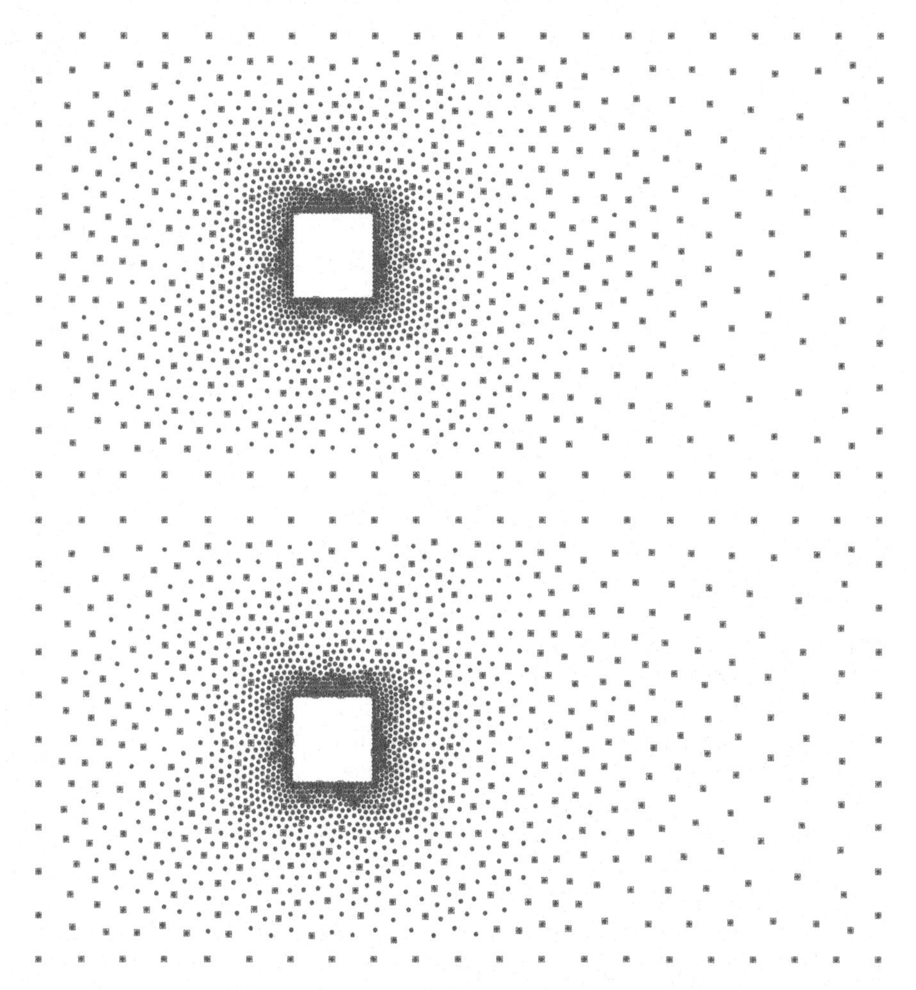

Fig. 4.3 The effect of the subsampling method is demonstrated on a cloud of 1722 points using a distance $\delta = 0.25$. In the original node numbering edges and corners are not preserved (top 434 points retained). The same algorithm applied on sorted points allows to preserve geometrical features (bottom 435 points retained)

4.4 Duplicates Check

The RBF fit process can be completed successfully if the original dataset is free of duplicates, i.e. of points that share the same position in the n-d space, if duplicates are introduced, the system (2.6) becomes singular because it has two rows and two columns identical for each duplicate pair. Solvability can be an issue even if two points are not coincident but very close each other. Duplicate search is a quite delicate task for many reasons: a metric for duplicate check is required, multiple duplicates may occur at the same position and inter-distance search may be time consuming for very large dataset.

The RBF method is meshless and so the duplicates check can be limited to the elimination of duplicate points without keeping track of removed point index. Nevertheless, there are scenarios in which some stages of the workflow require the management of points that are nodes of a mesh and the preservation of original indexes is of paramount importance to keep the topology of connected elements valid. This means that duplicates cannot be simply eliminated; entities that point to them (i.e. connected elements or cells) have to be updated replacing index of duplicate nodes to be deleted with index of coincident nodes to be kept so that the duplicates can be safely deleted without corrupting the mesh topology.

The duplicates elimination task is usually divided in two stages: duplicates search and duplicates merge.

In the first stage the cloud is visited. Each point is compared with all the others computing the inter-distance and every time that the distance is lower than a given tolerance radius the visited point has to be removed and is flagged with the index of the coincident node; if the coincident node is already flagged its flag is used, this action is repeated until an unflagged point is found. Such iterative flagging guarantees the elimination of multiple duplicates while preserving the information to reconstruct the topology, when required.

At the end of the first stage all the points flagged have to be removed. If the mesh topology has to be preserved, an efficient approach consists in completing the flag vector using the original point index as a flag for unflagged entries: at the end of such operation the flag vector contains the original index for kept points, a reference to the index of a kept point for the points that are flagged to be removed. The topology can be then updated visiting all the elements and replacing the point index of the vertex with the one pointed by the flag vector. Once all the elements are visited and all the nodes of each element are replaced with the corresponding flag, the duplicates are not used anymore and can thus be removed. An optional node/element renumbering may be considered for a compact index representation.

The process may be slow when processing a large datasets if all the inter-distance are evaluated. A substantial improvement of performances can be achieved using auxiliary data structures as binary trees, buckets or boxes. The idea is to search for duplicates just in the near field, i.e. among the points that are known to be close to the one to be checked.

The algorithm described in (4.7) of Sect. 4.3 can be effectively used for duplicates detection setting a small value of the distance $\boldsymbol{\delta}$. As already noticed, the performances of the direct approach usually are not an issue if the retained set is small; however, if the method is adopted for duplicate detection the final cloud will have a size very close to the original one and so performances could become an issue. This is why a substantially faster version (4.8) is proposed and demonstrated in this section.

$$
xg\_fast(\delta, x_{file}, g_{file}) := \left|
\begin{array}{l}
\text{for } i \in 1..\text{rows}(x_{file}) \\
\quad X_i \leftarrow \left(x_{file_i}\right)_1 \\
xg \leftarrow \text{augment}(X, x_{file}, g_{file}) \\
xg_{sort} \leftarrow \text{csort}(xg, 1) \\
x_{file} \leftarrow xg_{sort}^{\langle 2 \rangle} \\
\text{for } ic \in 1..3 \\
\quad g_{file}^{\langle ic \rangle} \leftarrow xg_{sort}^{\langle ic+2 \rangle} \\
x_{new_1} \leftarrow x_{file_1} \\
g_{new}^{\langle 1 \rangle} \leftarrow \left(g_{file}^{T}\right)^{\langle 1 \rangle} \\
idnew \leftarrow 2 \\
\text{for } i \in 2..\text{rows}(x_{file}) \\
\quad \left|
\begin{array}{l}
addi \leftarrow 1 \\
j \leftarrow idnew - 1 \\
\text{while } j \geq 1 \wedge \left|\left(x_{file_i}\right)_1 - \left(x_{new_j}\right)_1\right| < \delta \\
\quad \left|
\begin{array}{l}
addi \leftarrow 0 \text{ if } \left|x_{file_i} - x_{new_j}\right| < \delta \\
j \leftarrow j - 1
\end{array}
\right. \\
\text{if } addi = 1 \\
\quad \left|
\begin{array}{l}
x_{new_{idnew}} \leftarrow x_{file_i} \\
g_{new}^{\langle idnew \rangle} \leftarrow \left(g_{file}^{T}\right)^{\langle i \rangle} \\
idnew \leftarrow idnew + 1
\end{array}
\right.
\end{array}
\right. \\
out \leftarrow \begin{pmatrix} x_{new} \\ g_{new}^{T} \end{pmatrix}
\end{array}
\right.
\tag{4.8}
$$

A significant acceleration can be achieved sorting the dataset with respect to one coordinate (in this example the first one). This operation can be usually conducted very fast with standard libraries (for instance the quick sort function **qsort(·)**. In the sorted dataset each individual query to search for duplicates is limited to the neighbouring points with respect to the sorted coordinate. Mathcad uses the fast sort function **csort(·)** which allows to sort the content of a matrix according to a column. In the modified version of the algorithm, **xg_fast(·)**, the **X** vector is first created extracting the first coordinate of all the points to be processed. The **augment(·)** function is successively used to pair with **X** the input vector $\mathbf{x}_{\mathbf{file}}$ and the input matrix $\mathbf{g}_{\mathbf{file}}$ so that the resulting xg matrix is then sorted; sorted versions of $\mathbf{x}_{\mathbf{file}}$ and $\mathbf{g}_{\mathbf{file}}$ are extracted to be further processed. The **for** cycles used in the previous version is here replaced by a **while** loop that decreases the j variable at each loop getting exactly the same behaviour of the original method. The acceleration is obtained thanks to the second check after the and ($\wedge$) operator meaning that the

Table 4.1 Complexity test for duplicate detection

# points	Direct time (s)	Fast time (s)	# points kept direct	# points kept fast
1000	0.34	0.09	998	998
2000	1.38	0.36	1988	1988
5000	10.32	2.55	4945	4945
10,000	35.32	9.18	9800	9802

A random distribution of centres is generated in the cube of unit side. Duplicate detection distance is set to 0.01

loop is continued only in the near field, i.e. if the first coordinate of scanned points is within the range. The first coordinate sorting guarantees that the points are ordered in such direction and if a duplicate exists, it will be certainly detected. Satisfying the first coordinate criterion matching is not enough because the point could be very close or coincident with respect to the first coordinate; this is why the usual inter-distance condition is in any case considered.

The method has been tested using points distributed randomly in the unit cube (as in Sect. 3.6.2) and a summary of the performances is given in Table 4.1. It is worth to notice that the number of kept point is not the same for the large cloud. This is because the results can change for a different visiting order, unless duplicates are exactly overlapped (i.e. duplicates have inter-distance that is a perfect zero), but this is not the case for a cloud of points with random positions.

4.5 Data Normalisation

Dataset used in the RBF problem can be related to physical quantities; a common case is that RBF centres in 1D, 2D or 3D space are the coordinates of points. However RBF can be used for generic data fitting in nD spaces and a centre can represent a state in multi-parametric spaces. A good example is given in Chap. 11 where RBF are used to generate the Response Surface (RS) as a meta-model for optimum search. In this case the input state may represent quantities that are not homogenous and that may span different dimensions and length.

In such situation the Euclidean distance used for RBF interaction can lead to a wrong interpolation and a good practice is to move the problem in a new space after scaling and with an optional offset of the axes. Such an approach can be used to map the original dataset onto an n-dimensional hypercube centred in the origin and unit dimension sides. Data normalisation was necessary for the shape optimisation of the glider presented in Sect. 11.4, the input parameters of DOE points were normalised using different factors, **nf1** and **nf2** of (4.9), so that the RBF fit is performed in the dimension-less space and in this case the same transformation function has to be used to look into the RBF fitted function as defined in (4.10).

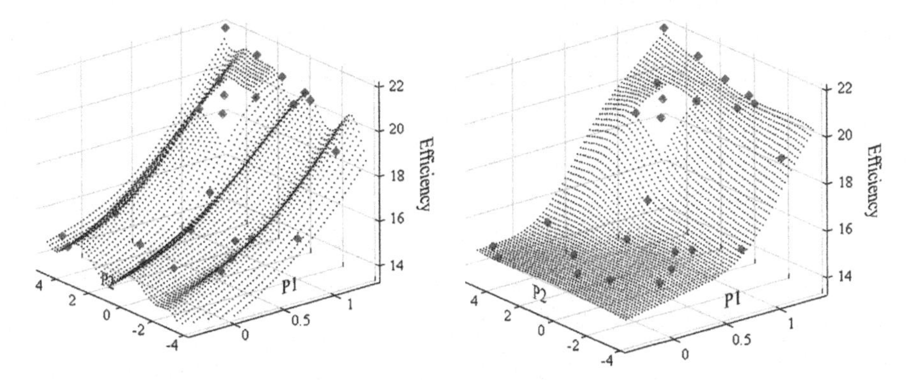

Fig. 4.4 The effect of the normalisation is demonstrated for the construction of the response surface. Without normalisation an oscillation is observed in the second parameter (left). After that the two normalisation factors are introduced (nf1 = 1.8, nf2 = 9) undesired oscillations disappear (right)

$$\begin{aligned} x_{kc} &:= \text{for } i \in 1 \ldots n_{DOE} \\ &\quad x_{k_i} \leftarrow \begin{pmatrix} \dfrac{DOE\_coarse_{i,1}}{nf1} \\ \dfrac{DOE\_coarse_{i,2}}{nf2} \end{pmatrix} \end{aligned} \tag{4.9}$$

$$s_{RBF}(x, ic, x_k, \gamma, \beta, i_\phi, R_{sup}) := \left[\begin{array}{l} x \leftarrow \begin{pmatrix} \dfrac{x_1}{nf1} \\ \dfrac{x_2}{nf2} \end{pmatrix} \\ \displaystyle\sum_{i=1}^{rows(x_k)} \left[(\gamma_{ic})_i \cdot \phi\left(|x - x_{k_i}|, i_\phi, R_{sup} \right) \right] + \beta_{ic} \cdot stack(1, x) \end{array} \right] \tag{4.10}$$

The wrong interpolation without normalisation is represented in Fig. 4.4 on the left. It is clear that a wavy behaviour of the function is captured because points have distances in the direction of the second parameter that are in the order of five times larger than the ones in the direction of the first parameter. The use of normalisation allows to get the desired smooth behaviour with a very good interpolation of both parameters.

4.6 Generation of Auxiliary RBF Clouds

The management of RBF problems, especially when full supported functions are used, could require the addition of special distributions. Points uniformly spaced on the surfaces of boxes, cylinders and spheres are commonly used for RBF based mesh morphing (see Sect. 6.2) to wrap the working domain with a distribution of points where the field is imposed to vanish or to accomodate a simple law (rigid movement, scaling). Algorithms to generate simple RBF clouds are given in this section.

4.6.1 Box

A simple method for generating an uniform distribution of points onto the surface of a box consists in the mapped generation of point in the volume of the box keeping just the ones that belong to the surfaces and neglecting the ones into the volume.

The approach is demonstrated in (4.11). The function $\mathbf{P_{box}}$ receives as input two points, $\mathbf{P_1}$ and $\mathbf{P_2}$, located at the corners of a box and the desired spacing $\mathbf{\Delta R}$. The counter **ID** is initialized to 1 and the vector **v** on the diagonal of the box is evaluated subtracting $\mathbf{P_2}$ and $\mathbf{P_1}$, integer number of intervals, **nx**, **ny** and **nz**, in the three directions are then computed dividing the components of **v** by the spacing $\mathbf{\Delta R}$ and truncated by the **ceil(·)** function. Actual spacing $\mathbf{\Delta x}$, $\mathbf{\Delta y}$ and $\mathbf{\Delta z}$ are then computed. The indexes **i**, **j**, **k** are finally used to span the box volume with three nested **for** cycles. The coordinate of the points to be added are computed only if at surface using the six conditions on the indexes and the "or" Boolean operator $\vee$.

$$
\begin{aligned}
P_{box}(P_1, P_2, \Delta R) := \;& ID \leftarrow 1 \\
& v \leftarrow P_2 - P_1 \\
& \begin{pmatrix} nx \\ ny \\ nz \end{pmatrix} \leftarrow \begin{pmatrix} ceil\left(\frac{v_1}{\Delta R}\right) \\ ceil\left(\frac{v_2}{\Delta R}\right) \\ ceil\left(\frac{v_3}{\Delta R}\right) \end{pmatrix} \\
& \begin{pmatrix} \Delta x \\ \Delta y \\ \Delta z \end{pmatrix} \leftarrow \begin{pmatrix} \frac{v_1}{nx} \\ \frac{v_2}{ny} \\ \frac{v_3}{nz} \end{pmatrix} \\
& \text{for } i \in 1..nx + 1 \\
& \quad \text{for } j \in 1..ny + 1 \\
& \quad\quad \text{for } k \in 1..nz + 1 \\
& \quad\quad\quad \text{if } i = 1 \vee j = 1 \vee k = 1 \vee i = nx + 1 \vee j = ny + 1 \vee k = nz + 1 \\
& \quad\quad\quad\quad Pb_{ID} \leftarrow \begin{bmatrix} (i-1)\cdot\Delta x \\ (j-1)\cdot\Delta y \\ (k-1)\cdot\Delta z \end{bmatrix} \\
& \quad\quad\quad\quad ID \leftarrow ID + 1 \\
& out \leftarrow P_b
\end{aligned}
\tag{4.11}
$$

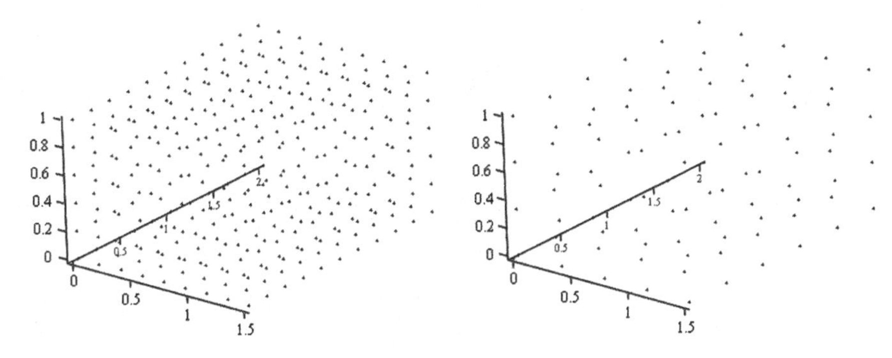

Fig. 4.5 Uniform points distribution onto the surface of a box with a corner in the origin and the other at (1.0, 1.5, 2.0) with spacing 0.2 (342 points left) and 0.4 (96 points right)

The method is demonstrated in the example of Fig. 4.5 where points of the surface of a box with a corner in the origin and the other one at (1.0, 1.5, 2.0) are generated for a spacing of 0.2 and 0.4. If a large spacing is prescribed, the algorithm produces in any case 8 points at the corners of the box.

4.6.2 Cylinder

A simple method for generating a uniform distribution of points onto the surface of a cylinder consists in the mapped generation of points in the volume of the cylinder keeping just the ones on the surfaces and neglecting the ones into the volume.

The approach is demonstrated in the algorithm of (4.12). The function $\mathbf{P_{cyl}}$ receives as input the radius $\mathbf{R_{cyl}}$, the desired spacing $\mathbf{\Delta R}$, two points, $\mathbf{P_1}$ and $\mathbf{P_2}$, defining the axis of the cylinder and a vector **vy** that is not parallel to the axis. The counter **ID** is initialized to 1, the vector **vz** on the axis is evaluated subtracting $\mathbf{P_2}$ and $\mathbf{P_1}$, unit vectors **vx** and **vy** are computed performing cross products and normalisation, integer numbers of interval, **nz** and **nR**, in the z and radial directions are then computed dividing the modulus of **vz** and the radius $\mathbf{R_{cyl}}$ by the spacing $\mathbf{\Delta R}$ and truncated by the **ceil**(·) function. The indexes **iz**, **iR**, $\mathbf{i\vartheta}$ are then used to span the cylinder volume with three nested **for** cycles. The coordinate of the points to be added are computed only if at surface using the three conditions on the indexes and the "or" Boolean operator $\vee$.

$$
\begin{array}{l}
P_{cyl}\left(R_{cyl}, \Delta R, P_1, P_2, vy\right) := \\
\quad ID \leftarrow 1 \\
\quad vz \leftarrow P_2 - P_1 \\
\quad vx \leftarrow vy \times vz \\
\quad vy \leftarrow vz \times vx \\
\quad vx \leftarrow \dfrac{vx}{|vx|} \\
\quad vy \leftarrow \dfrac{vy}{|vy|} \\
\quad nz \leftarrow \mathrm{ceil}\left(\dfrac{|vz|}{\Delta R}\right) \\
\quad nR \leftarrow \mathrm{ceil}\left(\dfrac{R_{cyl}}{\Delta R}\right) \\
\quad \text{for } iz \in 0..nz \\
\quad\quad \text{for } iR \in 1..nR \\
\quad\quad\quad \text{if } iR = nR \vee iz = 0 \vee iz = nz \\
\quad\quad\quad\quad R \leftarrow \dfrac{iR}{nR} \cdot R_{cyl} \\
\quad\quad\quad\quad n\theta \leftarrow \mathrm{ceil}\left(\dfrac{180 \cdot deg \cdot R}{\Delta R}\right) \\
\quad\quad\quad\quad \Delta\theta \leftarrow \dfrac{180 \cdot deg}{n\theta} \\
\quad\quad\quad\quad \text{for } i\theta \in -n\theta + 1..n\theta \\
\quad\quad\quad\quad\quad \theta \leftarrow \Delta\theta \cdot i\theta \\
\quad\quad\quad\quad\quad vr \leftarrow vx \cdot \sin(\theta) + vy \cdot \cos(\theta) \\
\quad\quad\quad\quad\quad {P_{cyl}}_{ID} \leftarrow vr \cdot R + P_1 + vz \cdot \dfrac{iz}{nz} \\
\quad\quad\quad\quad\quad ID \leftarrow 1 + ID \\
\quad out \leftarrow P_{cyl}
\end{array}
\tag{4.12}
$$

The method is demonstrated in the example of Fig. 4.6 where points of the surface of a cylinder with axis starting from the origin and ending at (2.0, 2.0, 2.0), radius 0.5 and vx (1.0, 0.0, 0.0) are generated for a spacing of 0.2 and 0.4.

4.6.3 Sphere

A simple method for generating points onto a spherical surface consists in the use of a uniform spacing in spherical coordinates. Unfortunately such spacing, whilst still uniform in the distance when traveling along meridians, is not uniform when traveling along parallels as the parallel radius decreases up to zero when the poles are reached.

An effective algorithm, which adapts the aforementioned concept, is given in (4.13). The centre $\mathbf{P_{centre}}$, the radius $\mathbf{R_s}$ and the desired spacing $\mathbf{\Delta R}$ are the inputs of the method, whereas a vector containing a cloud of points is the output. The method

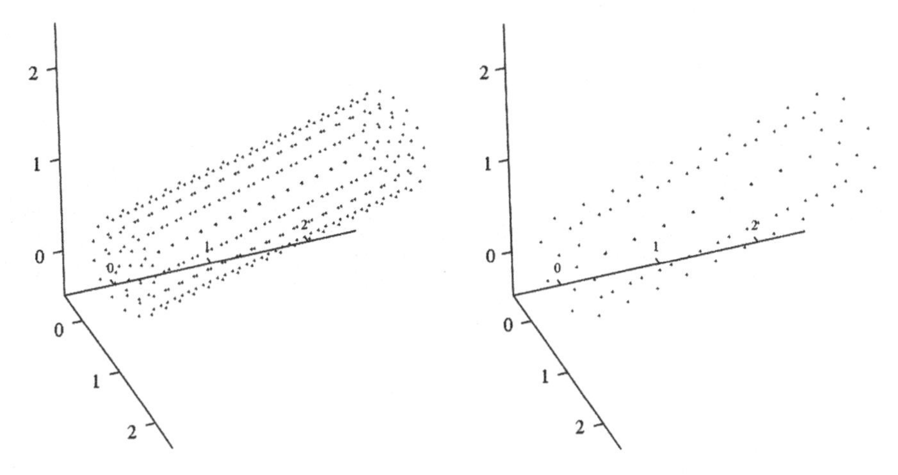

Fig. 4.6 Uniform point distribution onto the surface of a cylinder with axis from the origin to (2.0, 2.0, 2.0), radius 0.5 and with spacing 0.2 (340 points left) and 0.4 (88 points right)

is not ordered and, so, the index of the generated point is obtained by incrementing the variable **ID** and the number of division along the meridian $\mathbf{n\varphi}$ is computed using the **ceil**(·) function that rounds to an integer the ratio between the meridian half-length and the spacing. The number of divisions along the parallel is computed by dividing the current parallel circumference half-length by the spacing. The unit vectors are finally processed to transform polar coordinates into Cartesian ones and then a translation is added to account for the position of the sphere centre $\mathbf{P_{centre}}$.

$$
P_{sphere}(R_s, \Delta R, P_{centre}) := \left|
\begin{array}{l}
ID \leftarrow 1 \\
n\phi \leftarrow \mathrm{ceil}\left(\dfrac{90 \cdot deg \cdot R_s}{\Delta R}\right) \\
\Delta\phi \leftarrow \dfrac{90 \cdot deg}{n\phi} \\
\text{for } i \in -n\phi + 1 .. n\phi - 1 \\
\quad \left|
\begin{array}{l}
\phi \leftarrow \Delta\phi \cdot i \\
n\theta \leftarrow \mathrm{ceil}\left(\dfrac{180 \cdot deg \cdot \left|R_s \cdot \cos(\phi)\right|}{\Delta R}\right) \\
\Delta\theta \leftarrow \dfrac{180 \cdot deg}{n\theta} \\
\text{for } j \in -n\theta + 1 .. n\theta \\
\quad \left|
\begin{array}{l}
\theta \leftarrow \Delta\theta \cdot j \\
vr \leftarrow \begin{pmatrix} -1 \\ 0 \\ 0 \end{pmatrix} \cdot \sin(\theta) + \begin{pmatrix} 0 \\ 1 \\ 0 \end{pmatrix} \cdot \cos(\theta) \\
P_{s_{ID}} \leftarrow \left[vr \cdot \cos(\phi) + \begin{pmatrix} 0 \\ 0 \\ 1 \end{pmatrix} \cdot \sin(\phi) \right] \cdot R_s + P_{centre} \\
ID \leftarrow 1 + ID
\end{array}
\right.
\end{array}
\right. \\
out \leftarrow P_s
\end{array}
\right.
\tag{4.13}
$$

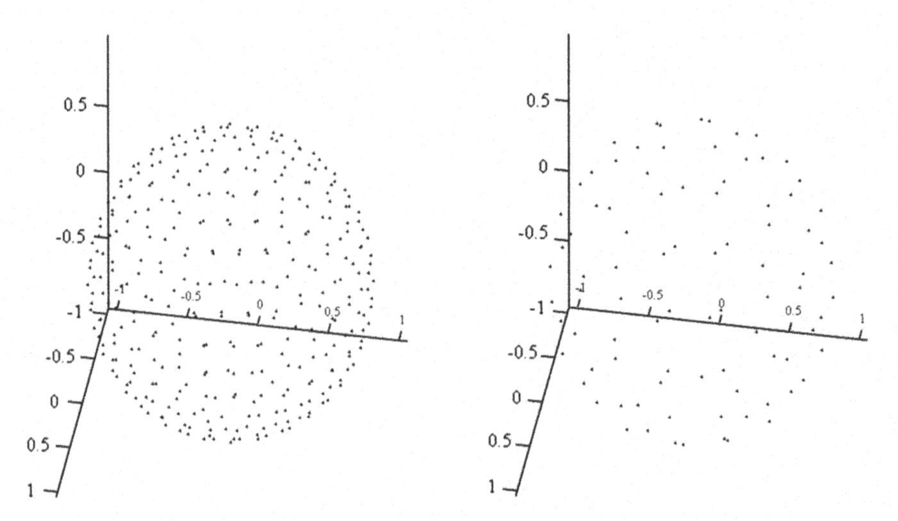

Fig. 4.7 Uniform point distribution onto the surface of a sphere of unit radius, centred in the origin with spacing 0.2 (340 points left) and 0.4 (88 points right)

The method is demonstrated in Fig. 4.7 where two different spacings (0.2 and 0.4) are adopted to generate a distribution of points (2580 and 652) onto the unit radius sphere centred in the origin.

References

Fasshauer GE (2007) Meshfree approximation methods with Matlab, vol. 6. World Scientific, Singapore

Rendall TCS, Allen CB (2010) Reduced surface point selection options for efficient mesh deformation using radial basis functions. J Comput Phys 229(8):2810–2820. https://doi.org/10.1016/j.jcp.2009.12.006

Part II
Engineering Applications Examples

Here a collection of practical applications of fast RBF in engineering is provided. For each presented application, the complete path is clearly and consistently exposed using the systematic approach defined in the first part of the book.

Chapter 5
RBF Implicit Representation of Geometrical Entities

Abstract In this chapter the implicit surface method based on RBF is explained and demonstrated. The classical method that consists of the definition of a scalar valued function in the 3D space that is null at on-surface points and positive (negative) at off-surface points that are positioned outside (inside) the surface is firstly introduced. The projection onto the surface is achieved by computing the gradient of the implicit function. The second one is a novel method that consists of the creation of a projection field (vector valued RBF) defined on the surface and at off-surface points that is zero at on-surface points whereas it is equal to the vector moving to the surface at off-surface points. The effectiveness of both methods is demonstrated with practical examples.

Shape transformation techniques can be broadly divided in two categories: parametric correspondence methods and implicit function interpolation methods (Turk and O'Brien 1999). Although parametric methods are commonly characterised by an higher performance in data processing and a lower memory request with respect to implicit function methods, they suffer from several problems including the occurrence of self-intersecting surfaces when changes in topology between objects need to be consistently handled. Otherwise, with regard to this specific aspect, techniques that make use of implicit function interpolation do not create self-intersecting surfaces.

An implicit function, in particular, can be considered as a relationship in which the dependent variable is not isolated on one side of the equation. Taking into account what just stated, an implicit surface is defined as a locus in space that satisfies the following equation:

$$s(x, y, z) = T \tag{5.1}$$

where T is the threshold of an isosurface. Due to the efficiency and the relative simplicity in modelling smooth and deformable surface through the use of implicit surfaces, these latter are extensively employed for modelling and treating complex objects in applications concerning, among others, constrained space deformation, surface reconstruction, transformation and modelling, computer assisted surgical planning, as well as mesh repairing and sketching (Jin et al. 2003).

M. E. Biancolini, *Fast Radial Basis Functions for Engineering Applications*,
https://doi.org/10.1007/978-3-319-75011-8_5

Adopting RBF, interpolated implicit surfaces were successfully used, for example, to reconstruct solids from scattered surface points and contours (Savchenko et al. 1995) and, successively, to build up a shape transformation tool (Turk and O'Brien 1999). As a matter of fact, describing interpolating implicit surfaces through RBF allows to directly specify surface points and surface normals with closed form solutions.

For what stated above, RBF can thus be effectively used in surface morphing and projection. In particular, this latter feature is the objective of the present section which aims at describing two RBF-based approaches for surface projection: the one hereinafter referred to as classic (Jin et al. 2003) and a novel one proposed by the author of the book. As demonstrated in Sect. 3.5.1 surface projection can be effectively accelerated using POU (the novel approach is the one demonstrated in the example); application of POU for the same purpose using the classic approach using fast RBF is presented by Tobor et al. (2006). Classic approach can be accelerated using FMM as demonstrated by Carr et al. (2001, 2003).

More complex applications of the novel RBF-based projection tool concerning stress analysis in the biomechanics sector are described in Sect. 7.2.

5.1 Point Projection onto the RBF Implicit Surface

In the following sections, a novel and high quality approach for projecting a point onto a tessellated surface, typically identified by a portion of the surface mesh of a geometrical model, is presented. This method is very useful and effective when a given mesh needs to be changed and adjusted onto a new configuration, herein called target surface, while preserving the original topology.

This novel approach will be compared to a classic one from which differs in the way the projection operator is set up. Both methods foresee at first the generation of an implicit surface interpolating the target mesh. Such an operation is carried out by means of RBF using the original mesh nodes and a set of offset nodes generated along the nodal normal vectors. This latter information is employed to define a scalar function f used to perform the point projection. In particular, the classic approach interpolates the scalar field through RBF imposing $f = 0$ at the points onto the surface and $f = \pm 1$ at the offset points, and applies the projection field by calculating the gradient of the function f. In a different way, the new approach, in which the three components of the projection field are fitted using an RBF field, envisages the setting of a non-null vector that connects each offset point with the original point of the target mesh and a null vector at the original points.

The formulation of both the outlined approaches will be provided using Mathcad software worksheets. Successively, these methods are compared considering a simple example concerning a sphere and, finally, the use of the novel approach, currently implemented in the commercial software RBF Morph, is applied to a more complex case concerning the creation of a fillet along a cube edge.

The working hypothesis for both test cases is that the target mesh is a triangular mesh in STL format. The normal of mesh nodes can be straightforwardly evaluated estimating it by averaging the normal of its neighbouring triangles exploiting the connectivity information provided by the input triangular mesh.

5.1.1 Processing of Target Mesh for Offset Points Generation

As already introduced, the definition of the offset points, required by both the presented approaches, is a crucial task and normal length validation is an issue to circumvent a bad surface reconstruction, namely the occurrence of overlapping normal vectors. To this end, the implicit surface is at first defined on the triangular elements of the target mesh whose nodal normals are initially unknown. To get such an information, the nodal normal at each node is calculated by averaging the normals of the connected elements. The element oriented areas can be determined as:

$$\begin{aligned}
\text{Area(iel)} :=\; & \text{for inod} \in 1..3 \\
& \quad \text{IDnod} \leftarrow \text{elements}_{\text{iel, inod+2}} \\
& \quad P_{\text{inod}} \leftarrow \text{submatrix(nodes , IDnod , IDnod ,2,4)}^{T} \\
& r31 \leftarrow P_3 - P_1 \\
& r21 \leftarrow P_2 - P_1 \\
& s3 \leftarrow r21 \times r31 \\
& e3 \leftarrow s3
\end{aligned} \tag{5.2}$$

The nodal normals are calculated in a two-stage process: first each element area is accumulated on the connected nodes, and the obtained values are then normalised to the same length according to what is following formulated:

$$\begin{aligned}
\text{Normals} :=\; & \text{comment} \leftarrow \text{"init"} \\
& \text{for } i \in 1..\text{rows (nodes)} \\
& \quad \text{Normals}_i \leftarrow \begin{pmatrix} 0 \\ 0 \\ 0 \end{pmatrix} \\
& \text{comment} \leftarrow \text{"accumulate oriented surfaces on nodes"} \\
& \text{for iel} \in 1..\text{rows (elements)} \\
& \quad \text{Nelement} \leftarrow \text{Area(iel)} \\
& \quad \text{for inod} \in 1..3 \\
& \quad\quad \text{IDnod} \leftarrow \text{elements}_{\text{iel, inod+2}} \\
& \quad\quad \text{Normals}_{\text{IDnod}} \leftarrow \text{Normals}_{\text{IDnod}} + \frac{\text{Nelement}}{3} \\
& \text{comment} \leftarrow \text{"normalise"} \\
& \text{for } i \in 1..\text{rows (nodes)} \\
& \quad \text{Normals}_i \leftarrow \text{versor}\left(\text{Normals}_i\right) \\
& \text{out} \leftarrow \text{Normals}
\end{aligned} \tag{5.3}$$

The offset points can now be generated using the distribution of the points x_{kb} on the surface by prescribing the values of the offset δ_n:

$$x_{kp} := \left| \begin{array}{l} \text{for } in \in 1..\text{rows}(x_{kb}) \\ \quad xkp_{in} \leftarrow x_{kb_{in}} + \text{Normalib}_{in} \cdot \delta_n \\ \text{out} \leftarrow xkp \end{array} \right. \qquad x_{km} := \left| \begin{array}{l} \text{for } in \in 1..\text{rows}(x_{kb}) \\ \quad xkp_{in} \leftarrow x_{kb_{in}} - \text{Normalib}_{in} \cdot \delta_n \\ \text{out} \leftarrow xkp \end{array} \right. \tag{5.4}$$

Note that the subscript "b" is used to flag the points processed on the boundary after sub-sampling (and related normal directions). RBF are very smooth and usually just a small subset of the surface points is able to represent the surface (the accuracy can be controlled calculating the distance between the surface and the non-retained point). The tool presented in Sect. 4.3 can be used for this specific task tweaking the resolution until a satisfactory result is obtained.

5.2 Classic Approach

According to the classic approach an interpolating implicit surface can be seen as a generalization of a thin-plate surface for scattered data interpolation. Adopting the notations used in (Jin et al. 2003), that in turn refer to the one earlier employed by Turk and O'Brien (Turk and O'Brien 2002), in a tri-dimensional space the scattered data interpolation can be formulated as described below. Assigned N distinct points $\{\boldsymbol{x}_{s_1}, \boldsymbol{x}_{s_2}, \ldots, \boldsymbol{x}_{s_N}\}\, in\, \mathbb{R}^3$ and their corresponding real values $\{g_{s_1}, g_{s_2}, \ldots, g_{s_N}\}$, determine a smooth unknown function $s(\boldsymbol{x})$ such that $s(\boldsymbol{x}_{s_i}) = g_{s_i}, i = 1, 2, .., N$. The smoothness of $s(\boldsymbol{x})$ is measured through the minimization of the energy functional:

$$E(s) = \int_{\mathbb{R}^3} \left| \frac{\partial^2 s}{\partial x^2} + \frac{\partial^2 s}{\partial y^2} + \frac{\partial^2 s}{\partial z^2} + 2\frac{\partial^2 s}{\partial x \partial y} + 2\frac{\partial^2 s}{\partial x \partial z} + 2\frac{\partial^2 s}{\partial y \partial z} \right| dx\, dy\, dz \tag{5.5}$$

This energy minimization problem can be solved using RBF through the analytical interpolant:

$$s(\boldsymbol{x}) = \sum_{i=1}^{N} \gamma_i \varphi(\|\boldsymbol{x} - \boldsymbol{x}_{s_i}\|) + h(\boldsymbol{x}) \tag{5.6}$$

where γ_i are real-valued weights, φ are the RBF and $h(\boldsymbol{x})$ is a one-degree polynomial accounting for the linear and constant portions of s (Turk and O'Brien 2002).

Choosing $h(\boldsymbol{x}) = \beta_1 + \beta_2 x + \beta_3 y + \beta_4 z$ as a linear polynomial can guarantee that the resulting surface is affine invariant to the assigned scattered points. Let

$\boldsymbol{x}_{s_i} = \left(x^x_{s_i}, x^y_{s_i}, x^z_{s_i}\right)$, so the sought weights γ_i and polynomial coefficients β_1, β_2, β_3 and β_4 can be determined by the interpolation conditions (5.7):

$$s(\boldsymbol{x}_{s_i}) = \sum_{j=1}^{N} \gamma_j \varphi\left(\left\|\boldsymbol{x}_{s_i} - \boldsymbol{x}_{s_j}\right\|\right) + h(\boldsymbol{x}_{s_i}), \quad i = 1, 2, \ldots, N \tag{5.7}$$

and the orthogonality conditions (5.8) (Carr et al. 2001):

$$\sum_{j=1}^{N} \gamma_j = \sum_{j=1}^{N} \gamma_j x^x_{s_j} = \sum_{j=1}^{N} \gamma_j x^y_{s_j} = \sum_{j=1}^{N} \gamma_j x^z_{s_j} = 0 \tag{5.8}$$

Let $\phi_{ij} = \varphi\left(\left\|\boldsymbol{x}_{s_i} - \boldsymbol{x}_{s_j}\right\|\right)$, $\boldsymbol{x} = (x, y, z)$, $h(\boldsymbol{x}) = \beta_1 + \beta_2 x + \beta_3 y + \beta_4 z$, then Eqs. (5.7) and (5.8) will lead to the linear system (5.9).

$$\begin{bmatrix} \phi_{11} & \phi_{12} & \cdots & \phi_{1N} & 1 & x^x_{S_1} & x^y_{S_1} & x^z_{S_1} \\ \phi_{21} & \phi_{22} & \cdots & \phi_{2N} & 1 & x^x_{S_2} & x^y_{S_2} & x^z_{S_2} \\ \vdots & \vdots & \vdots & \vdots & \vdots & \vdots & \vdots & \vdots \\ \phi_{n1} & \phi_{n2} & \cdots & \phi_{NN} & 1 & x^x_{S_N} & x^y_{S_N} & x^z_{S_N} \\ 1 & 1 & \cdots & 1 & 0 & 0 & 0 & 0 \\ x^x_{S_1} & x^x_{S_2} & \cdots & x^x_{S_N} & 0 & 0 & 0 & 0 \\ x^y_{S_1} & x^y_{S_2} & \cdots & x^y_{S_N} & 0 & 0 & 0 & 0 \\ x^z_{S_1} & x^z_{S_2} & \cdots & x^z_{S_N} & 0 & 0 & 0 & 0 \end{bmatrix} \begin{bmatrix} \gamma_1 \\ \gamma_2 \\ \vdots \\ \gamma_N \\ \beta_1 \\ \beta_2 \\ \beta_3 \\ \beta_4 \end{bmatrix} = \begin{bmatrix} g_{S_1} \\ g_{S_2} \\ \vdots \\ g_{S_N} \\ 0 \\ 0 \\ 0 \\ 0 \end{bmatrix} \tag{5.9}$$

The matrix of the system is symmetric and positive semi-definite, and its unique solution can be evaluated by direct LU decomposition. Adopt tri-harmonic spline $\phi(\boldsymbol{r}) = |\boldsymbol{r}|^3$ for 3D interpolation as done in Turk and O'Brien (Turk and O'Brien 2002), the interpolant gets the form (5.10).

$$s(x, y, z) = \sum_{j=1}^{N} \gamma_j \left(\sqrt{\left(x - x^x_{s_j}\right)^2 + \left(y - x^y_{s_j}\right)^2 + \left(z - x^z_{s_j}\right)^2}\right)^3 + \beta_1 + \beta_2 x + \beta_3 y + \beta_4 z \tag{5.10}$$

The interpolating implicit surface can be thus generated using the concept of 3D scattered data interpolation using boundary constraints and normal constraints (Turk and O'Brien 2002). In particular, the mesh nodes form a set of boundary constraints where the implicit function takes on zero, whilst the normals form a set of normal constraints where the function takes on one. The level set of the implicit surface forms a surface interpolating the vertices of the input triangular mesh. In such a manner, the resulting surface is gained by means of a variational interpolation problem.

In the considered work (Jin et al. 2003) the built up mathematical framework was finally used to map the new vertex added at the midpoint of each edge of the target mesh onto the corresponding points on the implicit surface, with the objective to obtain the polygonised mesh of the interpolating implicit surface. Such a procedure is carried out by Newton's iteration method (Hartmann 1998; Karkanis and Stewart 2001). The gradient of the function *f*(x,y,z) is $\nabla s = \left(\frac{\partial s}{\partial x}, \frac{\partial s}{\partial y}, \frac{\partial s}{\partial z}\right)$ where:

$$\begin{aligned}
\frac{\partial s}{\partial x} &= 3\sum_{j=1}^{N} \gamma_j \left(x - x_{s_j}^x\right)\left(\sqrt{\left(x - x_{s_j}^x\right)^2 + (y - x_{s_j}^y)^2 + \left(z - x_{s_j}^z\right)^2}\right) + \beta_2 \\
\frac{\partial s}{\partial y} &= 3\sum_{j=1}^{N} \gamma \left(y - x_{s_j}^y\right)\left(\sqrt{\left(x - x_{s_j}^x\right)^2 + (y - x_{s_j}^y)^2 + \left(z - x_{s_j}^z\right)^2}\right) + \beta_3 \\
\frac{\partial s}{\partial z} &= 3\sum_{j=1}^{N} \gamma_j \left(z - x_{s_j}^z\right)\left(\sqrt{\left(x - x_{s_j}^x\right)^2 + (y - x_{s_j}^y)^2 + \left(z - x_{s_j}^z\right)^2}\right) + \beta_4
\end{aligned} \tag{5.11}$$

The mapped point on implicit surface for a new added vertex $\boldsymbol{r}_0$ can then be calculated iteratively by (5.12).

$$\boldsymbol{x}_{k+1} = \boldsymbol{x}_k + \frac{s(\boldsymbol{x}_k)}{\|\nabla s(\boldsymbol{x}_k)\|^2} \nabla s(\boldsymbol{x}_k) \tag{5.12}$$

The above iteration runs until $\|\boldsymbol{x}_{k+1} - \boldsymbol{x}_k\|$ is less than a given tolerance.

Now the implementation of the classic method using the Mathcad software is provided following the main passages previously deepened. As above described, the identified scalar function s is set to a fixed value at the offset points generated using all the target mesh nodes, usually 1 at positive side and -1 at negative side, while it is forced to get the 0 value at their exact locations.

Starting from the Mathcad data already supplied for target surface processing (see Sect. 5.1.1), if normal vectors have different lengths, a better choice is to use the distance from the surface as the scalar value with sign depending on the spatial side. In this case, the RBF scalar function is fitted using all points, namely sub-sampled points on-surface and off-surface in both directions, prescribing the following values for the scalar functions at source points:

```
g_kb := | for in ∈ 1..rows(x_kb)      g_kp := | for in ∈ 1..rows(x_kb)      g_km := | for in ∈ 1..rows(x_kb)
        |    gkb_in ← 0                       |    gkp_in ← δ_n                     |    gkp_in ← -δ_n
        | out ← gkb                           | out ← gkp                           | out ← gkp
```
(5.13)

where the scalar at offset points is equal to the offset distance (with positive sign on the positive side and negative on the other) so that the scalar function has the physical meaning of distance from the surface.

Once the RBF problem is fitted, it is possible to evaluate the function f_{surf} that defines the implicit surface as follows:

$$f_{surf}(x) := \sum_{i=1}^{rows(x_k)} (\gamma_i \cdot \phi(|x - x_{k_i}|)) + \beta \cdot stack(1, x) \tag{5.14}$$

where the cloud is comprised of on-surface and off-surface points:

$$\begin{aligned} x_k &:= stack(x_{kb}, x_{kp}, x_{km}) \\ g_k &:= stack(g_{kb}, g_{kp}, g_{km}) \end{aligned} \tag{5.15}$$

Such a function is defined everywhere and can be evaluated, for instance, along a direction from the origin to a given point using ray tracing to find the surface (Fig. 5.1): when the function reaches zero an intersection between the ray and the surface is found and resulting point belongs to the implicit surface.

The target surface is represented by the isosurface of the scalar function *f* at 0. This means that the gradient of the function is directed in the normal direction as defined in Sect. 2.4. A projection algorithm can be defined exploiting this vector. Differentiating the function f_{surf}, the following closed form for the gradient is hence obtained:

$$gradf_{surf}(x) := \sum_{i=2}^{rows(x_k)} \left[\frac{\gamma_i \cdot \phi p(|x - x_{k_i}|)}{2 \cdot |x - x_{k_i}|} \cdot \begin{bmatrix} 2 \cdot [x_1 - (x_{k_i})_1] \\ 2 \cdot [x_2 - (x_{k_i})_2] \\ 2 \cdot [x_3 - (x_{k_i})_3] \end{bmatrix} \right] + \begin{pmatrix} \beta_2 \\ \beta_3 \\ \beta_4 \end{pmatrix} \tag{5.16}$$

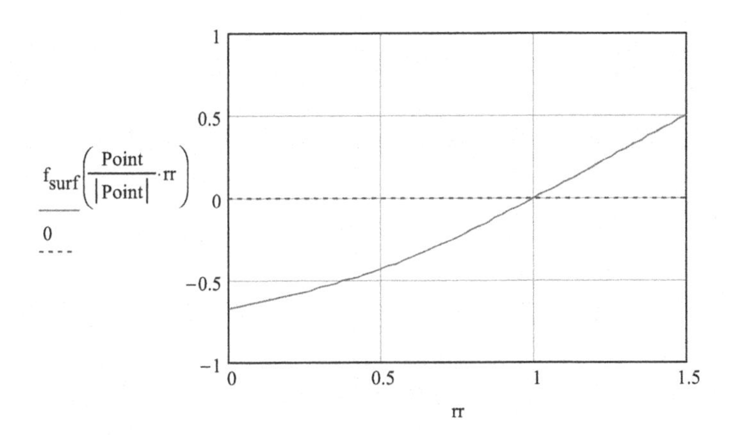

Fig. 5.1 The implicit function fsurf(·) can be computed everywhere in $\mathbb{R}^3$. In the plot the function is evaluated along a line from the origin defined by a given Point onto the target mesh. The scalar variable rr spans the space by multiplying the unit vector obtained dividing the vector connecting the point with the origin by its length

where $\phi p(r)$ is the first derivative of the radial function. The projection of a point onto the surface can be achieved using the Newton method adopting the point updating formula (5.17).

$$Pnew(P_{old}) := P_{old} - f_{surf}(P_{old}) \cdot \frac{gradf_{surf}(P_{old})}{gradf_{surf}(P_{old}) \cdot gradf_{surf}(P_{old})} \tag{5.17}$$

Whilst an iteration scheme can be defined prescribing the surface evaluation tolerance:

$$\text{project}(Px, \varepsilon) := \left| \begin{array}{l} P1 \leftarrow Px \\ it \leftarrow 1 \\ err \leftarrow \varepsilon \cdot 2 \\ \text{while } err > \varepsilon \wedge it < 30 \\ \quad \left| \begin{array}{l} P2 \leftarrow Pnew(P1) \\ err \leftarrow |P2 - P1| \\ P1 \leftarrow P2 \\ it \leftarrow it + 1 \end{array} \right. \\ out \leftarrow P2 \end{array} \right. \tag{5.18}$$

5.3 New Method: Projection Field

The same off-surface points calculated for the classic method are used in the new method. As previously introduced, the main difference is that a vector field is interpolated at each point using RBF. The vector that connects the current offset point to the corresponding point on the target surface is used as known term of the RBF field at each off-surface point, whereas a null vector is prescribed at each node of the target surface. The resulting field, which attracts points on the tessellated surface with an intensity proportional to the distance is given by the following expression:

$$g_{kfield} := \left| \begin{array}{l} Npoint \leftarrow rows(x_{km}) \\ \text{for } i \in 1..Npoint \\ \quad \left| \begin{array}{l} \left(G_{i+Npoint,1} \;\; G_{i+Npoint,2} \;\; G_{i+Npoint,3} \right) \leftarrow (0 \;\; 0 \;\; 0) \\ \left(G_{i,1} \;\; G_{i,2} \;\; G_{i,3} \right) \leftarrow \left(x_{kb_i} - x_{km_i} \right)^T \\ \left(G_{i+2 \cdot Npoint,1} \;\; G_{i+2 \cdot Npoint,2} \;\; G_{i+2 \cdot Npoint,3} \right) \leftarrow \left(x_{kb_i} - x_{kp_i} \right)^T \end{array} \right. \\ out \leftarrow G \end{array} \right. \tag{5.19}$$

The point updating equation changes in the following form:

$$P_{new\_field}(P_{old}) := P_{old} + \text{smoother}(\gamma_{field}, \beta_{kfield}, x_{kfield}, g_{kfield}, P_{old}) \tag{5.20}$$

where smoother is defined as follows:

$$\text{smoother}(\gamma, \beta, x_k, g_k, x) : = \sum_{i=1}^{\text{rows}(x_k)} \left[\begin{bmatrix} (\gamma_1)_i \\ (\gamma_2)_i \\ (\gamma_3)_i \end{bmatrix} \cdot \phi(|x - x_{k_i}|) \right] + \begin{pmatrix} \beta_1 \cdot \text{stack}(1, x) \\ \beta_2 \cdot \text{stack}(1, x) \\ \beta_3 \cdot \text{stack}(1, x) \end{pmatrix} \tag{5.21}$$

5.4 Implicit Surface Representation Test Cases

5.4.1 Spherical shell

The first test is conducted on a portion of a spherical shell centred in the origin and having a unity radius. Such a surface is depicted in Fig. 5.2 (left) together with the triangular elements of the mesh it is discretised into (right). A not uniform grid spacing has been intentionally used for generating the mesh to test the robustness of the surface representation method.

Fig. 5.2 Portion of the spherical shell (left) and its triangular mesh (right)

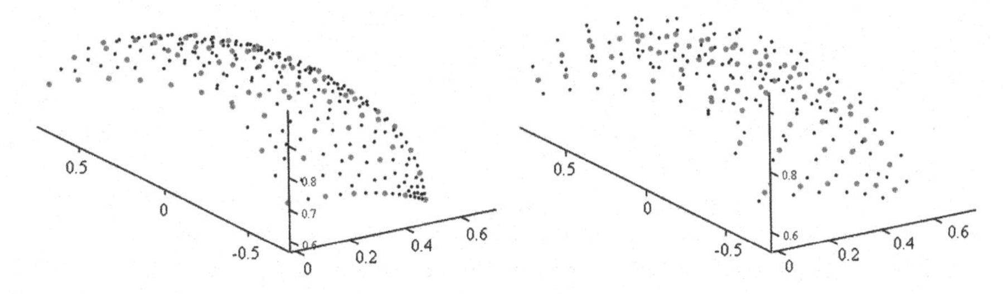

Fig. 5.3 Subsampled set of mesh nodes (left) and offset points generation (right)

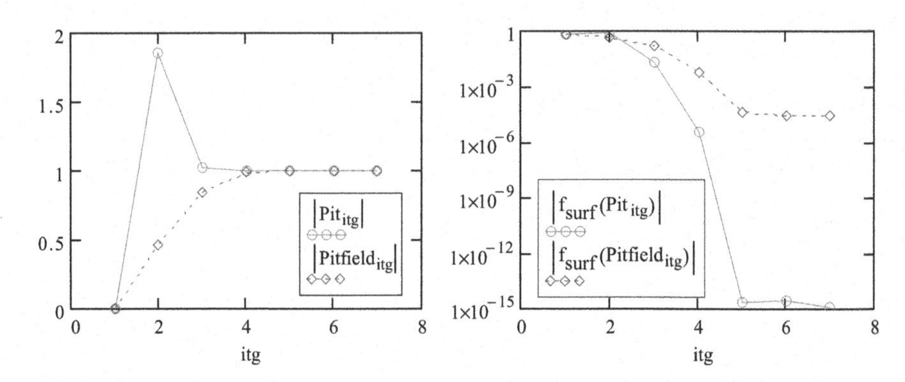

Fig. 5.4 Projection convergence plots computed by both methods: distance from the centre (left) and tolerance (right)

The acquired points distribution is shown in Fig. 5.3 in black colour, where the subsampled set, selected on a fixed distance basis, is highlighted using a different colour and a larger size (left). In the same figure the complete points distribution including off-surface points used for implicit surface definition, is depicted. In particular, the offset δ_n has been set equal to 0.05 (right).

The effectiveness of both methods has been investigated inspecting how a probe point is projected onto the target surface. As the original geometry was a sphere, the correctness of the final position can be easily checked. Projection convergence paths of a probe point initially located at the centre of the sphere is demonstrated by the plots of Fig. 5.4. The traditional method first moves the point beyond the sphere surface whilst the new method shows a smooth and monotonic progression toward the surface; the accuracy of the traditional method is higher as highlighted inspecting the error in the logarithmic plot.

5.4.2 Add a Fillet on a Cube Edge

The practical application of the implicit surface is demonstrated in the present test case that aims at creating a fillet on a cube with edges 1 m in length. The starting triangular surface mesh of the cube is shown in Fig. 5.5 (left) and it is coloured in pink colour.

In particular, the rounding of one of the sharped edges of the cube is gained using, as target, an STL file created with two plane surfaces blended by a curved surface with radius of 0.2 m. The mesh is shown in Fig. 5.5 (left) in yellow colour. The STL file, in particular, is supposed to be generated and suitably discretized by the user such to properly capture the target geometry. The final effect of the morphing action application is visualised in Fig. 5.5 (right).

The implemented projection field based on an RBF interpolation, foresees that the points and normals of the target tessellated surfaces to be morphed are used to

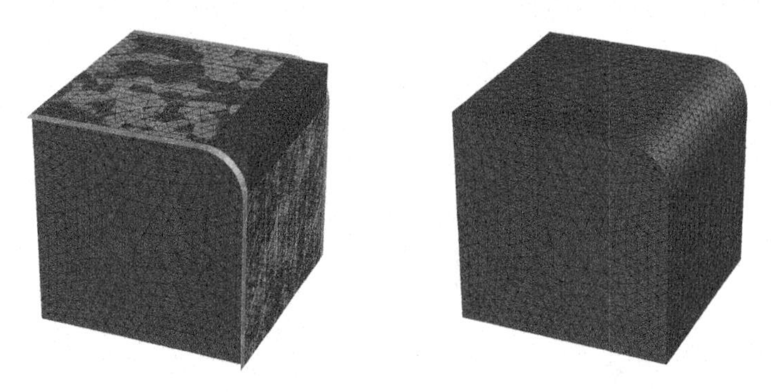

Fig. 5.5 The starting mesh of the cube and target mesh (left) and morphed mesh (right)

generate off-surface points with respect to both sides of such surfaces. The produced vector field to attract morphed points onto the target surface is defined by assigning, at each point, the vector that connects the off-surface points to the connected on-surface one, taking care that the RBF field is zero valued for these latter (points laying on the surfaces).

Figure 5.6 shows the location of both on-surface and off-surface points in proximity of the target tessellated surface (left) as well as the morphed surface mesh superposed to the target mesh (right).

Figure 5.7 (left) shows an overlay of two profiles: the profile of the analytic definition of the fillet and of the one obtained through projection, namely morphing the surface mesh. This latter profile is obtained by cutting the rounded tessellated surfaces with a plane parallel to the x-y plane of the cube mesh. An error check procedure, in which all the points are examined and compared with the analytic representation of the surfaces, is performed. The results of this verification process, depicted in Fig. 5.7 (right), allow to judge that the error is lower than 0.4%.

Fig. 5.6 Creation of on-surface and off-surface points (left) and morphed mesh superposed to target mesh (right)

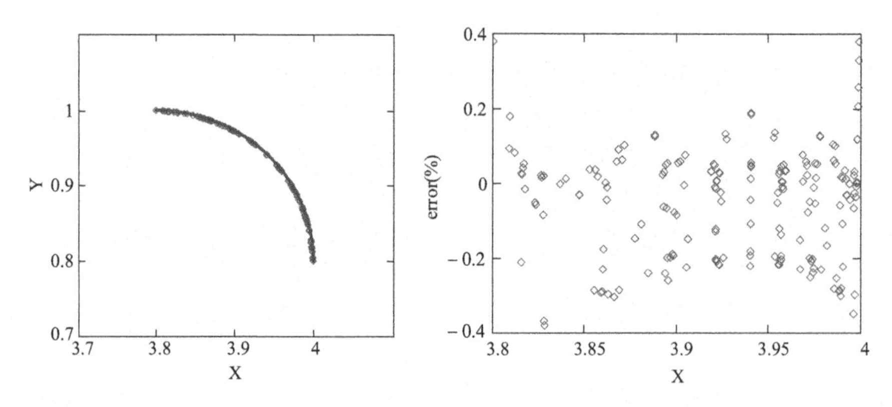

Fig. 5.7 Comparison between morphed points and the analytic definition of the fillet (left) and error distribution (right)

The error distribution evidences that an accurate tracking of the analytical fillet is achieved even at most critical areas where an abrupt change of curvature is imposed. Such a finding demonstrates the accuracy of the implemented projection algorithm.

References

Carr JC, Beatson R, Cherri J, Mitchell T, Fright W, McCallum B (2001) Reconstruction and representation of 3D objects with radial basis functions. In: Proceedings of the 28th annual conference on Computer graphics and interactive techniques, Los Angeles, CA, p 67–76

Carr JC, Beatson RK, McCallum BC, Fright WR, McLennan TJ, Mitchell TJ (2003) Smooth surface reconstruction from noisy range data. In: First international conference on computer graphics and interactive techniques, p 119

Hartmann EA (1998) Marching method for the triangulation of surfaces. Vis Comput 14(3): 95–108

Jin X, Sun H, Peng Q (2003) Subdivision interpolating implicit surfaces. Comput Graph 27: 763–772. https://doi.org/10.1016/S0097-8493(03)00149-3

Karkanis T, Stewart J (2011) Curvature dependent triangulation of implicit surfaces. IEEE Comput Graphics Appl 21(2):60–9

Savchenko VV, Pasko AA, Okunev OG, Kunni TL (1995) Function representation of solids reconstructed from scattered surface points and contours. Comput Graph Forum 14(4):181–188

Tobor I, Reuter P, Schlick C (2006) Reconstructing multi-scale variational partition of unity implicit surfaces with attributes. Graph Models 68(1):25–41

Turk G, O'Brien JF (2002) Modeling with implicit surfaces that interpolate. ACM Trans Graph 21(4):855–873

Turk G, O'Brien JF (1999) Shape transformation using variational implicit functions. In: SIGGRAPH '99 Proceedings of the 26th annual conference on Computer graphics and interactive techniques, ACM Press/Addison-Wesley Publishing Co. New York, NY, USA, pp 335–342, ISBN:0-201-48560-5, https://doi.org/10.1145/311535.311580

Chapter 6
RBF Mesh Morphing

Abstract This chapter concerns the use of RBF for mesh morphing in computer-aided engineering (CAE) applications. The main paradigms, including the typical morphing strategies, are provided with a specific focus on surface mesh morphing. The effective use of the reported concepts and numerical means is concisely showcased through some test cases whose RBF is just mentioned because its exhaustive description is one of the principal objectives of the three subsequent chapters pertaining the solution of CAE applications coming from different engineering sectors.

6.1 RBF Mesh Morphing Basic Principles

When RBF are used for mesh morphing the three components of a displacement field, that is typically assigned at a cloud of control points, here defined as RBF centres or source points, are interpolated in the space and used to update the nodal positions of the mesh nodes to be morphed.

$$\begin{cases} s_x(\boldsymbol{x}) = \sum_{i=1}^{N} \gamma_i^x \varphi(\|\boldsymbol{x} - \boldsymbol{x}_{s_i}\|) + \beta_1^x + \beta_2^x x + \beta_3^x y + \beta_4^x z \\ s_y(\boldsymbol{x}) = \sum_{i=1}^{N} \gamma_i^y \varphi(\|\boldsymbol{x} - \boldsymbol{x}_{s_i}\|) + \beta_1^y + \beta_2^y x + \beta_3^y y + \beta_4^y z \\ s_z(\boldsymbol{x}) = \sum_{i=1}^{N} \gamma_i^z \varphi(\|\boldsymbol{x} - \boldsymbol{x}_{s_i}\|) + \beta_1^z + \beta_2^z x + \beta_3^z y + \beta_4^z z \end{cases} \tag{6.1}$$

The field of (2.9), here rewritten in (6.1), is applied to process all the nodal positions to be updated according to Eq. (6.2).

$$\boldsymbol{x}_{node\_new} = \boldsymbol{x}_{node} + \begin{bmatrix} s_x(\boldsymbol{x}_{node}) \\ s_y(\boldsymbol{x}_{node}) \\ s_z(\boldsymbol{x}_{node}) \end{bmatrix} \tag{6.2}$$

M. E. Biancolini, *Fast Radial Basis Functions for Engineering Applications*,
https://doi.org/10.1007/978-3-319-75011-8_6

The definition of the RBF problem, i.e. the arrangement of the cloud of RBF centres and their input values, is the key enabler for RBF based mesh morphing and, considering its meshless nature and the great flexibility offered by the RBF mathematics, there are a variety of options (all working!) to impose a desired morphing action. An overview of RBF mesh morphing strategies was recently published by Cella et al. (2016), whilst a deeper presentation about the use of RBF mesh morphing in industrial applications can be found in Biancolini (2012).

Mastering RBF mesh morphing techniques requires some experience and a deep knowledge of the method so that the optimal strategy can be defined for a specific application. In this chapter some useful elements are provided; however, RBF tools are so flexible that new or improved paradigms can be defined for new applications. This great flexibility, that sometimes is a source of confusion for beginners, has to be considered as a great opportunity to exploit the still undiscovered potential of this young mathematical method for CAE applications. Beside the strategy adopted, it is important to say that a successful application of RBF mesh morphing is the one that allows to update (in a single passage or in a parametric fashion) the mesh according to the wanted shape modification while preserving at best the quality of the mesh after the deformation.

6.1.1 RBF Mesh Morphing Main Advantages and Drawbacks

There are many advantages of the RBF method that makes it very attractive in the area of mesh morphing and smoothing. Consider again Eqs. (6.1) and (6.2) that highlight the meshless nature of the approach as the deformed position depends on the original position to be moved only; this means that RBF are able to manage every kind of mesh element type (tetrahedral, hexahedral, polyhedral, etc.) and, consequently, the RBF fit can be efficiently used for both surface and volume mesh smoothing ensuring the preservation of their topology, namely the same number of nodes and cells with the same typology. Furthermore, due to the meshless characteristic, RBF mesh morphing may be exploited with all solvers (structural, fluid-dynamic, electromagnetic, etc.) providing that mesh nodes' position can be altered. As only grid points are moved regardless of the connected elements, RBF are also very suitable for parallel implementation and, thus, quite prone to be run on high performance computing (HPC) systems and embedded in HPC tools. In fact, once the RBF solution is known and shared in the memory of every calculation node of the cluster for instance, each partition has the ability to smooth its nodes without taking care of what happens outside the partition because the smoother is a global point function and the continuity at interfaces between partitions is implicitly guaranteed. In turn, the parallel calculation enables the morphing of large size models (many millions of cells) in a short time and with a low memory requirements as well because just the parameters of the RBF solutions need to be allocated, and not the large amount of data related to the morphed mesh.

There are indeed many other advantages that are related to the fact that mesh morphing preserves the original model topology. Providing that the deformation does not decrease the quality of the CAE model, RBF allow to prevent the noise due to remeshing. This is especially true if medium/low fidelity models are used (i.e. with a known fidelity limitation due to mesh refinement) because the designer is sure that a variation of the sought output (e.g. an objective function value) following the application of morphing is just caused by the shape variation rather than the uncertainty connected to the generation of a new mesh.

Keeping the same mesh topology allows to quickly compare CAE results computed on the morphed mesh; this could be useful for optimisation post-processing, so that areas where the computed field are changed by the new shape can be easily identified; it can be useful for statistical purposes as well, when variation of the CAE shape is used to represent all the items of a population of similar objects. Having the same topology is relevant also for advanced connection with CAE solvers. As a matter of fact, it is an effective way to interact with adjoint solvers that allows to drive the morphing according to sensitivities or to feed the morphing parameters derivatives according to fields. Furthermore, it allows to get the CAE model parametric with respect to structural modes, for FSI analyses, or other decomposition (as the proper orthogonal decomposition) for the definition of reduced order models (ROM). The synergy of RBF mesh morphing and ROM is demonstrated in the paper by Luboz et al. (2017).

Another positive characteristic of RBF mesh morphing is the precision. The exact control of nodes movement allows to get the exact preservation of surfaces making RBF better than free form deformation (FFD) methods (Sederberg and Parry 1986). This means that a node-wise control is possible and complex shapes, as the ones coming from other CAE results (adjoint, FEA, …), can be assigned.

As far as drawbacks are concerned, these basically are the high computing demand when large RBF datasets (source nodes) need to be processed, the limitation in modifications extent, the deterioration of the mesh quality introduced by morphing and, in some situations, the need of a tool to generate the CAD model of the morphed mesh.

Nevertheless, there are several strategies and techniques that allow to mitigate and give an answer to the afore-mentioned issues. Relating to high computing requests it is possible to exploit fast methods (see Chap. 3), to delimit the portion of the mesh processed by morphing through auxiliaries entities (see Sect. 6.2.1) and, if feasible, to use HPC solutions that guarantee the application of morphing in parallel on many partitions of the whole computational domain. With regard to mesh quality degradation and limited extent in modifications, the designer may benefit the most suitable RBF selecting it among those available (compact or global support and with high order) depending on the specific case to be tackled.

Referring to the generation of the CAD model of the morphed mesh, there are two well-established options that can be considered. The first is the possibility to use the RBF to morph the CAD model with the same RBF fields employed for the mesh providing that the surface are created according to the non-uniform rational B-splines (NURBS) formulation; the coordinates of the CAD model are warped according to the RBF field and, providing that enough resolution is guaranteed by the NURBS degree, the surfaces are moved onto the deformed shapes and can be then adjusted and trimmed using the CAD software. The second option is related to the preservation of mesh topology; after deformation the associativity with original underlying surfaces is maintained. Advanced surface reconstruction tools, which major weakness is in the set-up of automatic skin detection, can be exploited imposing the same topology for the geometrical model to be regenerated.

6.1.2 RBF Mesh Morphing Application Scheme

There are a variety of possible usage of RBF mesh morphing; whatever is the technique adopted, the designer actions are usually addressed to the modification of the shape of a surface mesh or geometry. Such a variation, in many cases, requires the opportune smoothing of the volume mesh cells surrounding the morphed surfaces as well. From a computational point of view, as already introduced in Sect. 6.1.1, this latter operation may be really critical because the worsening of the mesh quality has to be kept as low as possible to guarantee the expected accuracy in the numerical outputs and, in some scenarios, to prevent the non usability of the mesh itself or the occurrence of numerical instabilities during the computing stage.

Taking into account the features of the RBF described in Chap. 2, if compact supported RBF are employed and no polynomial correction is added, the morphing action is automatically limited in space by their radius and, so, the designer is commonly not forced to impose restraints to the motion of surfaces that are far from the controlled region; the affected volume is in fact enveloped by the outer surfaces of all the spheres centred in the sources and with the support radius. On the other hand, if the global supported RBF are used, the long distance interaction are not vanishing and, then, the designer has to take care of prescribing a null displacement at mesh morphing far field; this effect can be achieved by selecting fixed points at surfaces that bound the morphing domain or by wrapping the volume within which the morphing action is applied with simple geometries as box, cylinders and spheres.

Given that, a great flexibility and expressiveness in RBF mesh morphing use can be achieved only if the designer is provided with various paradigms for an effective surface mesh morphing, requirements that can be straightforwardly fulfilled because plenty of morphing strategies and solutions may be considered and implemented. These topics are deepened in the following paragraphs dealing with RBF mesh morphing strategies, RBF solution arrangement and paradigms for surface mesh morphing.

6.1.3 *RBF Mesh Morphing Strategy and RBF Solution Arrangement*

To morph a mesh of a numerical model in view of modifying the surfaces of interest, two main strategies may be adopted: the first strategy foresees the application of all the deformation field with a single RBF action, whereas the second envisages the sequential application of more RBF subdividing the complex deformation in a certain number of steps. Distributing the deformation in small steps should allow a better propagation of the deformation; however, as far as the experience of the author, the added complexity does not always guarantee a better quality of the final morphed mesh. However, there are several situations in which the subdivision of the morphing in small increments is the only way to achieve the sought shapes as in the cases of the simulation computed morphing actions presented in Chap. 9 (ice and snow growth, crack propagation, Biological Growth Method) and in Chap. 8 (adjoint sensitivity driven surface sculpting).

Apart from the particular embraced strategy, an RBF solution can be arranged according to a lot of ways. As a matter of fact, to generate an RBF solution the designer can use either a single RBF solution or a combination of two, or more, RBF solutions. In addition, a single RBF solution can be expressed in the respect of a multi-step (hierarchical) approach and each single step can be designed in function of a simple geometrical relationship or, alternatively, driven by physical data or by a target entity.

The just mentioned concepts constitute the surface mesh morphing paradigms which are the topic detailed in Sect. 6.3.

6.2 Volume Mesh Morphing Paradigms

As already introduced, it is of fundamental importance limiting the lowering of the volume mesh quality caused by the morphing action as much as possible. To this end, the bi-harmonic kernel $\varphi(r) = r$ is the basis that guarantees to keep the mesh quality degradation at minimum, so it is in general advisable to use it to accomplish this task.

Global supported RBF, as $\varphi(r) = r$, requires the definition of source points to set zero the morphing field. In some cases (more common for the mesh morphing of FEA solid mesh) the morphing action is limited adopting as fixed RBF points all the nodes on the surfaces that are not controlled and are not directly neighbours of controlled surfaces; usually there are surfaces directly controlled, surfaces left free to be deformed and surfaces that are, again, directly controlled, but to be preserved at original configuration.

For the case of CFD mesh, or for large assembly of FEA mesh, aforementioned approach could be difficult to implement and results in too large RBF dataset. Auxiliary entities, as boxes, cylinders and spheres, can then be used to generate

RBF points suitable to locally control the RBF mesh morphing field. Auxiliary points can be generated at new locations on the skin of such auxiliary entities or at original mesh nodes locations of the volume mesh selected by specific Boolean operations. Only the first option is presented in the next sections of this chapter, an example of the second one is demonstrated in Fig 9.52 of Sect. 9.7.2.

6.2.1 Delimiting Encapsulation Domains

The first type of auxiliary feature is the delimiting encapsulation domain. This is a geometrical entity that consists of a set of surfaces with a set of RBF centres laying over them. Such surfaces identify either the boundaries of a single volume primitive such as box, sphere or cylinder (see Sect. 4.6 for detailed algorithms suitable for auxiliary clouds generation), or the ones of a more complex volume generated through Boolean operations among volume primitives.

The entity created according to this approach has a two-fold objective: at first it aims at delimiting the action of morphing into its volume and, furthermore, to specify the RBF centres where the morphing action vanishes. In the case that no RBF centres are generated, the intent is just to impose the first type of constraint during morphing.

An example of delimiting encapsulation domain, shown in Fig. 6.1 (Biancolini et al. 2013), is set to restrict the morphing in a box-shaped volume surrounding the nacelle of an aircraft CFD model.

Properly tuning the position and the dimensions of the delimiting encapsulation domain as well as the density (number) of the source points, the user may gain the

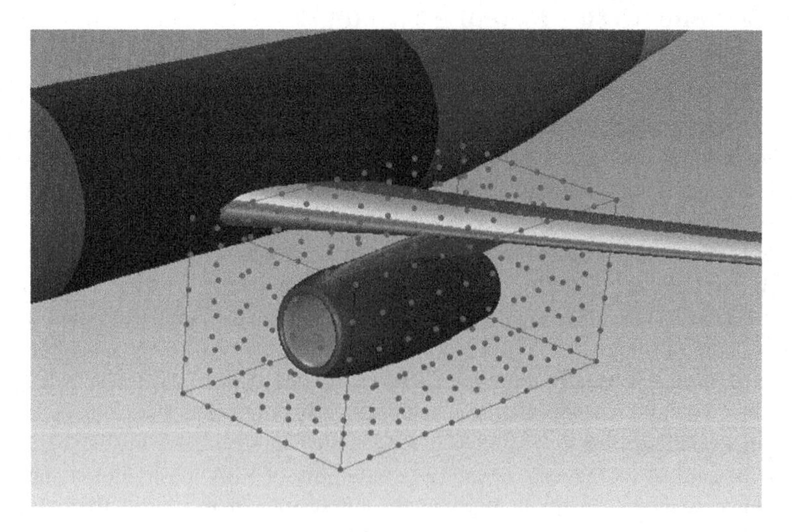

Fig. 6.1 Source nodes distribution over the surfaces of the delimiting encapsulation domain

wanted morphing result lowering a lot the computing request because only a portion of the whole volume mesh is processed to carry out the RBF fit and the successive morphing.

The size of such delimiting encapsulations has to be defined as a trade-off between performances and quality: large domains allow to distribute the morphing actions on more cells and to minimize the distortion, but more RBF points are used because of the larger encapsulations and because more surfaces of the original mesh are included; smaller domains allow to save performances as less RBF points are required but they could lead to bad quality of the deformed mesh.

6.2.2 Moving Encapsulation Domains

The second type of the auxiliary feature is the moving encapsulation domain. Similarly to the delimiting one, it can be generated through volume primitives but it is conceived to apply the same wanted motion to the mesh nodes wrapped by its boundary surfaces. In the case such a motion is null, the domain protects the included mesh nodes meaning that it maintains unchanged their position during morphing. Considering that to get the same result the user may be forced to employ the RBF centres extracted from the surface mesh that may have a very large number of nodes, also in this case the adoption of the moving encapsulation domain allows to relevantly diminish the computing requests.

The concept is explained by Fig. 6.2 that shows a moving encapsulation box-shaped domain created to apply a motion to a portion of the surfaces of an aircraft nacelle in order to perform a parametric CFD analysis. Only the pylon needs to be deformed by the RBF field and the entire nacelle (including all the interior mesh) undergoes a rigid movement. Just a few RBF centres on the box surface are used, together with the portion of the nacelle surface that is not wrapped by the box, to propagate the rigid movement of the nacelle in the volume mesh of the fluid. Adopting all the nodes of nacelle surface produces the same effect but with a substantial higher number of source points. The combined action of moving encapsulation (Fig. 6.2) (Biancolini and Gozzi 2013) and delimiting encapsulation (Fig. 6.1) allows to act with the RBF field just inside the delimiting and outside the moving; elsewhere simple actions (zero or rigid movement, scaling) are applied to the mesh.

The moving encapsulation domain can be utilized as an effective sculpting tool for surfaces as well. The idea is that the distribution of a moving encapsulation domain can be placed close to the surfaces to be sculpted or partially overlapped to them. An example of this further use is depicted in Fig. 6.3 where the intersection between the fuselage and the wing of a glider is reshaped using two cylinders as sculpting tools, and in Fig. 6.4 where the hull of a ship is reshaped adopting a distribution of boxes with different transversal scaling actions.

A complete RBF set-up to for an internal flow application is demonstrated in Fig. 6.5 (left) where one box-shaped delimiting encapsulation domain without

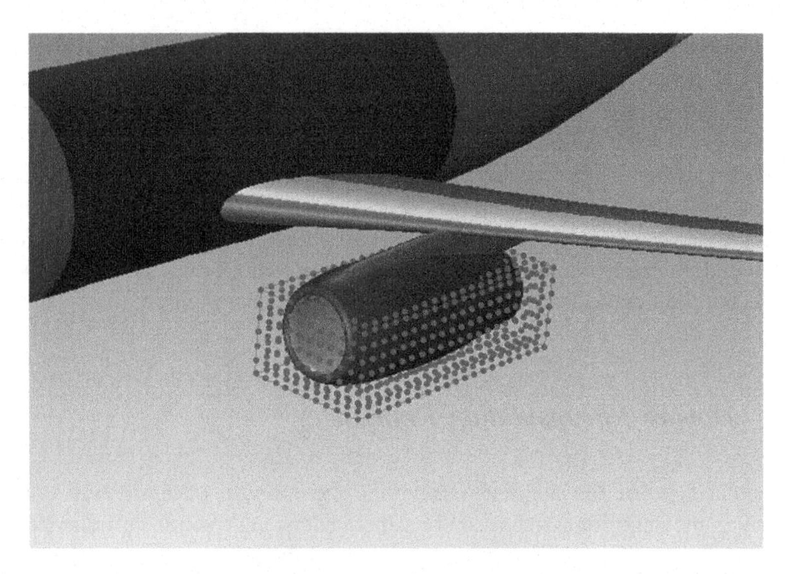

Fig. 6.2 Moving encapsulation domain to apply a displacement to a part of nacelle surfaces

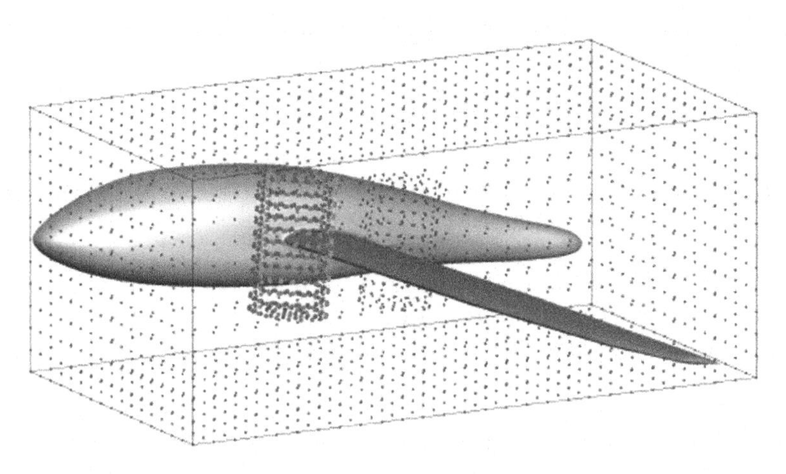

Fig. 6.3 The fuselage of the glider is reshaped using two moving cylinders as sculpting tools (Costa et al. 2014)

source points is created to apply the morphing action only to a part of the entire model mesh, whilst five moving encapsulation domains are employed to impose a wanted movement to the duct junctions in view of gaining the wanted deformation (Fig. 6.5 right). In particular, two cylinder-shaped moving encapsulation domains with zero motion are positioned close to the delimiting encapsulation domain to maintain unaltered the dimension and shape of the tube, whilst other three cylinder-shaped moving encapsulation domains are located at tube junctions to modify its shape causing a narrowing at the middle of the "U" turn.

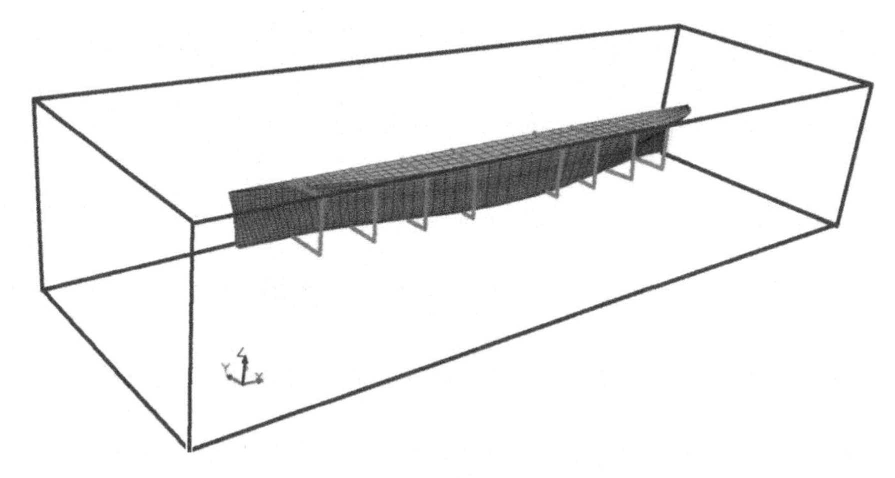

Fig. 6.4 Hull form is parametrised adopting scaling boxes (Pranzitelli and Caridi 2011)

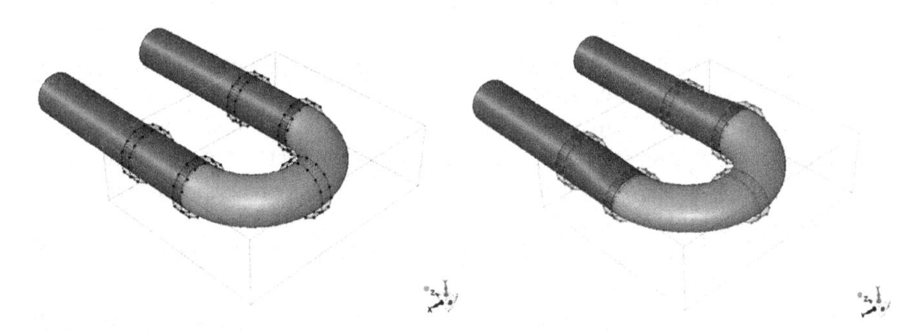

Fig. 6.5 Tube morphing performed using auxiliary entities only

6.3 Surface Mesh Morphing Paradigms

When defining a modelling paradigm, the context in which it will be used must be taken into account evaluating the technologies involved and then, as a consequence, the kind of interaction that should be guaranteed to the user.

Computer graphics (CG), for instance, is a field that allowed the development of many original approaches to modelling such as soft object animation (SOA) for the FFD techniques. Several algorithms were developed for CG with the aim of obtaining a geometry control oriented especially for the visualization of the entities of interest. For this reason, the modelling paradigms employed in three-dimensional CG are mostly interactive.

On the other hand, CAD systems are oriented to a more accurate control of surfaces and geometrical features as they are required to represent and manage geometries and information of engineering value. As such, over the years several approaches to modelling have been explored leading to the implementation of modelling and transformation algorithms that often exploit parametric geometries able to guarantee an accurate control over continuous geometries.

In CAE working environments the modelling paradigm has to deal with discretized geometrical representations (meshes) and, similarly to what developed for CAD systems, an accurate control of the wanted modifications is then required. Exploiting RBF, a continuous interpolation field describing translations and deformations can be achieved and, thus, the local quality requirements can be satisfied by accurately respecting the imposed boundary conditions. To generate this continuous field, a discrete number of RBF centres in the space must be generated and employed. Given that, the definition of such RBF centres shall be eased as much as possible by a suitable and effective set of tools.

Attempts for the definition of real time editing tools based on RBF were presented in the past by Botsch and Kobbelt (2005) where RBF handles and modelling tools were defined, and more recently by Levi and Levin (2013) where interactive tools for the animation of 3D objects are defined. An example based on augmented reality was presented by Valentini and Biancolini (2017a).

A basic set of tools required by any modelling paradigm is composed by the geometric modifiers, namely a certain number of transformation algorithms ruled by geometrical relationships that allow the user to apply the wanted modifications to the discretised geometries. In addition to a proper expressiveness, these modifiers should also guarantee a degree of control similar to what happens for CAD tools enabling a persistent morphing with it, if required (Cenni et al. 2016; Sieger et al. 2014). In addition to the geometric modifier, there are other important paradigms such as those working according to the hierarchical input of an RBF solution or those processing the data coming from an adjoint solution that enrich a lot the possible way to drive the morphing of a surface.

In the following sections, the most important paradigms for surface mesh morphing are described.

6.3.1 Translation Modifier

Translation is the most basic modifier. It is a geometric transformation that allows to move every selected point by the same amount in a given direction, making it a rigid transformation very easy to be achieved by adding a constant vector to each point included in the modification. Translation is a linear modifier, meaning that the order of application of several different displacements does not affect the final result.

Taking into account that the interpolation of a motion field in the space is driven by source points' displacements only, every RBF centre has three degrees of freedom. By defining a zero translation, this modifier can be thus used to apply a null movement type of constraint. In Fig. 6.6 an example of the application of the translation modifier is shown, where a set of source points on the ground is kept

Fig. 6.6 Translation modifier effect: baseline (left) and deformed geometry (right)

fixed by imposing a zero translation, whereas the cube top surface is translated in order to obtain the wanted deformation.

Several methods to help the user in applying a proper translation with the wanted precision are available. In the simplest manner the displacements can be defined by specifying three values relative to the principal axes. However, when managing complex geometries a more accurate and easier control can be gained by exploiting local reference systems that permits, for example, the translation of sloping surfaces using only one displacement value.

6.3.2 Rotation Modifier

Rotation is a non-linear geometric transformation that always guarantees at least one fixed point in space. While it is quite simple to be defined in a two-dimensional domain, space rotations are not commutative in meaning that the order used to define more than one rotation influences the final result if the centre of rotation changes.

In a three-dimensional space a single rotation around an axis causes the motion of each point which it is applied to, over its orthogonal plane. Such a motion can be described using the rotation matrix:

$$R = \begin{bmatrix} \cos\theta + u_x^2(1-\cos\theta) & u_xu_y(1-\cos\theta) - u_z\sin\theta & u_xu_z(1-\cos\theta) - u_y\sin\theta \\ u_xu_y(1-\cos\theta) - u_z\sin\theta & \cos\theta + u_y^2(1-\cos\theta) & u_xu_y(1-\cos\theta) - u_x\sin\theta \\ u_xu_z(1-\cos\theta) - u_y\sin\theta & u_xu_y(1-\cos\theta) - u_x\sin\theta & \cos\theta + u_z^2(1-\cos\theta) \end{bmatrix} \quad (6.3)$$

where ux, uy and uz define the components of the unitary vector u, around which the rotation angle is accomplished. Such components are expressed with respect to the three orthogonal axes x, y and z.

In Fig. 6.7 an example of the effect of the rotation modifier application is depicted. Since the right hand rule is valid, the result of the morphing is an anti-clockwise rotation when using positive angles with the rotation axis pointing towards the viewer.

Fig. 6.7 Rotation modifier effect: baseline (left) and deformed geometry (right)

To completely define a rotation, the user can supply the information relative to the axis of rotation and the desired angle. Alternatively, the rotation can be also imposed using principal/relative reference systems in order to easily apply the modifier to complex shaped geometries.

Being rotation a non-linear transformation, direct amplifications of the obtained shape variations would result in a distorted geometry. As such, special care must be taken when dealing with amplified rotations by correcting the geometry through the adopting of a scaling correction. Besides, to tackle the non-commutativity property of rotations, a strategy must be employed to expressively use them by appropriately blending multiple modifiers.

An example of how rotation modifier can be used to investigate the effect of wing slope of a sport car is given in Fig. 6.8.

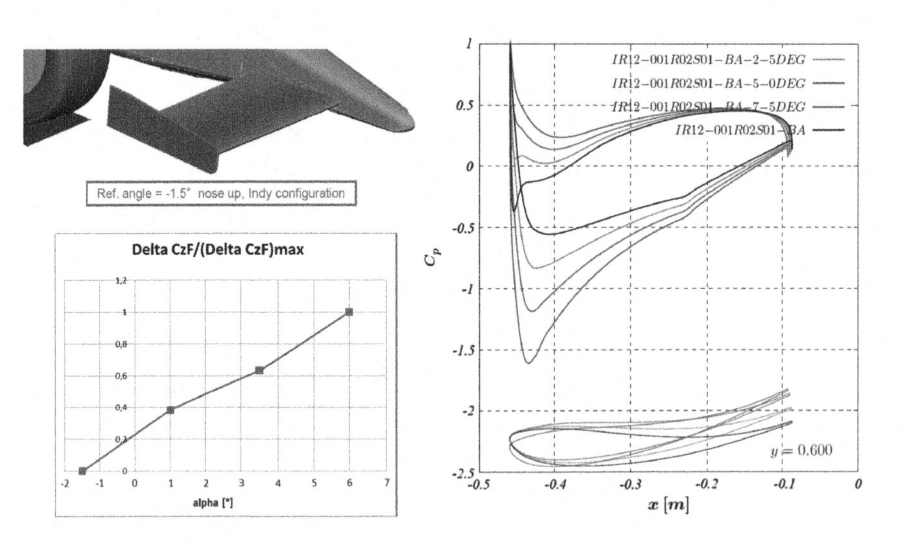

Fig. 6.8 Formula Indy IR6 front wing slope study. Rotation operator allows to investigate the effect of front wing angle in the range −1.5° to 6° (Invernizzi 2013)

6.3.3 Scaling Modifier

Scaling is a linear transformation that let the user shrink or dilate the selected centres about a given point. By using a scaling factor it is possible to reduce or enlarge the geometry along the orthogonal axes using, respectively, a negative or positive value. A unitary scaling maintains source points' position unvaried, obtaining the same result got using a zero translation.

A simple scaling transformation around the axes origin can be achieved by multiplying the point vector for a scaling matrix:

$$S = \begin{bmatrix} S_x & 0 & 0 \\ 0 & S_y & 0 \\ 0 & 0 & S_z \end{bmatrix} \tag{6.4}$$

For a scaling transformation about a given point S_o, source points can be translated of the vector:

$$-S_o = \begin{Bmatrix} -S_{ox} \\ -S_{oy} \\ -S_{oz} \end{Bmatrix} \tag{6.5}$$

in order to put S_o on the top of the axes origin. The matrix S, above defined, can be then used to properly apply the scaling order and a new translation of the vector:

$$S_o = \begin{Bmatrix} S_{ox} \\ S_{oy} \\ S_{oz} \end{Bmatrix} \tag{6.6}$$

can be finally accomplished to bring the point S_o back to the original place.

Figure 6.9 shows an example of the application of the scaling action of the cube top surface mesh nodes about a given point, where the points in red were fixed using a zero displacement modifier.

Fig. 6.9 Scaling modifier effect. Baseline (left) and deformed geometry (right)

6.3.4 Surface Offset Modifier

Another useful surface based shape modifier is the surface offset one. It consists in moving nodes belonging to a surface along the normal direction for a specified amount. An example of the application of the offset modifier is shown in Fig. 6.10 (Groth et al. 2015). The size of the rod lightening hole is increased moving all the points along the local normal direction.

It is worth to mention that the offset modifier can be straightforwardly implemented and employed both when the CAD features are exploitable and if only the surfaces mesh is available. In fact, in the first scenario the CAD features are exploitable and, then, the nodal normals can be easily retrieved just querying the CAD modeller such information, whilst in the second scenario they can be calculated directly from the mesh averaging the normals of the neighbouring surface elements. Curve offset modifier can be defined similarly for plane problems or, in spatial problems, specifying the working plane to be used of the computation of local normal direction to the curve.

6.3.5 Projection Modifier

Translation, rotation, scaling and offset are the three fundamental shape modifiers allowing the user to completely set up a proper shape modification. However, a more expressive degree of freedom is achieved employing advanced methods exploiting geometry definitions and features.

When dealing with numerical simulations, a common workflow foresees the generation of the grid using a geometrical model as a starting point. As such, several CAE systems provide the associativity between mesh and the underlying geometry. This bi-directional association helps the user in selecting straightforwardly mesh entities at different levels of detail by basing selections on geometrical features as surfaces or edges. Furthermore, this information can be exploited to set advanced shape modifications by processing surface data such as normals, distances between surfaces and points and parametric coordinates.

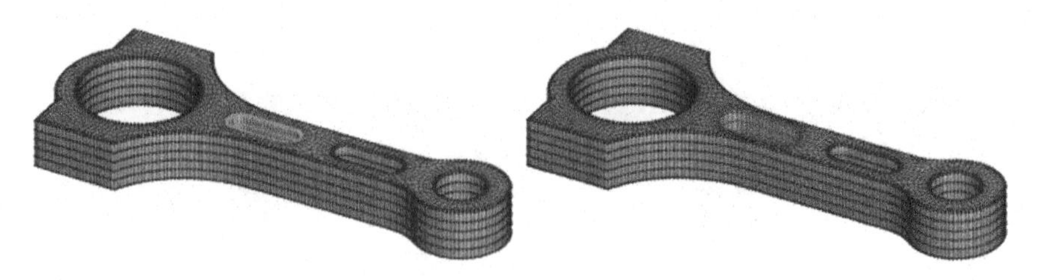

Fig. 6.10 Offset modifier. Baseline (left) and deformed geometry

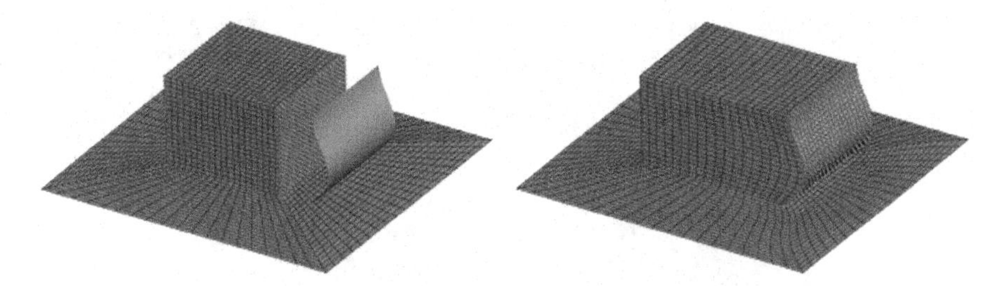

Fig. 6.11 Projection modifier. The target surface (left) is used to project the selected points to the final position (right)

Figure 6.11 depicts an example of the projection modifier. To obtain the final geometry, the opposite face was kept constrained using source points with zero displacement, whilst the floor was kept free to move in order to absorb deformations.

For what described, in the case that geometrical information can be retrieved by the CAD modeller, the projection modifier implementation is not a difficult task to accomplish.

Actually, the underlying geometric representation is not always available. This happens when dealing with dead meshes or when using software not provided with geometric associativity. In such working conditions alternative methods must be employed.

According to what proposed by several authors (Carr et al. 2001, 2003; Xiaojun et al. 2005), RBF provide an useful set of properties making them very attractive to describe implicitly geometrical surfaces. The method, here summarized and fully described in Chap. 5, consists in the definition of a scalar space function which zero isosurface approximates the faceted mesh. The scalar function is defined by generating a cloud of on-surface points onto the faceted mesh (with the function zero valued) augmented with a cloud containing a certain amount of off-surface points, usually computed with a surface offset operation, positive or negative valued depending on the offset side.

6.3.6 *Free Form Deformation Using RBF*

A new shape for a surface can be obtained according to direct RBF modelling, i.e. by directly controlling the displacement of a certain number of RBF control points. Points can be located onto the surface, as in the example of Valentini and Biancolini (2017a), or close to the surface using regular space arrangement.

In the first case a certain number of RBF sources is uniformly distributed onto the surface, as in the example of Fig. 6.12 (Valentini and Biancolini 2017b) where nine points are picked onto the surface defining for each one a displacement orthogonal to the surface itself.

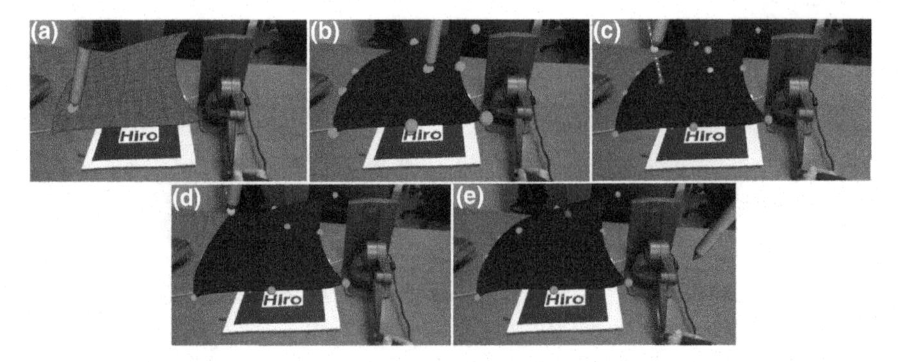

Fig. 6.12 Demonstration of interactive sculpting using the RBF4ARTIST platform

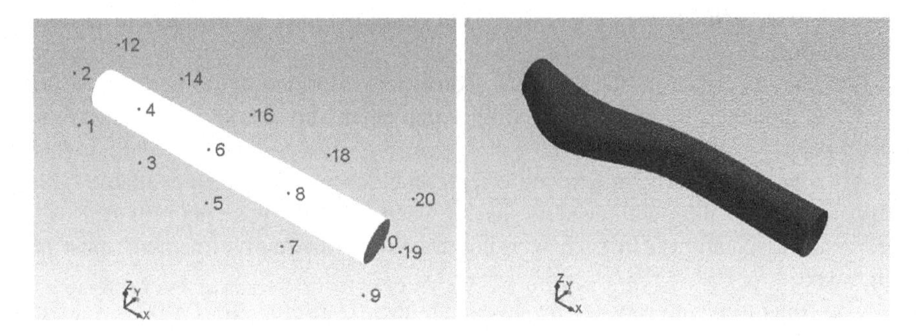

Fig. 6.13 Source points generated over the surfaces of an encapsulation domain (RBF Morph 2017a)

A second approach, similar to FFD deformation boxes, consists in arranging a certain number of RBF points on a volume surrounding the geometry. In the example shown in Fig. 6.13 twenty RBF centres are firstly positioned along the edges of a box-shaped encapsulating domain according to an equispaced distribution, and successively the source points with ID 3 and 4 are moved so as to gently deform the tube.

6.4 Multi-step Paradigm (Hierarchical Approach)

As already specified RBF define a smooth interpolating and extrapolating field in the space and, depending on the chosen basis function, each source point of the RBF problem may influence the field even at large extents. This behaviour of RBF is often neither required and nor desired. This happens, for instance, when complex RBF set-ups in which different modifications must be applied to different mesh areas. By defining a single set of source points, the RBF centres positioned in the different areas would interact with each other resulting in an unwanted interpolating behaviour. To avoid this problem, and to increase at the same time the user degree of freedom, an hierarchical approach for modelling can be adopted.

Hierarchical RBF mesh morphing approach was firstly introduced as a feature of the RBF Morph Fluent Add On software where the two-step morphing approach is a common modelling practice (RBF Morph 2017b). The final mesh morphing step to be applied at the volume grid is in this case controlled applying desired reshaping information to a certain number of surfaces and propagating the field with the bi-harmonic RBF that allows to get the minimum distortion of volume cells; at the same time some of the controlled surfaces can receive as input an already computed and stored RBF (saved in a file in this specific implementation). Such local RBF are defined to gain a specific control of an individual surface; high order RBF can be used to have the desired continuity (Wendland C0, C2 and C4 are available) and just a few points can be used to have the desired local control adopting for instance Free Form sculpting explained in Sect. 6.3.6. In Fig. 6.14 the concept is demonstrated; the top surface, originally flat, is sculpted by moving up a point at the centre face and keeping the border fixed; desired effect on surface is then applied to the volume mesh surrounding the cube in the second step where a linear RBF ($\varphi(r) = r$) is adopted to propagate the morphing action.

The multi-step method, which effectiveness was proven in years of industrial and research usage of the technology, has been better established in the ACT Extension of RBF Morph where the hierarchical multi-step feature is directly available in the RBF definition tree and many steps are allowed before propagating the final deformation to the computational mesh without the need of storing on file intermediate RBF solutions. In Fig. 6.15 an example of surface reshaped using the hierarchical approach of RBF Morph ACT software is shown. The edge highlighted with green circles (top left) is controlled using three source points, two fixed at the end of the edge and one moving at its centre (their final position is visualized in

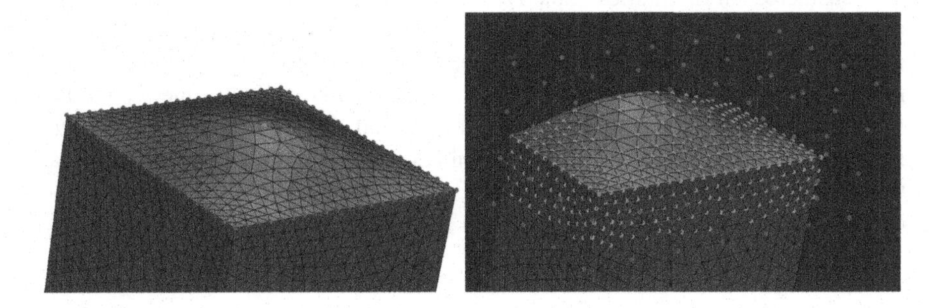

Fig. 6.14 The top surface of the cube is reshaped with a 80 points cp-c4 RBF (left) in the first step, in the second step the achieved shape is applied to the top surface, other surfaces are fixed and the morphing in the volume is limited by a delimiting encapsulation

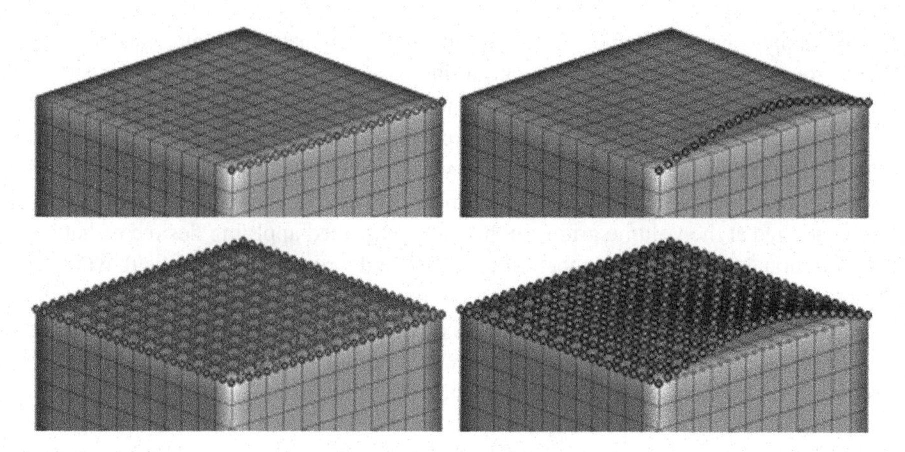

Fig. 6.15 Hierarchical set-up. The target of the first problem (top) is employed to define the source of the second problem (bottom)

orange). The result obtained using a cubic spline, highlighted with black circles (top right), is then employed in a second RBF set-up as a source displacement to mould the top surface (green circles) where new fixed edges are also added (orange circles along the three edges). The final result of the morphed surface is shown through black circles (bottom right). Such a morphing action is then propagated in the volume mesh which is assigned as final target adopting a linear RBF ($\varphi(r) = r$). Nesting the set-up tree allowed in this case to apply a three-step RBF set-up: (3) the volume is controlled by points at surfaces, (2) one of the surfaces is reshaped according to the movement of points at its edges and (1) one of the edges is reshaped according to input at points along it.

Two industrial examples of hierarchical approach are demonstrated in Figs. 6.16 (Biancolini et al. 2013b) and 6.17 (Biancolini 2017); two aeronautical applications are considered: in the first example two-step mesh morphing is adopted to have a parametric shape of an aircraft wing with respect to sweep angles and the parametric CFD grid is exploited to investigate the effects on aerodynamic performances by morphed variations of the high fidelity CFD model; in the second the fillet at the root of a turbine blade is reshaped and the high fidelity FEA model morphed accordingly to investigate a possible reduction of the maximum principal stress peak occurring at blade root which is responsible for fatigue failure of the component.

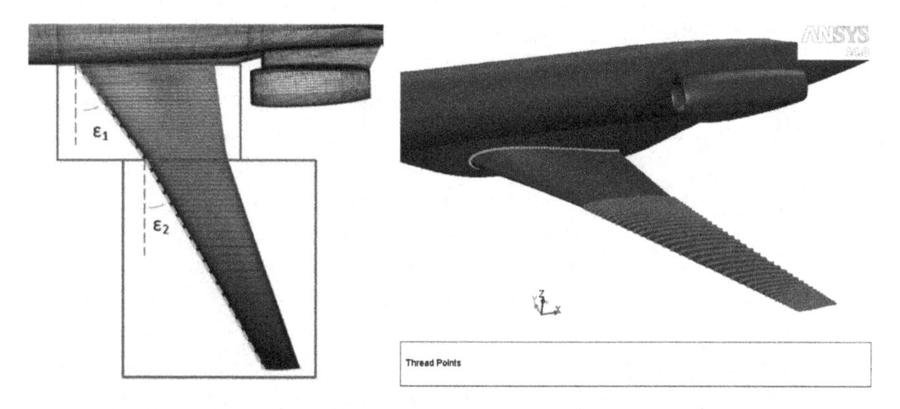

Fig. 6.16 Example of hierarchical morphing for the P1xx aircraft model. Sweep angles (left) are controlled with a two-step approach. The first step for controlling the first sweep angle (right) is defined to apply the desired shape to the wing (red points are moved fore and aft, green points are fixed)

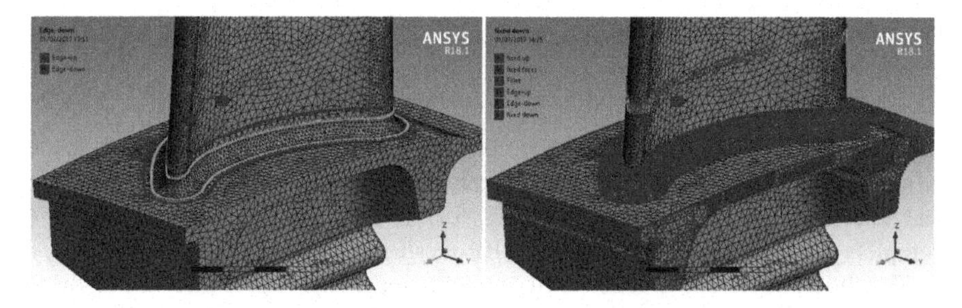

Fig. 6.17 The fillet of the blade is reshaped by controlling the positions of edges with translation and offset operations (left), desired shape is then propagated to the volume mesh (right)

6.5 Multi-physics Data Driven Paradigm

RBF can be efficiently used for steering the nodal displacement field of a surface using the data coming from the results of a multi-physics computing or the outputs of an adjoint solution suggesting the shape modifications to apply to decrease or increase an objective function.

For many of the afore described applications the adoption of the hierarchical approach is advisable because the exact control of surface movements is ensured by an RBF solution that is then passed ad input in another RBF solution mainly aiming at controlling the volume mesh smoothing.

When tackling multi-physics simulations such as ice and show accretion that produce a modification of the surfaces' shape that can be predicted in advance or also directly during a CFD or CSM computing, RBF mesh morphing is an effective way to update the shape of the covered component where the accretion develops. In

Fig. 6.18 Mesh morphing can be used to update the high fidelity CFD mesh according to computed ice profile

this case the accreted surface can be simulated processing the local thickness of ice or snow so as to generate the displacement field needed to define the RBF set-up and solution. An example is given in Fig. 6.18 (Biancolini and Groth 2014) and a complete description of this usage of RBF mesh morphing is given is Chap. 9.

The fluid-structure interaction (FSI) computational analyses, namely the study of the deformation of a body under the external loads due to a fluid flowing around it, can be performed through the use of RBF mesh morphing according to three main approaches: direct assignment of a motion law to the wall surfaces of the component to be studied; mode-superposition method, which concerns the calculation of a certain number of natural modes of the deformable components, the successive setting-up of the corresponding RBF solutions and, finally, the application of morphing during the CFD computing so as to update the actual shape of the structure; the standard two-way method envisaging the exchange of data between the CFD and the CSM model. An example of the second approach is given in Fig. 6.19 where the embedding of a modal base into the CFD high fidelity model of a sport car is sketched (Invernizzi 2013); a detailed description of the method is provided in Chap. 10.

The mesh sensitivities resulting from an adjoint solution can be exploited to drive morphing and to perform automatic shape optimizations as deepened in Sect. 8.3 that demonstrates how, after an optional filtering actions, a new shape of a surface can be obtained after a certain number of sequential morphing actions. The local field to sculpt a surface so that the performances of a part are increased, can be computed in some cases according to local solution data, as in the case of the

Fig. 6.19 Natural modes computed by FEA are used for the FSI study of the Dallara Ir5 car which aerodynamic performances are computed accounting for front wing flexibility

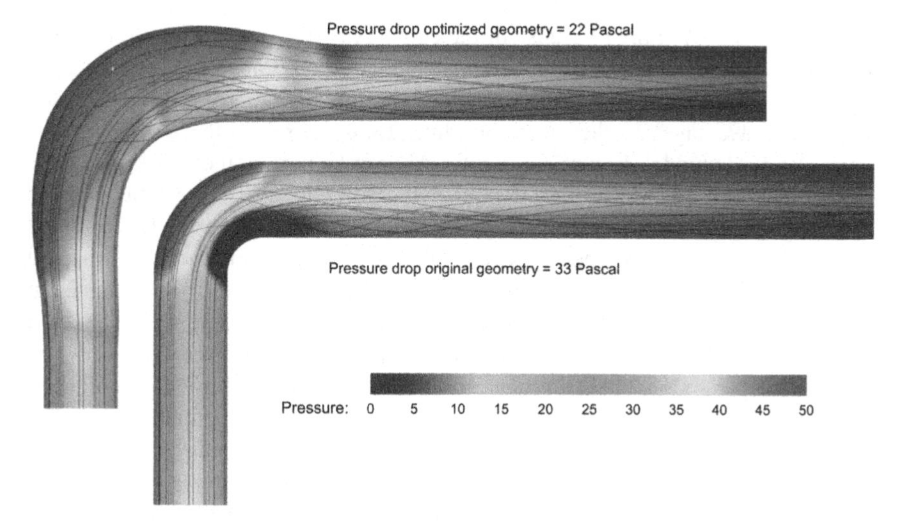

Fig. 6.20 A new shape sculpted according to the adjoint sensitivities is computed to reduce the pressure drop at the 90° bend in a pipe which initial cross section is circular

biological growth method explained in Sect. 9.7. An example of how adjoint sculpting can reduce the pressure loss in a 90° bend is given in Fig. 6.20.

6.6 CAD Driven Morphing

RBF mesh morphing can be better driven if a CAD representation of the underlying geometry representation is available; hybrid workflow, in which the mesh based modelling action of RBF mesh morphing and advanced parametric CAD modelling tools are combined.

Parametric CAD are able to regenerate quickly a design variation for the shape of a component. In CAE applications the workflow required to generate numerical models could be tricky as the geometrical model needs to be processed in several steps including meshing as well as the application of boundary conditions, which are not easily automated. Mesh morphing can be an effective way to update just the nodal positions of the final mesh; if a persistent associativity between the parametric CAD and the numerical model is required the morphing task should be automated so that the mesh is morphed from the baseline using the updated CAD representation as a target.

This application of mesh morphing is not mature enough yet, but there are scenarios in which it can be effectively used as demonstrated in (Sieger et al. 2014) where the use of RBF mesh morphing for adapting an existing mesh onto a new isotopological brep variation is demonstrated.

There are many design conditions in which a full associativity with the parametric CAD is not required because of the complexity of using mesh morphing but also because a parametric CAD is not available. An hybrid approach, enjoying the use of both CAD-based and mesh morphing based modifications can be integrated to join the advantages of the mesh morphing techniques with the flexibility of a parametric CAD model. In particular, the CAD can be used in the first stage of the hybrid procedure to modify only a part of the component (i.e. a single surface) exploiting the parameterization of the geometrical entities of interest, whereas in the second stage the mesh morphing is employed to attach the mesh on the updated geometry of the parametric CAD (the surface projection method of Sect. 6.3.5 can be adopted) as demonstrated in the example of Fig. 6.21 where the fillet of a complex assembly is updated according to a parametric single face.

Auxiliary parametric CAD entities can be adopted not only for the definition of geometrical targets suitable for geometrical operations as in the previous example, they can be used, in combination with parametric meshing, as a source of new RBF points suitable to drive the mesh morphing process. An example is shown in Fig. 6.22 showing how a parametric clew can be imprinted onto a flat rectangular plate.

Parametric surfaces are generated in the CAD system and then meshed onto the current design position; in the example the rectangular face is divided in three regions: the inner surface that is used to define the flat area of the bump to be embossed, an intermediate surface that is left free to be deformed to accommodate the bump with a transition, and the outer surface that is kept flat and fixed. Nodes generated on the auxiliary faces are used as RBF sources assigning an offset normal to the surface for the inner surface and fixed positions for the outer one; nodes on the back faces of the plate are kept fixed. The input parameters are the clew depth (that is the extent of the offset) and the parametric faces position. It is worth to notice that in this case the auxiliary geometry acts as a stencil.

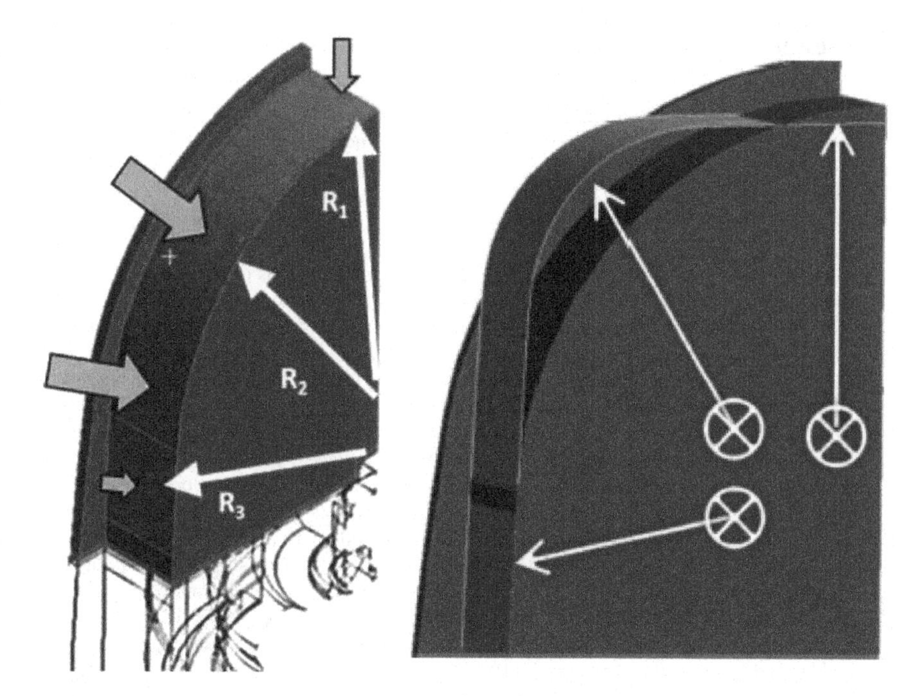

Fig. 6.21 Shape modification in the CAD-mesh hybrid approach (Cenni et al. 2016). Pressure loads related to geometry makes direct morphing complex to be implemented (left), the parametric surface is used to drive the morphing (right)

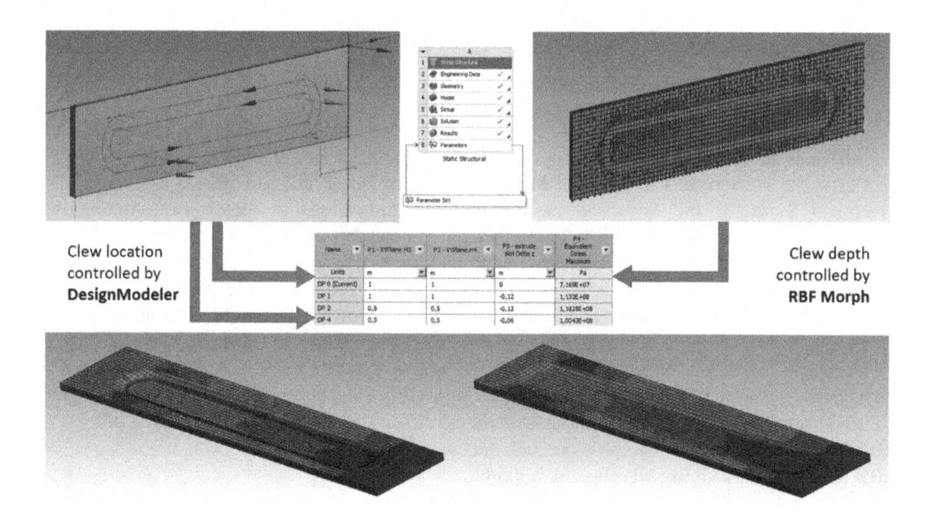

Fig. 6.22 Example of workflow with the use of auxiliary CAD

References

Biancolini ME (2012) Mesh morphing and smoothing by means of Radial Basis Functions (RBF): a practical example using Fluent and RBF Morph. In: Handbook of research on computational science and engineering: theory and practice, p 34. https://doi.org/10.4018/978-1-61350-116-0.ch015

Biancolini ME (2017) Ridurre le concentrazioni di tensione con il mesh morphing. Analisi e Calcolo n 81 Luglio/Agosto

Biancolini ME, Gozzi M (2013). Aircraft design optimization by means of radial basis functions mesh morphing. In: ANSYS user group meeting Italia 2013, Salsomaggiore 1

Biancolini ME, Groth C (2014) An efficient approach to simulating ice accretion on 2D and 3D aerofoils. In: Applied Aerodynamics Conference 2014: Advanced Aero Concepts, Design and Operations, Proceedings of a meeting held 22–24 July 2014, Bristol, UK. Royal Aeronautical Society (RAeS). ISBN 9781510802698.

Biancolini ME, Cella U, Mancini M, Travostino G (2013) Shaping up: mesh morphing reduces time required to optimize an aircraft wing. ANSYS Advantage 7:32–34

Botsch M, Kobbelt L (2005) Real-time shape editing using radial basis functions. In: Marks J, Alexa M (eds) Computer graphics, vol 24, no 3. Eurographics 2005

Carr JC, Beatson R, Cherri J, Mitchell T, Fright W, McCallum B (2001) Reconstruction and representation of 3D objects with radial basis functions. In: Proceedings of the 28th annual conference on computer graphics and interactive techniques, Los Angeles, CA, pp 67–76

Carr JC, Beatson RK, McCallum BC, Fright WR, McLennan TJ, Mitchell TJ (2003) Smooth surface reconstruction from noisy range data. In: First international conference on computer graphics and interactive techniques, p 119

Cella U, Groth C, Biancolini ME (2016) Geometric parameterization strategies for shape optimization using RBF mesh morphing. In: Advances on mechanics, design engineering and manufacturing, Lecture notes series in mechanical engineering, pp 537–545, Sep. https://doi.org/10.1007/978-3-319-45781-9_54

Cenni R, Cova M, Bertuzzi G (2016) A CAD-MESH mixed approach to enhance shape optimization capabilities. In: CAE conference, Parma, Italy, 17th–18th Oct 2016

Costa E, Biancolini ME, Groth C, Cella U, Veble G, Andrejasic M (2014) RBF-based aerodynamic optimization of an industrial glider. In: 30th international CAE conference, Pacengo del Garda, Italy

Groth C, Chiappa A, Giorgetti F (2015) Ottimizzazione strutturale di una biella automobilistica mediante mesh morphing. AIAS – Associazione Italiana per l'Analisi delle Sollecitazioni 44° Convegno Nazionale, Università di Messina, 2–5 Sept, AIAS 2015, p 596

Invernizzi S (2013) In: Advanced mesh morphing applications in motorsport. Automotive simulation world congress 2013, 29–30 Oct 2013, Frankfurt am Main, Germany

Levi Z, Levin D (2013) Shape deformation via interior RBF, Pubmed. US National Library of Medicine National Institutes of Health. https://doi.org/10.1109/tvcg.2013.255

Luboz V, Bailet M, Boichon Grivot C, Rochette M, Diot B, Bucki M, Payan Y (2017) Personalized modeling for real-time pressure ulcer prevention in sitting posture. J Tissue Viability

Pranzitelli A, Caridi D (2011) An optimization study of a ship hull. ANSYS webinar 7, 9 June 2011. http://www.rbf-morph.com/wp-content/uploads/2016/03/ansys_webinar_leeds.pdf

RBF Morph (2017a) RBF Morph—tutorials documentation

RBF Morph (2017b) RBF Morph—users guide

Sederberg TW, Parry SR (1986) Free-form deformation of solid geometric models. In Evans DC, Athay RJ (eds) Proceedings of the 13th annual conference on computer graphics and interactive techniques (SIGGRAPH '86), New York, pp. 151–160

Sieger D, Menzel S, Botsch M (2014) RBF morphing techniques for simulation-based design optimization. Eng Comput 30(2):161–174

Valentini PP, Biancolini ME (2017a) Interactive sculpting for engineering purposes using augmented-reality, mesh morphing and force-feedback. IEEE consumer electronics magazine

Valentini PP, Biancolini ME (2017b) Modellazione interattiva integrando realtà aumentata, morphing geometrico e interfacce aptiche, Analisi e Calcolo N. 78 Gennaio/Febbraio

Xiaojun X, Michael Y, Wang QX (2005) Implicit fitting and smoothing using radial basis functions with partition of unity, In: Proceedings—ninth international conference on computer aided design and computer graphics, CAD/CG 2005, vol 2005, pp 139–148

Chapter 7
Advanced RBF Mesh Morphing for Biomechanical Applications

Abstract In this chapter, examples of advanced mesh morphing for biomechanical applications are presented demonstrating the benefits of meshless deformation for situations in which an underlying CAD geometry is not available. Virtual surgery using CFD simulation of the hemodynamics is faced showing how a morphing interaction can provide the insight in the estimation of the evolution of a pathology or to plan the insertion of a cannula in a virtual environment where patient shapes and medical devices shapes coexist and can interact. The use of mesh morphing for the remodelling is then demonstrated on a femur and a tibia bones. FEA method is adopted for the structural assessment and mesh morphing can be used as a tool to transform a baseline (library) geometry onto the patient one. The approach is useful not just as a simplified mesh generation tool (adapting is faster than remeshing), but rather because it allows to enable the direct comparison between models preserving the one-to-one nodes correspondence.

Chapter 6 of the book shows the use of mesh morphing through the description of many practical engineering applications that are relevant for several industrial fields. The resolution of such applications typically starts from the CAD representation of the baseline shape of the model to be studied, that is supplied some way to the user.

The ability to introduce shape parameters capable to update both surface and volume meshes, can be exploited even when such a CAD representation is not accessible at all. This is the case of biomechanical applications where the baseline models of the investigated morphologies, usually available just as tessellated surfaces, consist of the result of an acquisition process of clinical images. Nevertheless, CAE workflows similar to those characterizing other industrial applications design are adopted to perform biomechanical numerical studies such as the analysis of the flow field using CFD codes, as in the case of hemodynamics, or the stresses analysis using FEA tools, as in the case of bone modelling and prosthetic components design.

This chapter addresses the role of mesh morphing in the field of biomechanical CAE simulations. In this context, RBF mesh morphing can be effectively used in predictive medicine workflows where a patient-specific numerical model is taken as reference to understand the physics of interest by means of simulation-driven

M. E. Biancolini, *Fast Radial Basis Functions for Engineering Applications*,
https://doi.org/10.1007/978-3-319-75011-8_7

techniques. Once a trusted baseline model is set up, the effect produced by shape modifications of the geometry can be explored in view of predicting the evolution of a pathology or to evaluate the effectiveness of a surgical correction. The interaction of the human body's parts with mechanical components' geometry can be addressed as well, as in the case of the dimensioning and positioning of prosthetic parts in the system or in the case of the simulation of a surgical operation.

Mesh morphing can be also employed for the automatic generation of a new patient-specific numerical model by suitably adapting the mesh. Such an operation enables to save the long amount of time of manual modelling mandatorily required for the generation of a reliable numerical model suitable for CFD or CSM simulations.

Examples dealing with CFD applied to hemodynamics and CSM applied to human bones are respectively described in the following sections.

7.1 Hemodynamics

The engineering applications involving biological fluids have highly transversal requirements because they deal with the numerical model pre-processing from clinical images, complex flow conditions, fluid rheological properties, structure motion and deformation, as well as the visualization and post-processing of the computational results.

In this section, two applications are detailed to demonstrate how mesh morphing can be used in this field. The first one, described in Sect. 7.1.1, pertains a workflow in which mesh morphing is utilised to investigate the effect of morphological variations of vessels (Biancolini et al. 2012). The second one, deepened in Sect. 7.1.2, showcases how mesh morphing can assist a surgical operation by predicting the effect of the positioning of a cannula in the aorta (Gallo et al. 2014). In both applications, the details of the RBF mesh morphing set-up referring to the ANSYS Fluent add-on version of the RBF Morph software are provided. Exploiting such information, the reader can define its own custom implementation using the basic RBF tools provided in the first part of this book, taking the set of concepts that will be identified and deepened hereinafter as a reference framework. Those concepts broadly are the placement of a cloud of source points to define the controlled locations, the assignment of a given displacement to each source point, the computation of the RBF solution and, finally, the position update of all the nodes of the computational domain which morphing is applied to.

It is worth to specify that the ability of mesh morphing to update the shape of the CFD domain can be also exploited to accomplish with success other applications not covered in this section. For example, an approach similar to the one presented in Sect. 7.2.2 can be followed to adapt a baseline template mesh to quickly generate the CFD model referring to a particular patient, whilst advanced fluid structure interaction (FSI) techniques, covered in Chap. 10, can be used to model the effect of wall movements using experimental data or, alternatively, numerical data computed by CSM models. Furthermore, thanks to the implicit surface approach, presented in

Chap. 5, experimental data can be managed by means of RBF mesh morphing to handle specific computational analyses in which the shape modifications to be applied are known in advance, as done for the applications discussed in Sect. 9.4. In the study by Capellini et al. (2017), 3D surface models defined from healthy subjects and patients with ascending thoracic aorta aneurysms (aTAA), selected for surgical repair, have been generated. A representative shape model for both healthy and pathological groups has been identified showing how mesh morphing allows to enable the parametric simulation of the aTAA formation.

7.1.1 Patient Specific Image-Based Hemodynamics: Parametric Study of the Carotid Bifurcation

The study outlined in this section is based on a workflow (Biancolini et al. 2012) that comprises the usage of three fully validated software to effectively fulfil the requirements related to hemodynamics: the vascular modeling toolkit (VMTK) for the pre-processing step that takes clinical images as input and gives anatomic models of cardiovascular districts as output, the RBF Morph add-on version for ANSYS Fluent to impose changes to the vascular anatomy through mesh morphing and the finite volume-based CFD tool ANSYS Fluent to solve the governing equations of the fluid motion.

The study is focused on a patient-specific carotid bifurcation, being this anatomical site of major interest in hemodynamics due to its relationship to atherosclerosis. In specific, the analysis aims at quantifying the impact that the shape of the carotid bulb and the bifurcation angle have on the resulting flow patterns. As a matter of fact, these latter characteristics provide a very useful information because they are thought to be tightly related to the focal development of atheromatous plaques at that site and can be put in relation to the outputs of remodelling surgical procedures.

Starting from the acquired carotid bifurcation, in order to perform a parametric study of its shape two relevant modifications are considered to enable the continuous change of the vessels through mesh morphing. In particular, the first one consists of the change of the external carotid artery (ECA) angle, whilst the second one concerns the deformation of the carotid bulb shape. Those mesh variations are emphasised through the green curves depicted in Fig. 7.1 on the left and right side respectively.

The morpher tool allows to set up and store an arbitrary number of shape modifiers. It is important to notice that only the information required for mesh updating (i.e. the RBF sought coefficients) are stored, and then there is no need for saving the morphed mesh. This approach makes the CFD model parametric allowing the desired set of amplifications for the taken shape modifications are applied at calculation stage, just before the CFD computations.

The morphing target for the RBF set-up, namely the set of points that will be updated using the RBF field, is defined employing a cylinder which allows to

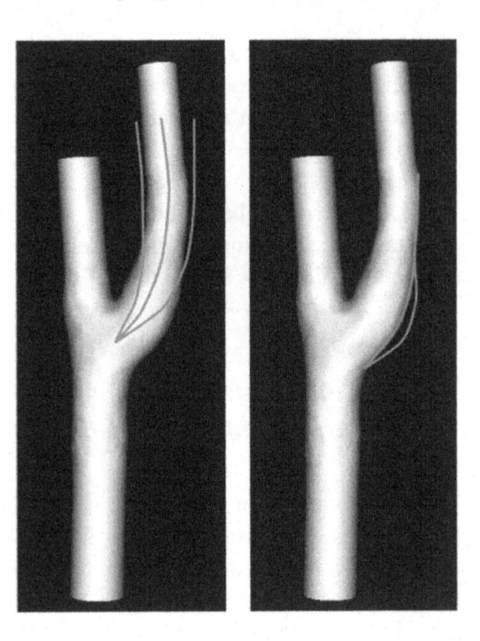

Fig. 7.1 Shape modifications of interest for the carotid bifurcation: ECA (left) and bulb (right) shape

delimit the morphing action just in the region around the ECA surface that has to be modified. Points of the branch root for which a zero displacement is imposed, are extracted from the vessel surface mesh nodes exploiting the intersection between the vessel itself and the cylinder already introduced to delimit the morphing volume target. Such an operation can be finalised taking all the nodes of the surface mesh that fall in the cylindrical morphing domain, and excluding the ones that belong to an auxiliary cylinder that is centred as the first one, but a little bit smaller. The resulting cloud of source points, visualised in red colour in Fig. 7.2 (left), is the portion of the surface mesh nodes in the clearance between the two employed cylinder-shaped domains.

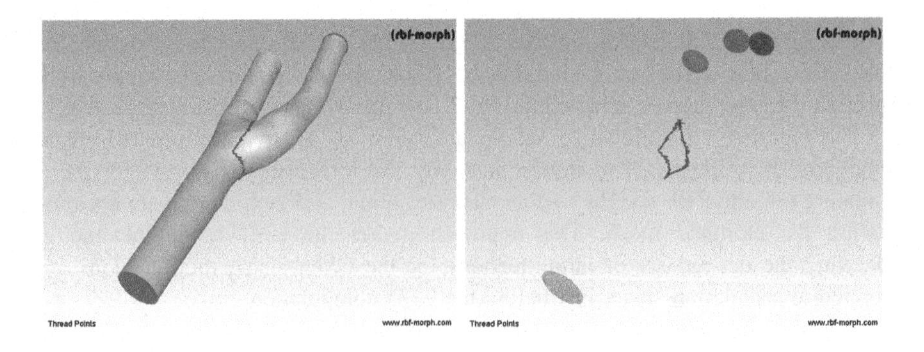

Fig. 7.2 RBF set-up for ECA angle variation

The bending of the vessel is set defining all the mesh nodes on the ECA outlet surface as RBF source points, and applying on them a rigid rotation field around an axis located at the root of the vessel. The rotation axis, visualised in blue, and the preview of the movement of the defined RBF points for one degree rotation are shown in Fig. 7.2 (right). It is important to notice that a rigid rotation produces a non-linear deformation field. The employed morpher tool has the ability to suitably manage such a type of motion and, according to that, rotations are then amplified with respect to the baseline value used for the set-up, without the need for re-computing the sought coefficients of the RBF problem. Given that, if the reader wants to build up its own custom workflow using the basic theory of RBF as support, the sought coefficients has to be recomputed with the desired angle and the corresponding input field unless such nonlinear behaviour is included in the workflow.

The effect of the first shape modifier can be explored using the RBF field for updating the nodes' position on the vessel mesh, as represented in Fig. 7.3 for two values of the ECA angle. As the morphing preview of a part of the whole numerical model, identified by a set of surfaces of the ANSYS Fluent case, is computed in a short time, it is worth to mention that this feature of the morpher tool is very useful to inspect the consistency of the RBF problem set-up and, if needed, to accordingly troubleshoot it. The same RBF field is successively used for the update of the volume mesh performed by smoothing all the nodes inside the cylindrical morphing domain defining the delimited target volume.

As the volume mesh has to be processed at some extent or entirely, to run simulations on the morphed variants, a specific attention should be paid on computing performances as well as the ways allowing to shorten the morphing time. In the case of parallel computing running on partitioned mesh, that is the common calculation environment used in HPC workflows for CFD, the same morphing field can be used to smooth the various portions of the mesh. Apart from this peculiar capability of a morpher tool due to the meshless characteristic of the technique it is based on, a further enhancement in terms of computing performance is gained considering the imposition of a local action of the RBF field. As earlier detailed, if a

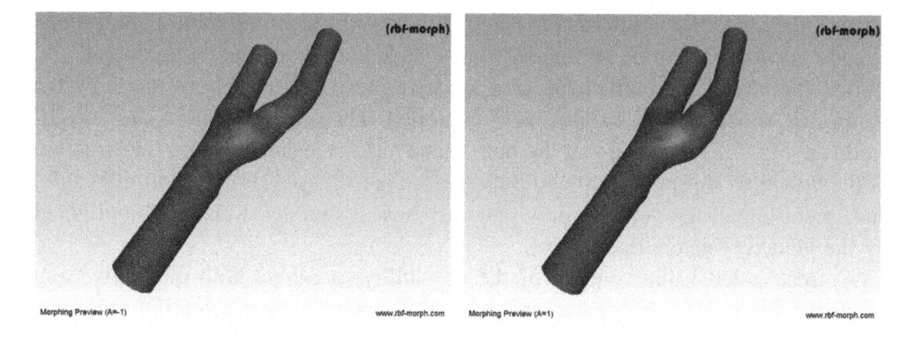

Fig. 7.3 The effect of the shape modification (ECA angle) is previewed at 2 values for the amplification (−1, 1)

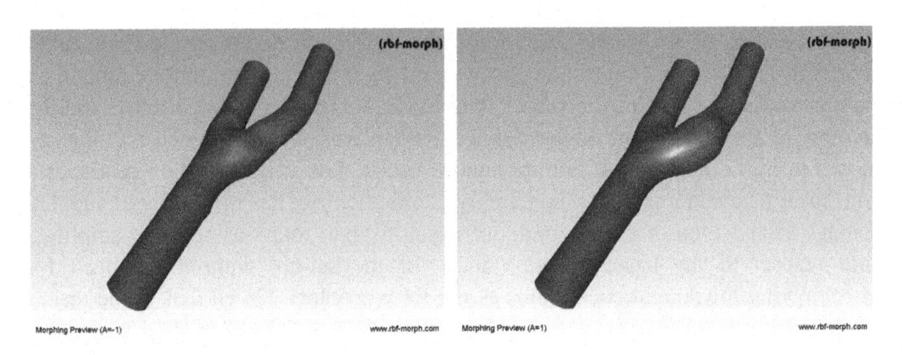

Fig. 7.4 The effect of the shape modification (bulb shape) is previewed at 2 values for the amplification (−1, 1)

delimiting target mesh domain is assigned, the volume mesh nodes that fall outside it are left unchanged in their original position with a relevant reduction of the computing demand. It is thus advisable that a similar approach is followed by the reader in the RBF set-up of a custom implementation.

Similarly to what done for the first shape modifier, the second shape modifier is defined delimiting the action of morphing with a sphere surrounding the carotid bulb. Also in this case a second auxiliary sphere, smaller than the first one, is used to select the vessel points at the boundary of the morphing domain (target mesh). A set of surface mesh nodes that fall within the clearance between the two concentric spheres is so used to define fixed source points at the boundary of the bulb area to be reshaped. The bulb shape is then controlled adding individual source points on the bulb surface, each one with a displacement normal to the surface, defined so that the combined action gently deforms the bulb surface in the intended fashion. The preview of the morphing action of the second shape modifier is depicted in Fig. 7.4 for two amplifications opposite in sign.

The described RBF mesh morphing approach can be thus used to define a parametric model of the carotid bifurcation that allows to explore new shapes obtained by combining the two shape modifiers just detailed. Other shape modifications, such as, for instance, the vessel bending in planes different from the one already accounted, can be straightforwardly added and consistently combined.

The outputs of the bulb shape case are here presented to demonstrate how the parametric model can be exhaustively exploited. Three amplifications are considered (see Fig. 7.5) showing that the bulb shape has a relevant effect on the structure of the flow field. In specific, the bulb dimension is increased (from left to right side) and, accordingly, the velocity flow patterns show a significant change highlighted by the velocity vectors distribution.

A great added value consists of the possibility to extract both qualitative and quantitative hemodynamics parameters via the user-defined functions (UDF) function of the employed CFD solver (ANSYS 2017a, b) with the purpose to get the physical insight and deeply understand how they are related to the defined shape parameters.

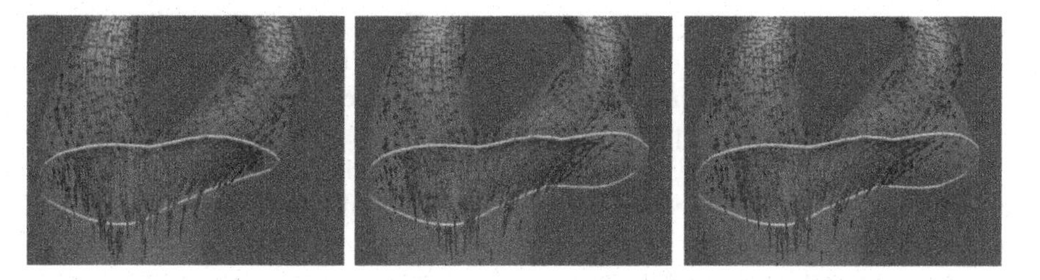

Fig. 7.5 Velocity vectors variation due to the effect of the bulb dimension increase

7.1.2 A Virtual Test Bench for Hemodynamic Evaluation of Aortic Cannulation in Cardiopulmonary Bypass

The cardiopulmonary bypass (CPB) study described in this section (Gallo et al. 2014), demonstrates how patient tissue geometry and CAD representation of surgical devices can be handled in the same multi-physics simulation workflow.

CPB is a common practice in cardiac surgery. According to this, venous blood is directed from the venous system to a heart-lung machine and then returned to the aorta through an inserted arterial cannula (Fig. 7.6 left). The altered flow conditions induced by the cannula cause the mechanical activation of platelet aggregation pathways with the risk of thromboembolic events. Thus, the study of the effect of cannulation and its influence on the described mechanisms are of high relevance from a clinical point of view and they may get a significance such to allow the optimization of therapeutic performances.

To this end, a CFD model is built up to numerically assess the flow field generated by the arterial cannula insertion. The general purpose CFD software ANSYS Fluent is used on a hybrid hexahedral-tetrahedral computational mesh

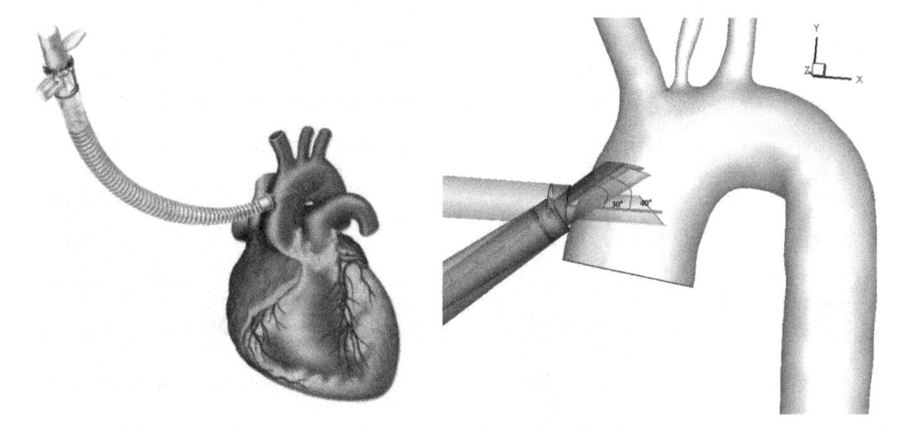

Fig. 7.6 Cardiopulmonary bypass arrangement (left) and cannula penetration and angle (right)

generated using the ANSYS ICEM CFD tool. The fluid domain is divided into about 11 million cells. A mesh morphing technique based on RBF, is used to explore the clinical use of parameters that influence the surgery performances.

As an example of the parameters that can be analysed with the developed benchmark, the cannula insertion angle (Fig. 7.6 right) is considered with the objective to evaluate its effect on flow dynamics related thromboembolic risk and, accordingly, on supra-aortic arteries perfusion. The desired effect is the tweaking of the cannula insertion angle. Supposing that a slight accommodation of the angle is imposed with respect to the initial position, the deformation of the tissue around the insertion hole has to be virtually reproduced by the morphing action. If such a local deformation is not desired, it is possible to preserve the original wall shape using the surface projection capability offered by the implicit surface method (see Chap. 5) of the morpher tool that enables to restore the surface onto the original shape after the first morphing application.

The first step of the RBF solution set-up consists of the definition of the morphing domain extent that is accomplished creating a delimiting cylinder able to wrap the portion of the geometry to be deformed (blue outlined cylinder in the right side image of Fig. 7.7), namely the cannula-aorta intersection district. The external shape of the aorta is then protected imposing a zero movement to the RBF source points extracted from surface mesh nodes of the aorta surface falling inside that cylinder (green points in the left side image of Fig. 7.7). In order to accommodate the cannula rotation, a small buffer over the vessel wall at the intersection with the cannula is left free to be deformed. Such a buffer is obtained excluding from the previous set, the mesh nodes inside a cylinder centred at the insertion hole. The nodes that belong to the mesh of the cannula (red in Fig. 7.7 left) are extracted, assigned as source points and controlled using a rigid rotation. The rotation of the remaining part of the cannula is gained by directly wrapping the volume mesh into a cylinder (red outlined cylinder in Fig. 7.7 right) that prescribes the same rotation of the controlled part of the cannula surface.

The arrangement of the defined RBF source points cloud allows to apply a rigid rotation to the cannula, supposed to behave as a perfectly rigid mechanical part, to deform the wall of the vessel at the intersection (the grey surface between green and red points in Fig. 7.7 left) and to accommodate the volume mesh inside the vessel

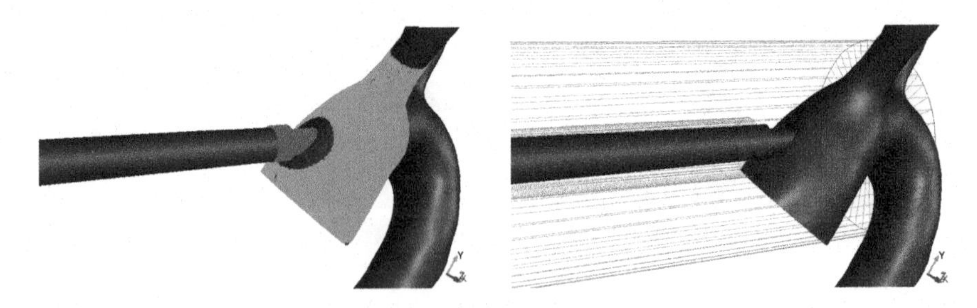

Fig. 7.7 RBF set-up for cannula angle variation (Color figure online)

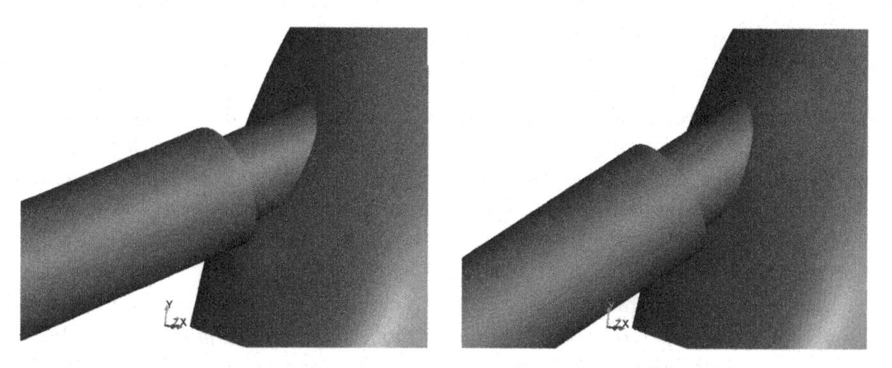

Fig. 7.8 The effect of shape modification is previewed for two amplifications opposite in sign

with respect to the rotation assigned to the protrusion of the cannula inside the aorta.

The effect of mesh morphing action is previewed in Fig. 7.8 where two values of the angular variation (five degree in negative and positive directions) are used. A gentle deformation of the buffer is noticed. The full morphing of the volume mesh can be implemented processing all the nodes in the morphing domain. It is worth to notice that two precautions are implemented to reduce the impact of the numerical cost of the RBF calculation. The first foresees the generation of a cylindrical domain surrounding the cannula insertion to delimit the morphing target. The second envisages the use of a rigid rotation applied to the portion of the cannula far from the intersection by wrapping it with a second cylindrical domain. According to what already introduced in the previous chapters, the morphing field can be applied in parallel if the mesh is partitioned.

The CFD results, shown in Fig. 7.9, demonstrate that the cannula insertion angle greatly influences the perfusion of carotid artery. In fact, the particle trajectories visualization put in evidence the presence of complex flow patterns in the aortic arch that strongly affect the platelet activation state (PAS).

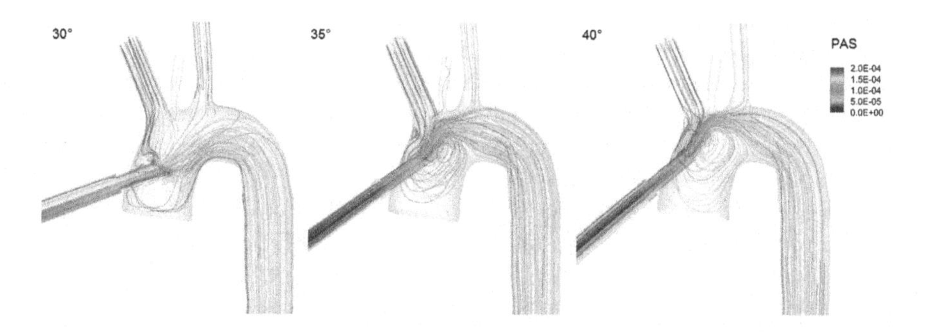

Fig. 7.9 Trajectories of platelet-like particle sets colour-coded with respect to local PAS value

7.2 Human Bones Stress Analysis

As already exposed at the beginning of this Chapter, the development of subject-specific models plays a relevant role in computational biomechanics. The digital modelling, having the purpose to reproduce mechanical and geometrical features of an anatomical region of interest from biomedical images, is a key step in the field of in silico evaluations aimed at diagnostic, therapeutic or surgical planning purposes.

Since the CSM capabilities have reached a detail in numerical prediction such to satisfy the accuracy required by in vitro models, many techniques have been developed and are now available in the market for generating high-fidelity CSM models of bones from computed tomography (CT) data.

The standard procedure for producing an accurate CSM model of a bone first foresees the segmentation of the CT dataset to obtain a surface tessellation of the bone segment contour. The shape obtained is successively mathematically parameterised, usually employing non-uniform rational B-splines (NURBS), and finally meshed using a dedicated software. The main drawbacks of such an approach are the low automation and flexibility, and the fact that the method is user-intensive and time consuming. The low flexibility limit, in particular, arises because the approach does not permit fast mesh adaptation and transposition between patients. This limit makes very difficult, and even impossible in many practical cases, to enable fast comparison among patients and to perform statistical analyses.

In this field mesh morphing can be used as a tool for shape registration as it allows the analyst to deform a template entity, represented by a CAD model or by a surface or volume mesh, onto a target one. This result can be reached by adapting a template mesh onto a subject-specific geometry extracted from magnetic resonance (MR) or CT images.

The morphing of the models of bone segments from CT data looks to be a promising approach to overtake the limits of the traditional approach, as it allows to quickly generate the patient mesh using adaption, to conduct sensitivity studies (e.g. on prosthesis design or positioning as demonstrated in Sect. 7.1.2), to easily compare results sets from two or more meshes with same topology as well as to improve the speediness and automation of the subject-specific CSM model generation (Grassi et al. 2011).

Starting from different working scenarios, a detailed workflow based on mesh morphing for a human tibia and a human femur are respectively demonstrated in the following sections. In particular, in the case of the tibia the template (source) is a volume mesh and the target is a surface mesh, whereas for the femur case both the template and target are provided as a CAD model.

The complete set-up proposed in this section consists of three principal steps. The first step (Step1) concerns the definition of an RBF field to gain a rough approximation of the target entity. The morphing action is performed by applying a transformation that maps the points of the source surface mesh, identified by landmark points, onto the homologues ones that belong to the target surface mesh.

The second step (Step2) deals with the definition of a second RBF field to project the rough computed shape obtained at the end of the first step, onto the target entity. Such an operation is carried out using, as target, a surface mesh that is fine enough to capture the shape variation. In the described test cases these latter consists of an STL file that, respectively, is directly available and generated from a STEP file.

The third step (Step3) finally foresees a sequential application of the two generated morphing fields for the final update of the CSM volume mesh.

It is worth to notice that in Step3 the amount of user interactions that enables the final movement of the CAE mesh is kept at minimum. The problem here faced, that is the mapping of similar shapes onto non matching meshes, is one of the most relevant and challenging application of the advanced mesh morphing.

Besides, the reader should also keep in mind that the burden of surface connection and projection faced in Step1 and Step2 can be avoided when isotopological shape variations are available in advance.

7.2.1 Tibia

The aim of the test case dealing with the tibia is to update a volume mesh CSM model for the Abaqus solver identifying the template (reference) item, onto a new shape, constituting the target, available as a surface mesh. This latter, as already introduced, is provided in the STL format generated through CT data.

The morphing strategy envisioned is comprised of three main steps already introduced in the previous section. Step1 foresees the use of an RBF field to match a small set of landmark points so that the position as well as the overall dimension of template and target are made similar. In Step2 an RBF projection field is set to finely project the nodes of the template mesh onto the target one. The two RBF solutions are applied in sequence in Step3 for finally updating the entire volume mesh of the bone through morphing.

In Step1 the landmark points on the template geometry are used as sources points, whilst those on the target are used for the calculation of the morphing field. The location of landmark points is of a paramount importance because the better is their positioning on the target, the better will be the result. Given that, such an arrangement is advisable to be improved with the support of an expert of anatomy and bone morphology and with the aid of haptic devices (Valentini and Biancolini 2017). Figure 7.10 shows the twelve landmark points generated on both the template (left) and target (right) mesh. The creation of these points, depicted in yellow colour, is performed through the Femap pre-processor tool disabling the visualisation of mesh edges.

The coordinates of the landmark points are then exported according to a specific format so that they can be imported and processed as source points by the morpher tool (in this example the GUI of the ANSYS Fluent add-on release of the RBF Morph technology). Such points, depicted in Fig. 7.11, enable to finalise the RBF set-up of the morphing action envisaged by the Step1 of the described procedure.

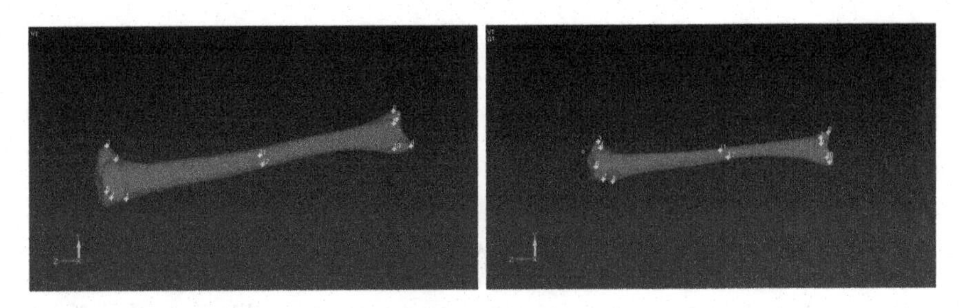

Fig. 7.10 Tibia case: landmark points generation (Step1)

Fig. 7.11 Tibia case: landmark points import in the RBF set-up (Step1)

The smoothing action at this stage is a simple transformation such to impose the correct bone length, that basically consists of an RBF translation that implicitly includes a proper mesh scaling as well.

At the end of Step1, the volume mesh turns out to be very similar to the target one as visible in Fig. 7.12, where an overlay of the final target (yellow) and the preliminary shape (grey), obtained using just twelve RBF source points, allows to understand how close the surface meshes are.

It is worth to specify that, such a step has the potential capability to make the morphing problem independent from the reference system used for the target. In fact, although in this test case a simple translation applied through RBF morphing is

Fig. 7.12 Tibia case: target and morphed mesh after the Step1 (Color figure online)

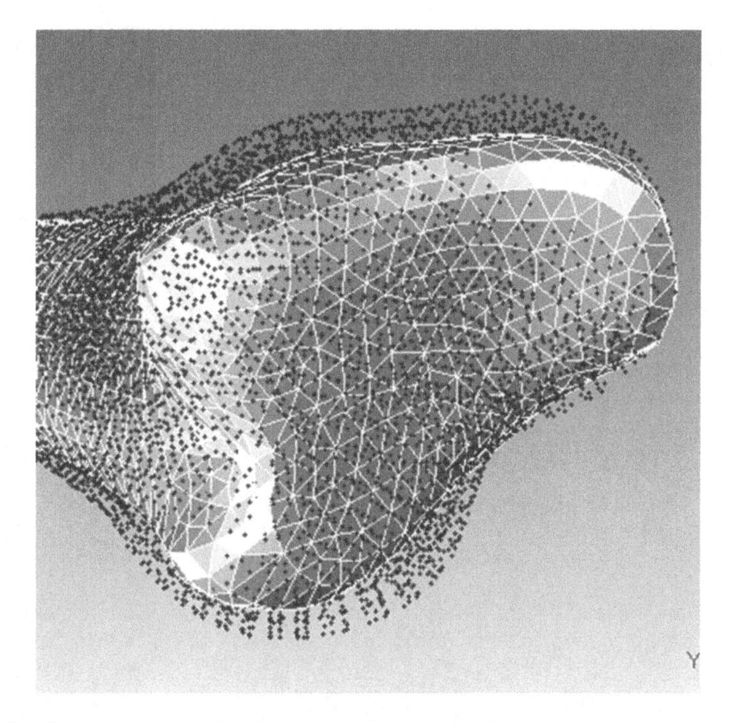

Fig. 7.13 Tibia case: cloud of RBF points to enable the RBF projection (Step2)

enough to gain the sought results of Step1, a similar operation can be also accomplished if rotations are required because the rigid movement to account for different reference frames is automatically introduced by the morpher tool. In any case, after the Step1 accomplishment, the faceted surfaces are close enough to allow the projection defined by the RBF solution generated in the successive Step2.

As already underlined, the updated volume mesh is close enough to the target faceted mesh and, so, the projection algorithm can now be used to project surface nodes of the solid mesh onto the target. This operation is executed using the implicit surface method (see Chap. 5) allowing to transform the target faceted mesh into a smooth implicit surface, and processing points and normal vectors connected to the STL file in view of generating a projection field valid around the target surface (the so called "STL target" feature of RBF Morph tool). In the present study, the STL file contains 13,328 nodes. To keep apart only the points at a distance more than 1 mm, a coarsening parameter is set to regularize and reduce the cloud of points to 10,705 items. When the offset points are generated, the size of the cloud, depicted in Fig. 7.13, comprises 32,115 items.

The obtained RBF field is then used for computing the updated position of the nodes at the surface of the volume mesh to be morphed, which is composed by 4897 points. Figure 7.14 shows the position of the volume mesh and the target STL mesh (yellow) respectively before (left) and after (right) applying the morphing dictated by the RBF solution generated in Step2.

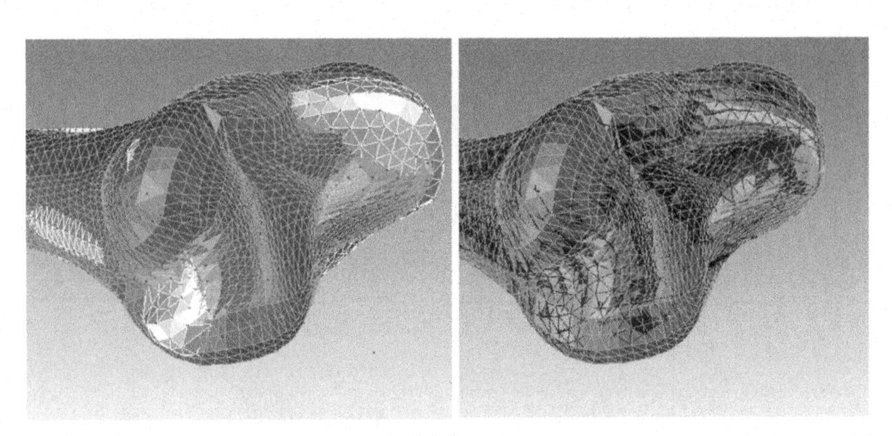

Fig. 7.14 Tibia case: RBF projection (Step2)

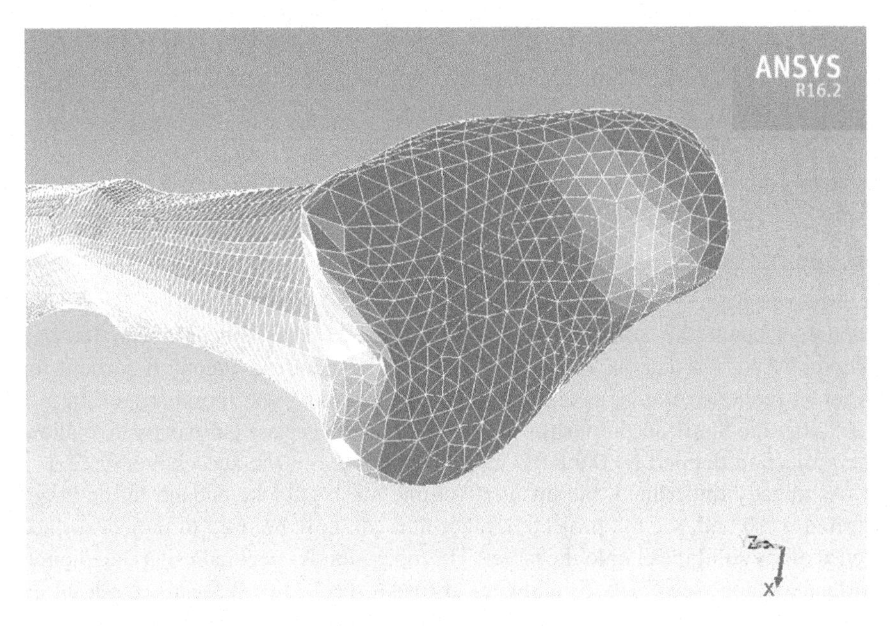

Fig. 7.15 Tibia case: final volume morphing (Step3)

In the last step, the sequential application of the RBF solution of the Step1 and Step2 allows to update the positions of all the nodes inside the volume according to the morphing action imposed at the surface mesh (Fig. 7.15).

7.2.2 *Femur*

The practical application dealing with a human femur that is demonstrated in this section, is representative of a quite common scenario: an high fidelity CSM model

Fig. 7.16 Femur case: the geometrical models are available as a BREP of a multi-patch of NURBS

of a reference geometry is available, or alternatively created, and its adaption to a new shape, characterised by the same morphology (same number and typology of elements), is addressed.

In this second bone biomechanics test case, the starting data for the CAE analysis pre-processing phase consists of the BREP reconstruction of both the template and the patient-specific geometry that, respectively, identify the reference model (source) and the target of morphing. These models, represented by a network of curved NURBS patches, are shown in Fig. 7.16 where the smaller bone, visualized in red, is the template that has to be morphed onto the larger (patient) bone, visualised in ivory.

The details of the two shapes are summarised in Table 7.1. Size, surface and volume of the two bones significantly differ. Moreover, meshes are not matching and the face count is different as well.

The BREP representations are imported in the Femap pre-processor to generate, on the one hand, the volume mesh of the source in CDB format to perform the stress analysis with the ANSYS Mechanical software and, on the other hand, the surface mesh of the target in STL format with the desired density. Actually, this latter operation is often not mandatory because typically the models are already supplied as very fine faceted meshes, and they can thus be used as they are or decimated according to a sampling algorithm as the one described in Sect. 4.3.

As previously introduced, the Step1 of the morphing procedure consists of the placement of a certain number of landmark points on the original and target entities. Thirteen landmark locations are firstly picked on the surface mesh of the source CSM model of the bone, and the corresponding points are secondly picked on the

Table 7.1 Femur case: geometrical details of the reference and target models

Feature	Template/source	Patient-specific model/target
Length (m)	0.369	0.458
Surface area (m^2)	0.04685	0.07560
Volume (m^3)	0.3324	0.7245
BREP multi-patch size (–)	286	282

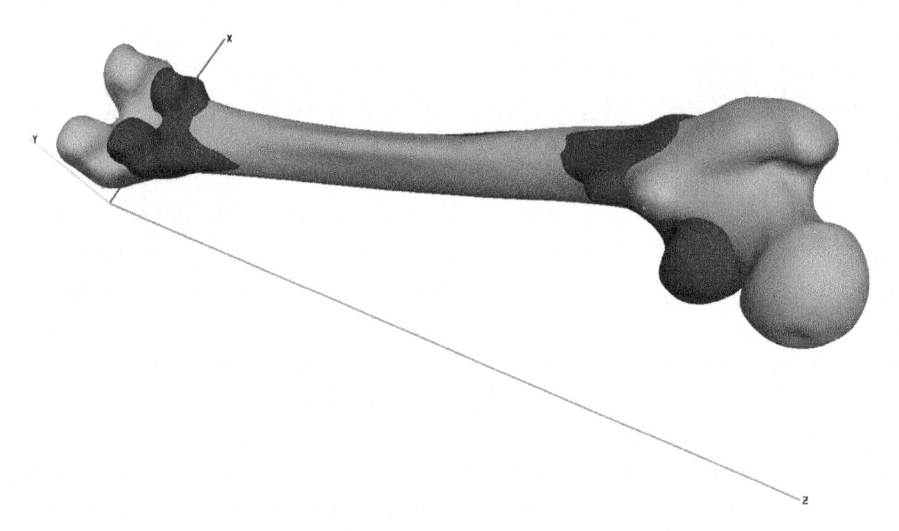

Fig. 7.17 Femur case: landmark points positioning (Step1)

surface mesh of the patient bone. Those sets of points are shown in Fig. 7.17, where both the source and target meshes are depicted disabling the visualisation of the edges by means of the stand-alone version of the RBF Morph software.

An RBF field allowing to move the original surface mesh very close to the target one is accordingly generated. The morphing results is depicted in Fig. 7.18 where the template mesh, coloured in yellow, is very close to the target one, coloured in ivory.

Due to the meshless feature of the approach, the warping field can be applied to any kind of volume mesh of the original geometry. The STL projection tool of the

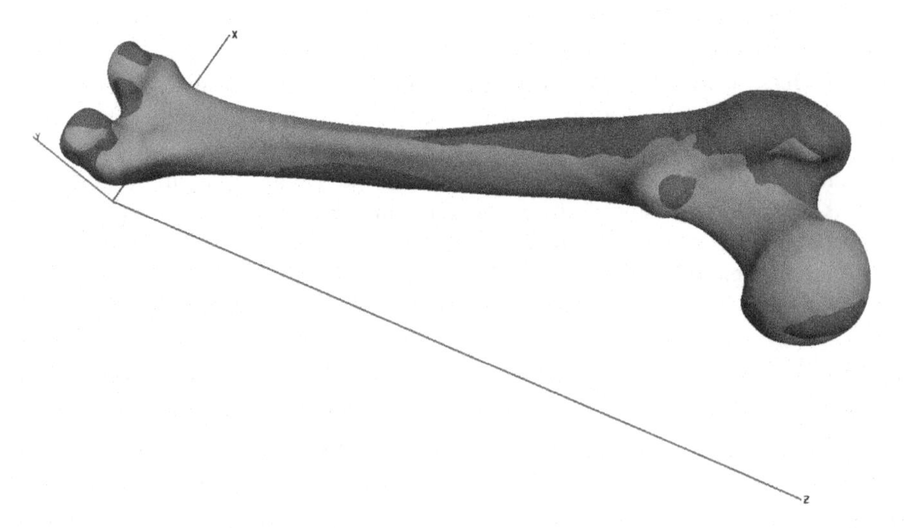

Fig. 7.18 Femur case: RBF projection (Step1)

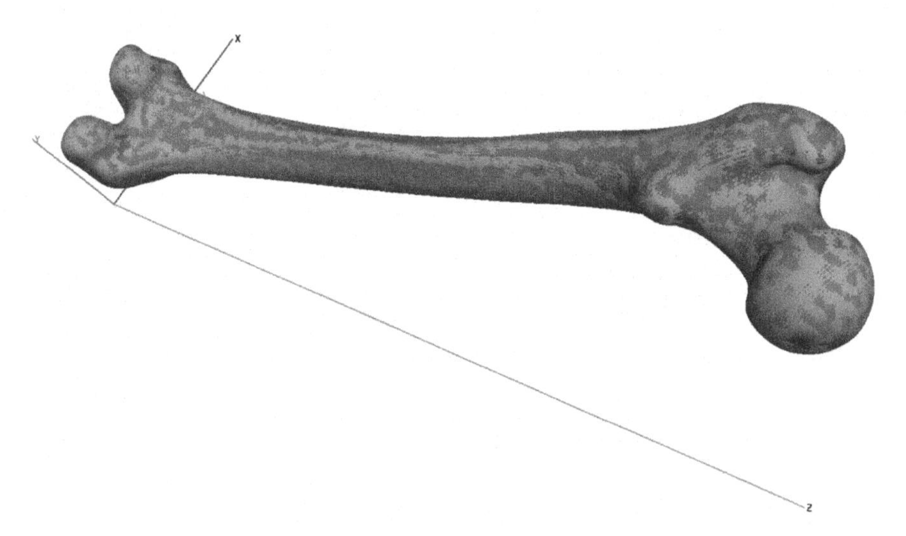

Fig. 7.19 Femur case: RBF volume morphing (Step2)

morpher tool is then used to complete the morphing. Figure 7.19 depicts the final result of the RBF mesh morphing which enables to accurately adapt the source to the target by morphing the volume mesh as well.

As already described, the RBF solutions of the first two steps are applied in sequence to get the final morphed configuration of the bone mesh. Two details of the final morphed mesh, obtained applying in sequence the RBF solution of the Step1 and Step2, superposed to the target CAD are depicted in Fig. 7.20.

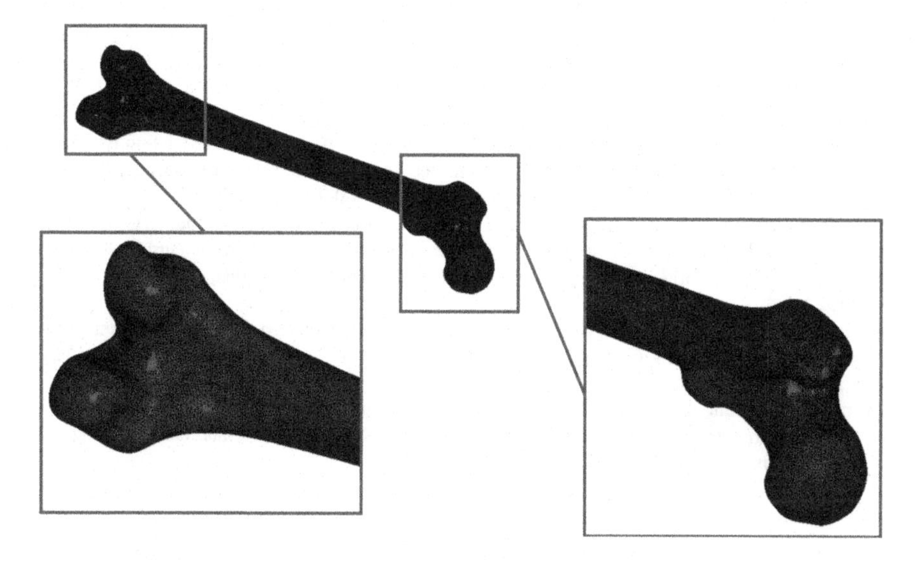

Fig. 7.20 Femur case: final morphed mesh (Step3) superposed to the target (starting) CAD

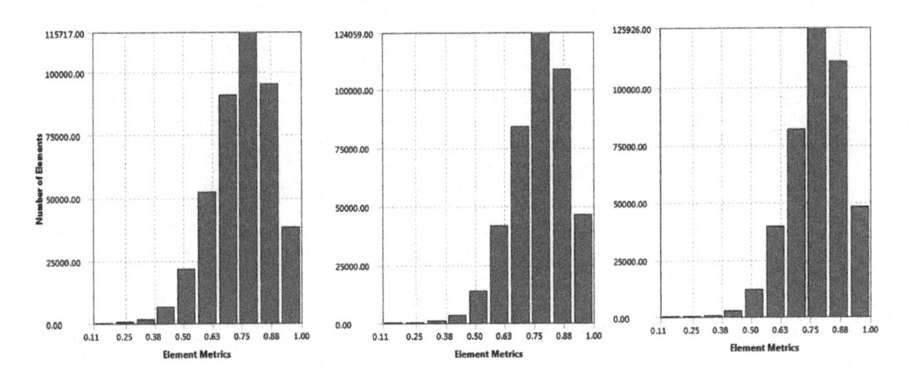

Fig. 7.21 Femur case: statistics of the various meshes

Figure 7.21 shows the elements metrics characterising the tetrahedral volume mesh of the starting mesh (left), and of the morphed mesh at Step1 (middle) and at Step2 (right). These data are expressed in terms of cell quality and, as visible, a slight worsening of the mesh quality accompanies the morphing of the bone discretised model.

References

ANSYS Inc. (2017a) ANSYS Fluent user's guide

ANSYS Inc. (2017b) ANSYS Fluent advanced add-on modules

Biancolini ME, Ponzini R, Antiga L, Morbiducci U (2012) A new workflow for patient specific image-based hemodynamics: parametric study of the carotid bifurcation. Comput Model Objects Represented Images III: Fundam Methods Appl. May 5–7, 2012: CRC Press, Rome, Italy, ISBN: 9780415621342

Capellini K, Costa E, Biancolini ME, Vignali E, Positano V, Landini L, Celi S (2017) An image-based and RBF mesh morphing CFD simulation for parametric aTAA hemodynamics. In: Giuseppe Vairo (ed) Proceedings of the VII meeting Italian chapter of the European society of biomechanics (ESB-ITA 2017). ISBN: 978-88-6296-000-7

Gallo D, Biancolini ME, Ponzini R, Antiga L, Rizzo G, Audenino A, Morbiducci U (2014) A virtual test bench for hemodynamic evaluation of aortic cannulation in cardiopulmonary bypass. In: 11th world congress on computational mechanics. Barcelona, Spain, July 20–25, 2014

Grassi L, Hraiech N, Schileo E, Ansaloni M, Rochette M, Viceconti M (2011) Evaluation of the generality and accuracy of a new mesh morphing procedure for the human femur. Med Eng Phys 33(1)

Valentini PP, Biancolini ME (2017) Interactive sculpting for engineering purposes using augmented-reality, mesh morphing and force-feedback. IEEE consumer electron mag (in Press). https://doi.org/10.1109/MCE.2017.2709598

Chapter 8
Adjoint Sensitivities and RBF Mesh Morphing

Abstract Optimizations based on adjoint sensitivity data are presented in this Chapter. RBF are adopted to set up an advanced filtering tool suitable for removing the noise usually observed when shape sensitivities data are computed using CFD so as to enable the adjoint sculpting method where surfaces are updated according to the information provided by the flow solution to get the desired performances (as drag reduction or pressure loss control). Advanced mesh morphing is used to propagate, once properly filtered if needed, the shape data known at surfaces into the full volume mesh required for the calculation. The concept is demonstrated for FEM as well showing how a bracket and a T beam can be updated to control a target displacement. The adjoint preview approach, which consists of the computation of derivatives with respect of shape variations known in advance is then detailed. A collection of fluid shape optimizations, taking into account both internal and external flows, is provided at the end of the Chapter.

Optimization plays a central role in many engineering and scientific fields. For centuries the ability to design better systems has been a driving force that shaped the world as we know. However, this process is slow and time-consuming and requires financial and human resources. In the modern competitive scenario, being able for companies to keep up the pace and to offer innovative design solutions with advanced, efficient and cutting edge technologies, is of vital importance. In this view, the current trend is to base design process on the use of computer-aided engineering (CAE) methods and tools such as computer-aided design (CAD), CFD, CSM, and of computer-aided manufacturing (CAM) methods.

The recent industrial success of production methods based on additive manufacturing (AM), characterised by a large degree of freedom in terms of the possible shapes that can be achieved, gave a new impulse to design technologies. Being nowadays optimal shape design (OSD) a requirement in various fields of engineering, lots of algorithms and methods have been developed in the last four decades for the solution of numerical optimization problems. Each technique comes at a price. Zero order methods (examples demonstrating how RBF can be used in this field to set-up surrogate models are provided in Chap. 11), the most common

M. E. Biancolini, *Fast Radial Basis Functions for Engineering Applications*,
https://doi.org/10.1007/978-3-319-75011-8_8

ones and with the longest tradition, have the advantage of an easy implementation but their drawbacks appear especially when dealing with multi-objective optimization (MOO) or when an high number of parameters needs to be considered (Vanderplaats 2005; Arora 2016). To alleviate these problems, several techniques have been developed as, for instance, the generation of the best distribution of points on the parameter space or procedures that make the population to evolve using evolutionary algorithms (EA) (Ashlock 2005; Montgomery 2012).

In this context, gradient-based methods have emerged as a powerful tool able to drive design, exhibiting good performances and speed also in the optimization scenarios characterised by an high number of design variables or objective functions. Depending on the problem complexity, two main approaches can be adopted to feed useful information to gradient-based methods, namely direct or adjoint. The direct approach allows to obtain sensitivities, namely the derivative of the objective function with respect to design parameters, with a low cost when more responses than design parameters are present. On the other hand, when dealing with a larger number of design parameters with regard to objective functions, adjoint methods are preferable. Advanced approaches, obtained by employing both gradient-based and zero order techniques, also represent a possible solution. The formers are indeed very accurate and can converge to a local minimum also when dealing with multiple parameters, whilst the latter are useful in identifying the most promising design globally in the design space.

Envisioning a node-based morphing approach, where the position of the numerical grid nodes are the parameters, the adjoint solver can be chosen to calculate the shape derivatives required to guide optimization. Sensitivities with respect to design parameters can be a key enabler in a powerful and cost efficient optimization process, but they require an effort to be seamlessly integrated in the common industrial design procedure. Strategies to process such complex and rich sensitivity data with the aim to obtain an optimized system integrating the workflow in a standard CAE framework are then needed.

Two workflows are proposed in the present Chapter. The first one, named adjoint sculpting, is particularly suitable for a full automatic optimization and the second one, termed adjoint preview, works fine for the evaluation of the imposed shape variations and their use in gradient-based optimizations.

8.1 Adjoint Sensitivity Background

Details of the adjoint solution background and framework are out of the scope of this book. What is important to say is that once the adjoint solution has been computed for a given observable F (recalling that for each observed objective a companion adjoint run is required), we get the shape sensitivity with respect to the observed quantity. Such sensitivity can be available as a vector field at all the mesh nodes (in the volume and at surface), as a vector field at surface or as a scalar function at surface. If a scalar is available, the direction will be given by the local normal at surface. Otherwise the component of the vector field can be used. Equation (8.1) gives an example of such sensitivity information sF, for the case of the full volume mesh.

$$sF_{in,ic} = \frac{\partial F}{\partial x_{in,ic}} in = 1\ldots nodes\ ic = x, y, z \quad (8.1)$$

At a first glance such sensitivity data could be used in a straightforward fashion to get a new shape of the part considering that the maximum variation of the observable is in the direction of the gradient and so, as demonstrated in (2.31), the position of the nodes could be optimised simply by updating them in the gradient direction, as explicated in Eq. (8.2) where the increment λ has to be defined small enough to keep the local linerarization valid.

$$x_{in,ic} = x_{in,ic} + \delta x_{in,ic} = x_{in,ic} + \frac{\partial F}{\partial x_{in,ic}} \lambda \quad (8.2)$$

Unfortunately, the sensitivity field is usually noisy and the direct update is not possible unless the analyst has already decided the deformation field. If the field is known in advance, the evaluation of its effect on the observable can be accomplished using Eq. (8.3). Where the total variation of the observable is a function of the intensity λ^* of the field δx^*.

$$\delta F(\lambda^*) = \sum_{in=1}^{nodes} \sum_{ic=1}^{3} \lambda^* \frac{\partial F}{\partial x_{in,ic}} \delta x_{in,ic}{}^* \quad (8.3)$$

It is worth to observe that in Eq. (8.3) we have assumed that a given field can be linearly amplified using the parameter λ^*. If a base of deformation vector is provided, such a concept can be applied for each shape according to Eq. 8.4 that is based on a linear relation between the deformation field and the shape parameter, or according to Eq. 8.5 which assumes that the variation of the deformation is not a linear function.

$$\delta F\left(\lambda_{ishape}{}^*\right) = \sum_{in=1}^{nodes} \sum_{ic=1}^{3} \lambda_{ishape}{}^* \frac{\partial F}{\partial x_{in,ic}} \delta x_{ishape}{}^*_{in,ic} \quad (8.4)$$

For the generic case of a nonlinear parameter, the variation of shape intensity due to the variation of the parameter has to be computed using the local derivative at current status of all shape parameters λ^* for a given variation of the individual parameter $\delta\lambda_{ishape}{}^*$.

$$\delta F\left(\lambda_{ishape}{}^*\right) = \sum_{in=1}^{nodes} \sum_{ic=1}^{3} \frac{\partial F}{\partial x_{in,ic}} \frac{\partial x_{ishape}{}^*(\lambda^*)_{in,ic}}{\partial \lambda_{ishape}{}^*} \delta\lambda_{ishape}{}^* \quad (8.5)$$

Equation 8.5 is usually expressed in differential form (8.6) and allows to directly compute the sensitivity with respect to the shape parameter. The variation of the shape $\frac{\partial x_{ishape}{}^{*}(\lambda^{*})_{in,ic}}{\partial \lambda_{ishape}{}^{*}}$ due to the changing of the parameter are usually known as mesh deformation speeds.

$$\frac{\partial F(\lambda^{*})}{\partial \lambda_{ishape}{}^{*}} = \sum_{in=1}^{nodes} \sum_{ic=1}^{3} \frac{\partial F}{\partial x_{in,ic}} \frac{\partial x_{ishape}{}^{*}(\lambda^{*})_{in,ic}}{\partial \lambda_{ishape}{}^{*}} \tag{8.6}$$

The gradient concept of (8.2) can be adapted acting on shape parameters instead of all mesh locations (8.7).

$$\lambda^{*} = \lambda^{*} + \delta\lambda^{*} = \lambda^{*} + \frac{\partial F(\lambda^{*})}{\partial \lambda_{ishape}{}^{*}} \lambda \tag{8.7}$$

A reference industrial implementation of the adjoint based design is provided by the discrete adjoint solver of ANSYS Fluent. It allows to get the sensitivity for all the nodes of the volume mesh that are exposed to the user for post-processing and custom usage, and can be exploited with the proprietary free-form deformation (FFD) morpher that allows to reshape according to computed sensitivities. The coefficients of the FFD (whose number depends on the number of divisions along the sides of the deformation boxes) are used as parameters according to Eq. 8.6 and computed according to Eq. 8.7 performing a volume integration of the morphing field projected onto the sensitivity field. Such an approach makes the effect of the sensitivity available at the surface with an implicit filtering applied as the shape variation complexity that can be represented is defined in advance by FFD resolution.

The concept of the parameter sensitivity is demonstrated in Fig. 8.1 where three shape parameters are introduced to reshape a cube immersed in a wind tunnel.

The variations that are typical of an automotive development consist of the modification of the angle of the boat tail, the nose and the roof. It is worth to see how the sensitivity prediction is tangent to the full computation (ten CFD runs for each parameter) and that the linear approximation validity range is verified just for small variations of the parameter itself.

8.2 Role of Fast RBF in Adjoint Based Design

Fast RBF give a feasible and comprehensive solution for the implementation of adjoint based optimisation workflows mainly for two reasons: first an RBF filter can be defined (see 2.7 to learn more about approximation of noisy data) using regression so that sensitivity shape information are properly smoothed, and second an RBF mesh morpher (see Chap. 6) can be adopted to update the full numerical grid (a volume mesh in 3D) according to adjoint sensitivity driven shapes (at surfaces in 3D). The effectiveness of the coupling between adjoint and RBF have been demonstrated

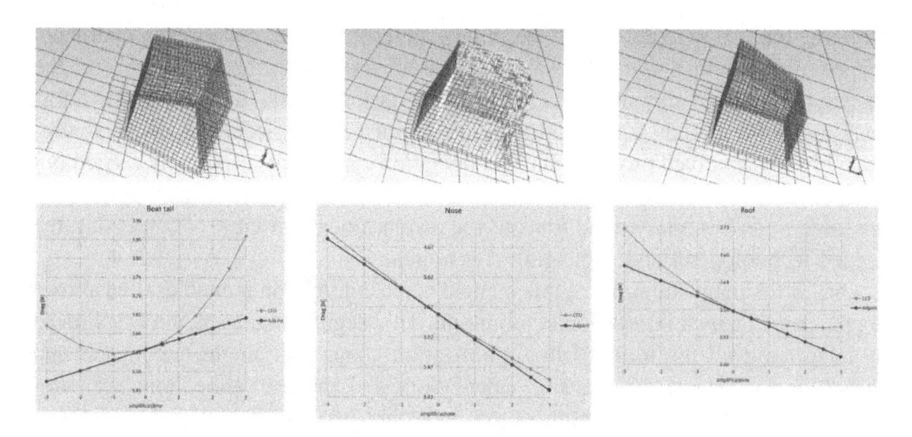

Fig. 8.1 Example of adjoint computed sensitivities for three shape parameters imposed using mesh morphing. Derivatives estimated by adjoint are compared with full exploration (10 analyses for each parameter). ANSYS Fluent adjoint solver and RBF Morph software are used for the CFD calculations and for the parametrisation of the mesh

in the past with RBF Morph Fluent Add On and ANSYS Fluent adjoint solver (Biancolini 2014). The gained experience has been then extended to the adjoint implementation of OpenFOAM within the framework of the EU FP7 project RBF4AERO (Papoutsis et al. 2018) and, recently, by coupling a basic adjoint implementation for linear structural analysis with ANSYS Mechanical CSM solver and the RBF Morph ACT Extension (Groth et al. 2018). The aforementioned examples are developed according to two workflows that are described in detail in the present Chapter. The first one, named adjoint sculpting, is particularly suitable for a full automatic optimization and the second one, termed adjoint preview, works fine for the evaluation of the imposed shape variations and their use in gradient-based optimizations. A complete description of the coupling of adjoint solver and RBF mesh morphing covered in this section can be found in Groth (2017).

8.3 Adjoint Sculpting

Defining a proper parametrization in real world applications is not an easy task, and engineering expertise in understanding the physics ruling systems and their response is generally required to properly foresee the most relevant variations. Although in several industrial fields such as automotive and aeronautics, some parameters on which is worth to act are well known because their influence in given conditions was theoretically and experimentally validated, it is common practice to simulate complex systems and physics for which the analyst encounters restrictive difficulties in suitably predicting the behaviour.

An answer to this problem is given by the use of adjoint sculpting, whose main idea is to exploit the mesh sensitivity data obtained through an adjoint solver as a sculpting tool, so as to deform the baseline geometry in order to obtain an optimized shape. By adopting a meshless node based morphing methodology for the accomplishment of the shape parametrization task as RBF, the sensitivity information can be directly employed to move surface nodes according to what suggested by the involved physics. Such an operation can be thus accurately carried out in an optimization process by controlling the volume deformations and applying, at the same time, boundary conditions to the displacements in view of respecting the needed technological and packaging constraints.

One of the main strengths of this method is the automation it enables, that allows to reduce the analyst interaction at minimum. This degree of automation is achieved by integrating all the tools of the optimization chain in a single, automated and closed loop in which data transfer, calculations and shape modifications are automatically managed by the optimization logic and performed until a convergence criteria is met. A weak point of shape optimization processes is the difficulty that arises in many design scenarios in providing meaningful shape variations to the optimization logic. Obtaining a valid shape parametrization is indeed one of the most delicate tasks, since it directly affects the final shape and, furthermore, the improvement obtainable with the optimization by influencing the quality of the morphed mesh and constraining the range of achievable shapes. As a matter of fact, the quality of the deformed mesh depends on both the technology used to morph the elements and the choices that the analyst makes when applying the boundary conditions for parametrization.

Although RBF-based methods allow to achieve an high degree of freedom when building shape parameters by prescribing potentially 3N nodal displacements for a 3D mesh described by N nodes, they can suffer of mesh dependent results and not smooth derivatives (Haftka and Gürdal 1992; Mohammadi and Pironneau 2009). Given that, to avoid in an adjoint sculpting based optimization workflow ill-posed problems and the consequent errors in numerical results (optimal shapes), the mesh sensitivity data need to be adequately filtered and smoothed prior their application in view of reducing the noise-related problems and guaranteeing a smooth final shape. Such data manipulation operations can be accomplished through RBF method introduced in Sect. 2.7 and detailed in Sect. 8.3.3 of this chapter.

8.3.1 RBF Morphing Set-Up for Adjoint Sculpting

As already introduced, according to the adjoint sculpting approach the geometrical description of the interested system is automatically modified during the optimization process exploiting the mesh sensitivities provided by an adjoint solver, with the final aim to obtain the required variation in the objective function.

Unfortunately, using this functioning principle the optimization may likely lead to industrially unfeasible or unusable shapes, especially when dealing with complex geometries and boundary conditions. Such an event represents an unacceptable limit when packaging or functional constraints must be respected.

To avoid this unwanted effect, exploiting the meshless property of RBF and their ability to interpolate given displacements everywhere in the space, the strategy adopted in the adjoint sculpting workflow foresees an RBF set-up, accomplished once before running the sculpting workflow, divided in two parts defining the fixed and moving source points separately. To generate the latter, sensitivity information are extracted only from the surfaces or areas of interest. This geometrical definition is automatically applied to the numerical results at each optimization iteration after the filtering application, moving the source points only where required. This set of RBF centres is extracted and generated from scratch at each iteration following, in a reliable manner, the shape suggested by the adjoint solver data.

On the other hand, the RBF set-up dealing with the fixed part consists of a set of motionless source points defined to constrain the motion of the surfaces or the portion of the geometries that must remain undeformed. Volume smoothing can be controlled in this RBF set by defining fixed points in the volume mesh and limiting the morphing action inside a portion of the whole domain. The analyst has to take care of leaving a free deforming buffer area between moving and fixed surfaces to accommodate deformations. In such a way, fixed and moving sets are blended together at each iteration obtaining a new RBF problem in the respect of the constraints defined in the fixed set while applying the shape deformations suggested by the mesh sensitivities in the moving set.

In Fig. 8.2 an example of moving RBF set is shown. The sensitivity information is available on the whole Ahmed body (Ahmed and Ramm 1984) surface (top) but

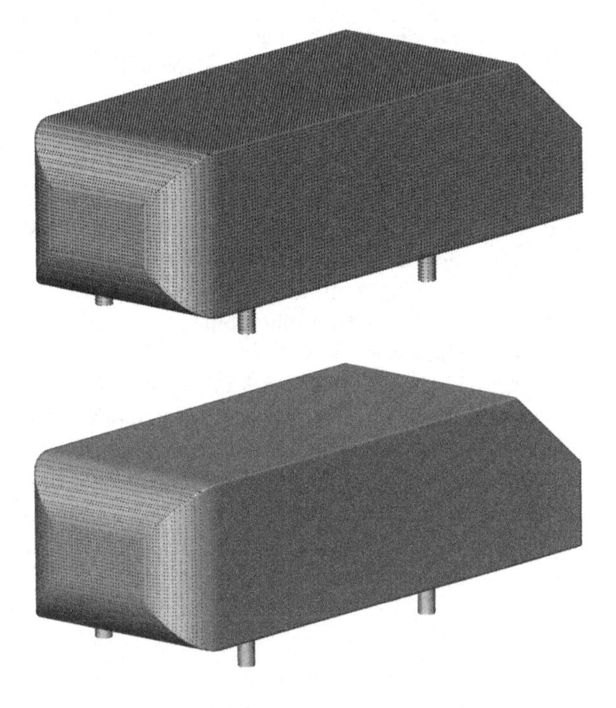

Fig. 8.2 Adjoint sculpting RBF moving set on the Ahmed body model surface mesh

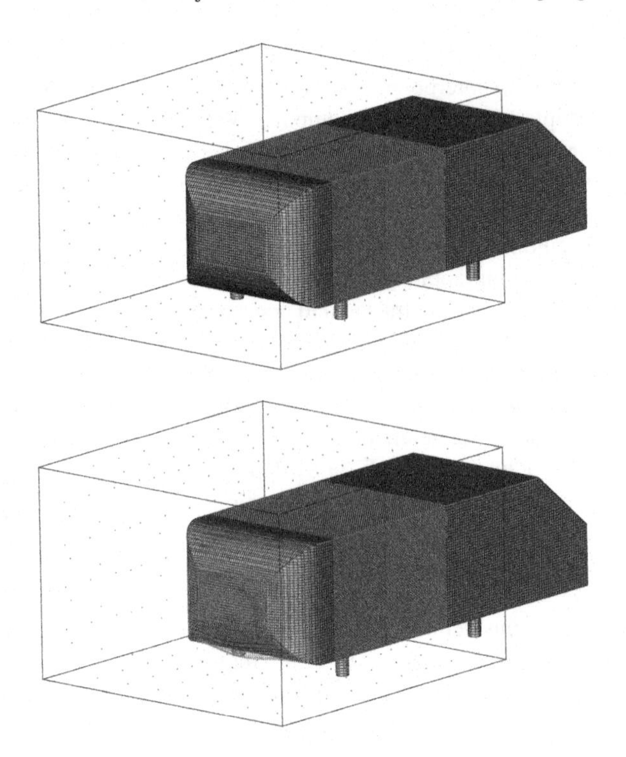

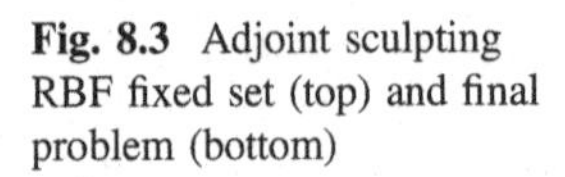

Fig. 8.3 Adjoint sculpting RBF fixed set (top) and final problem (bottom)

only the nose area has to be deformed. For this reason, the data available on the surface mesh nodes defining the nose are taken into account for surface smoothing (bottom). This selection is also kept for the successive optimization iterations in which the nodal positions, as well as the applied nodal displacements, are different.

Referring to the same RBF case, the moving source points, depicted in green colour in Fig. 8.3, are selected through a box-shaped domain wrapping the volume around the nose (delimiting domain). The remaining areas of the Ahmed body model are kept fixed by applying a zero motion to the RBF centres extracted from the surface mesh nodes of the body till a certain extent, using a delimiting domain and leaving a buffer zone to absorb deformations imposed by the moving set. As stated in Chap. 6, the delimiting domain aims at reducing the computational cost of the RBF application.

In the right image of Fig. 8.3 the achieved set-up, obtained by blending fixed and moving sets, is shown for a given iteration. In such a configuration, moving source points are visualised in both the initial (green) and final position (orange).

The baseline and morphed Ahmed body geometries obtained using this approach are shown in Fig. 8.4.

Fig. 8.4 Ahmed body geometry before (top) and after (bottom) adjoint sculpting iterations

8.3.2 Adjoint Sculpting Workflow

The adjoint sculpting workflow is envisaged to be an automatic and evolutionary workflow for shape optimization where one, or more solvers, feed sensitivity information to the optimization logic in order to drive geometry reshaping toward an optimal configuration. Several calculations, data transfers and transformations are required in order to fulfil this task. The adjoint sculpting workflow is shown in Fig. 8.5 through the block diagram where the tasks interested by the use of RBF are distinguishable.

The baseline discretized model is firstly employed to carry out the numerical calculations at the end of which the objective functions of interest are calculated and extracted for later use. The achieved numerical results are sent to the adjoint solver with the grid, in order to extract sensitivity data for the N performance measures chosen so as to generate N sensitivity maps for the whole domain. At this point, the subset of nodes interesting the deformable areas are automatically extracted generating the N moving sets, one for each performance measure and all insisting on the same points. The obtained data are then filtered, as detailed in the following section, to avoid noisy information that could lead to ill-posed problems.

At each iteration, the N different smoothed moving sets are blended to the same fixed set defining the constrained surfaces' areas. These N RBF problems are solved and stored for later use. Because of the chosen RBF set-up, assigning fixed boundaries and deforming buffers, only a portion of the model is actually used to apply morphing. This means that the sensitivity maps calculated for the achieved shape parametrisations are different from the original ones. For this reason, the original sensitivity maps are used to calculate the new expected observable variations by comparing them with the obtained mesh velocity fields, one for each shape parametrization. These N observable variations are sent to the optimization logic that calculates the required amplifications in order to achieve the optimal configuration.

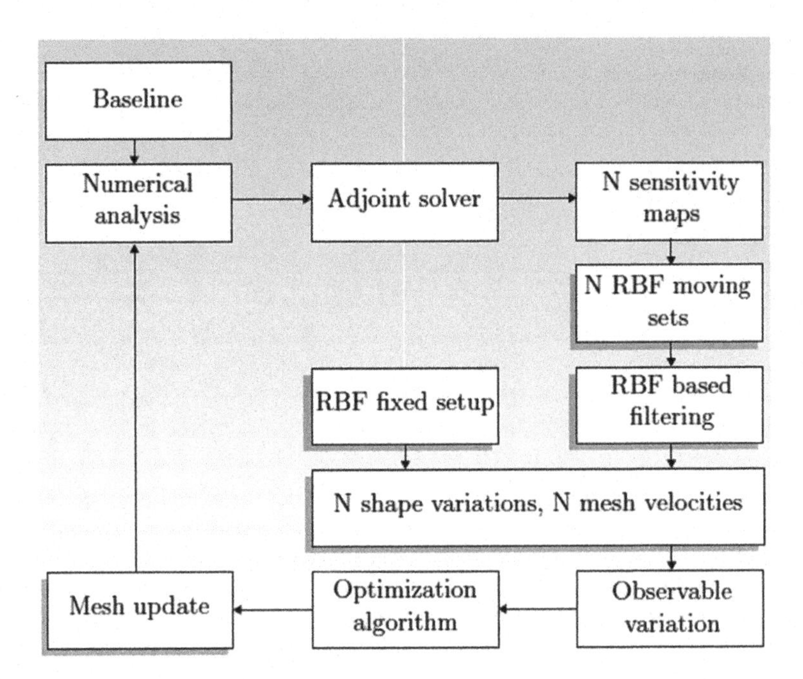

Fig. 8.5 Adjoint sculpting workflow

8.3.3 Filtering Tool for Sensitivity Data Processing

In this section the results of the application of a filtering tool on a simple case according to two approaches are shown. These approaches are referred to as sampling-length and regular-grid method respectively. The former, in particular, has been developed to enable adjoint sculpting optimizations through the RBF4AERO platform (RBF4AERO, 2013) by means of which several test cases, addressed in this book, were carried out.

To identify a reference morphed configuration, at first the effect of the application of non-filtered noisy sensitivities data is shown. These data, processed using a Mathcad worksheet and visualized by means of red dots in Fig. 8.6 (left), concern the steady state solution of an external aerodynamic analysis on a cube-shaped body having the edges of length 0.2 m, whose surface nodes are depicted with black dots. If all points are kept, the surface faithfully follows the original data as depicted in the same figure (right).

The first of the afore-cited approaches pertains to the use of a sampling algorithm that makes use of a length defined by the user to filter sensitivity data. Such a parameter is just related to the wavelength beyond which retaining adjoint output data. Consequently, this approach can run on whatever type of surface mesh because its functioning is just based on the distance between surface mesh nodes. The specified sub-sampling value tunes how aggressive the smoothing action is required to be by sub-sampling the nodes used for the least square regression:

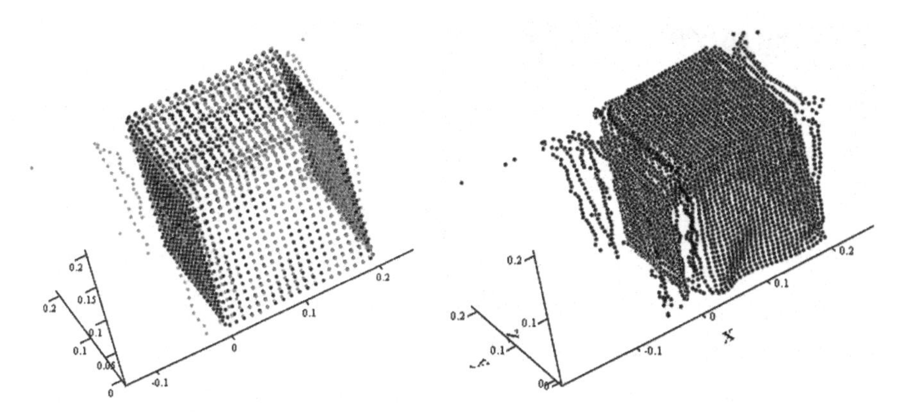

Fig. 8.6 Non-filtered noisy sensitivity data on a cube (left) and the corresponding morphing field (right)

the larger the value of the sub-sampling radius, the faster the fitting time and the smoother result. The method detailed in Sect. 4.3 can be used for this specific task.

The second approach foresees the direct creation of the distribution of points to be used; an auxiliary mesh, positioned as the one to be morphed and characterised by a regular element distribution and a resolution typically much coarser than the high-fidelity one used for the computational analyses, can be employed to generate the cloud of source points. In the case of regular shapes the regular cloud can be directly generated using the tools of Sect. 4.6.

Whatever is the approach, an RBF regression that uses the properly spaced cloud as source and the full cloud as the regression target, has to be defined according to the approach detailed in Sect. 2.7.2; system (2.42) is solved using Mathcad according to (2.43) building the constraint matrix at fit according to (2.38) with all the fit points, i.e. the full dataset, and constraint matrix at sources according to (2.8), i.e. using the coarse distribution.

The effect of the noise filter for the cube example is demonstrated using a fine structured mesh generated in the Femap pre-processing tool to ease the comprehension of the effect of morphing.

The first method, i.e. based on the sampling of a subset of the original mesh nodes a sources, is presented in Fig. 8.7 that shows the morphed configuration of the cube for three values of the sampling, namely 0.065, 0.05 and 0.025 m respectively. In specific, on the left column of images the morphed cube are depicted, while on the right the sampled points, visualized through large squares in red colour, and surface mesh nodes, visualized through small dots in black colour, are shown. It is well visible how decreasing the sampling value the morphing induces an increasingly rough shape.

The second method, in this case based on a regular distribution of points on the cube used as sources, is presented in Fig. 8.8 that visualises the morphed configurations as well as the source points distribution using 3, 4 and 5 points along each edge of the cube respectively.

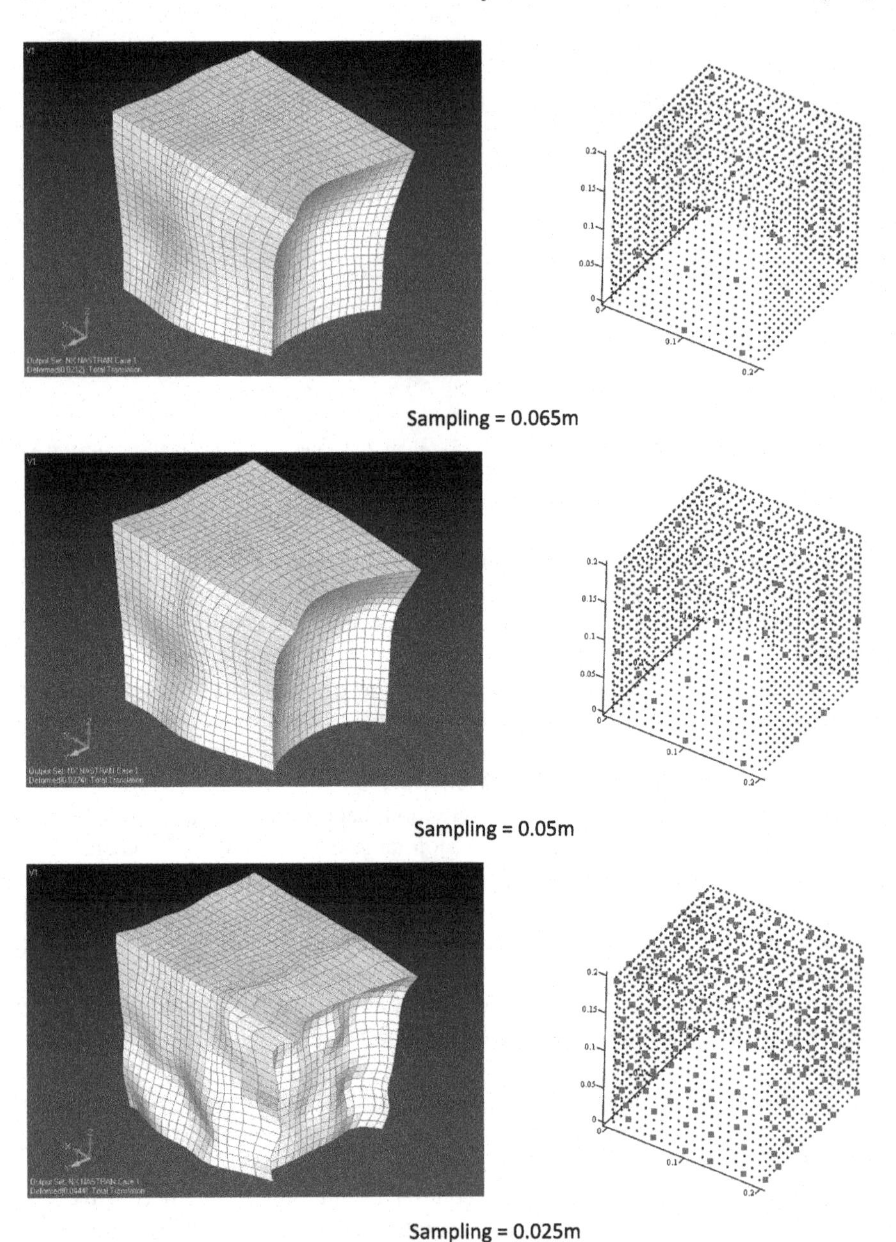

Fig. 8.7 Results obtained through the sampling-length based filtering tool approach

The regular grid allows to better control the filter but requires the extra effort of generating an auxiliary cloud; for a simple shape as the cube this is not an issue as the auxiliary cloud can be easily be parametric. In industrial geometry this task could be more complex instead, and multiple variations of the auxiliary mesh are

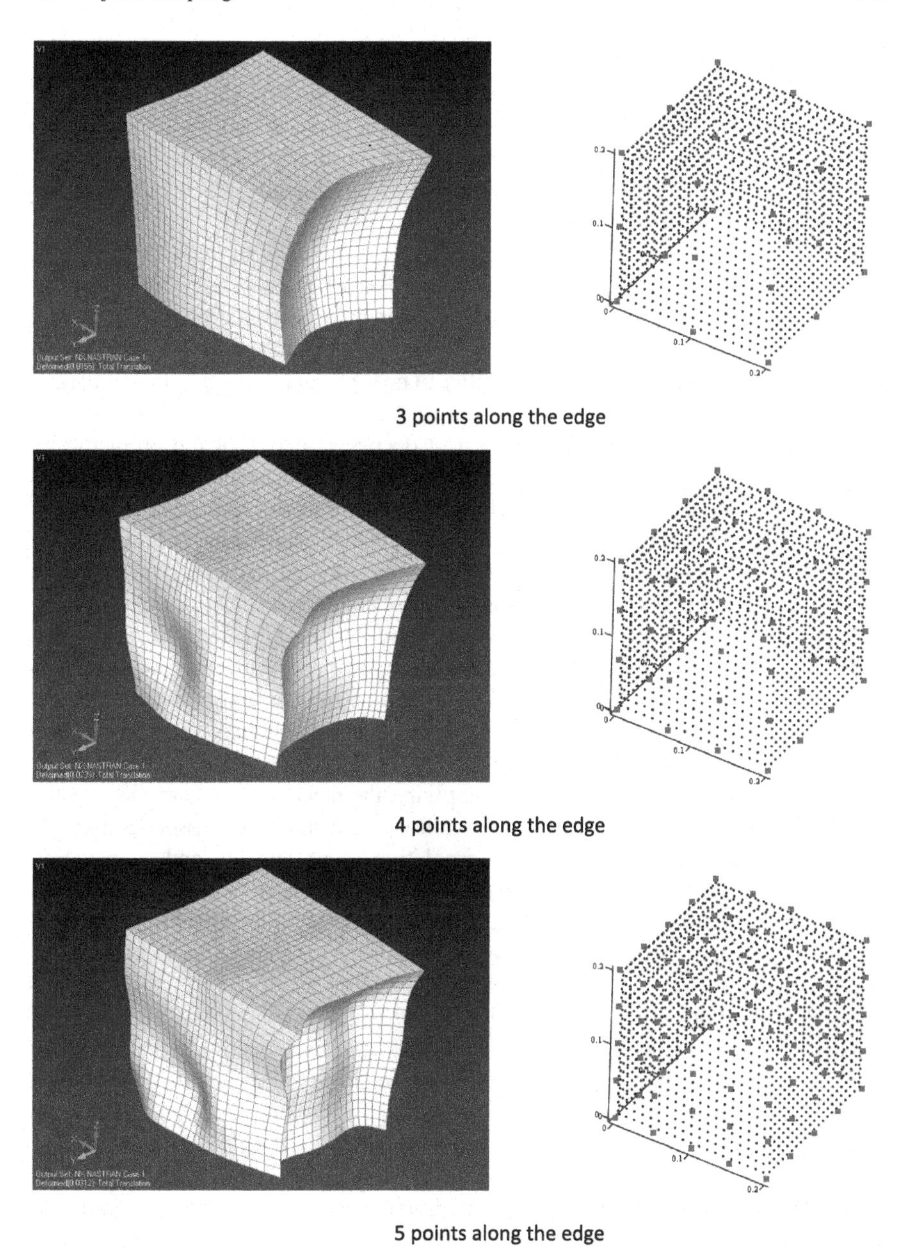

Fig. 8.8 Results obtained through the regular-grid based filtering tool approach

required if the analyst wants to tune the spacing in advance. Sub-sampling method work without further input and the filtering effect can be tweaked just acting on a single variable in a full automated way.

8.4 Adjoint Preview

The adjoint preview method foresees that the shape variations are imposed and defined directly by the analyst that, exploiting the sensitivity data supplied as output by an adjoint solver, can obtain an optimal shape by blending such modifications with the proper amplification.

Given what stated, the core of this workflow is the need to have a meaningful shape parametrization of the numerical grid. For what previously introduced, a good shape parametrization is the outcome of two fundamental elements: a valid interpretation from the analyst of the possible design changes able to guarantee a good impact on optimization and the ability to properly apply them to the numerical grid.

In order to fulfil the first requirement, a deep knowledge of the problem and physics involved in the optimization is strongly needed, fact that makes the engineering expertise and experience essential. It is worth to say again that the lack of such skills can be sometimes alleviated because either shapes changes are dictated by given constraints or suggested by commonly adopted variations to design in some fields. In addition, there are tools that can help the analyst in understanding how the system behaves under the given boundary conditions and how it would react with regard to a variation in some design parameters. With regard to this latter scenario, the sensitivity maps given by an adjoint solver, suggesting the analyst where and how to modify the shape in order to obtain the required variation of the defined objective function, are an example.

Relating to the second requirement, applying the desired shape variations to the numerical grid can be a difficult process not only because the chosen technology must guarantee a good quality in interpolating displacements and propagating deformations, but also because it must provide an expressive and elastic paradigm for defining it. This degree of freedom is particularly felt in technical fields such as the AM one, where the complex shapes achievable by production methods must be re-playable by the design stage as well.

Mesh morphing based parametrisation allows to easily compute the sensitivity of a shape parameter with respect to the observed objective function; sensitivity raw data can be used for this purpose without the need of filtering because the overall effect is averaged by projecting the sensitivities on the morphing field due to the parameter. In Othmer et al. (2011) an example based on the FFD mesh morphing implementation of ANSA software, OpenFoam and the adjoitn solver of NTUA is given. The key concept exploited in the adjoint preview is provided by Eq. (8.6), which tell us how to compute the derivative with respect to the shape parameters and steepest descent algorithm of (8.7) or its advanced adaption (that, for instance, could account for constraints).

As already exposed in Chap. 6, an expressive paradigm is guaranteed by the accurate control given by RBF and can be influenced by the well-posedness of the parametrization problem, making once again the importance of the user skills of central importance. On the other hand, elasticity must be made available to the analyst through a set of tools and a modelling logic suitable to help her/him in properly defining shape parameters. These tools should assist the user in generating geometry variations by guaranteeing a simple but efficient interaction, with the aim of obtaining the wanted shape as final result.

Being the adjoint preview method based on user defined shape variations, the modelling paradigm acquires a central role, making it fundamental in obtaining not only the wanted parametrization but also the optimal shape (see Sect. 6.3).

8.4.1 Adjoint Preview Workflow

Figure 8.9 shows the basic elements of the block diagram of the adjoint preview workflow where the tasks involving the usage of RBF are highlighted. The baseline mesh is employed to carry out the numerical analysis to calculate the N objective functions of interest. The results and mesh are sent to the adjoint solver in order to provide a sensitivity map for each objective function.

Using the M different shape variations generated, it is quite straightforward to calculate the M mesh velocities obtained for each geometrical change by multiplying each nodal velocity for the sensitivity of each node with respect to the objective function. In this way, the $N \cdot M$ observable variations can be calculated. As the observable variations define the influence that each shape parameter has on the variation of the objective function, the most effective parametrizations can be evaluated and, thus, the number of the shape parameters to be employed to finalise the optimization can accordingly be reduced, if needed. Using only a subset of the M original shape variations, namely the most efficient ones, zero order methods, whose use would be otherwise unfeasible, can be run avoiding the risk of getting trapped in local minima.

Choosing to continue the optimization process using gradient methods, at each optimization cycle the observable variation can be employed to calculate the search direction and the amplification of each shape parameter, obtaining an MOO. Gradient-based methods can be employed to calculate the optimal amplification of

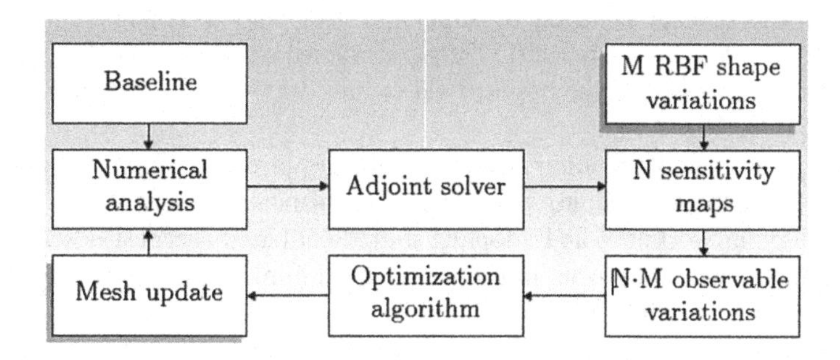

Fig. 8.9 Adjoint preview workflow

each shape parameter and MOOs can be performed. RBF can be eventually employed to update the numerical mesh accordingly. New optimization cycles can be performed until a satisfactory convergence in terms of objective functions or the maximum number of iterations is completed.

When employing the adjoint preview method, packaging or functional constraints are implicitly respected being expressed directly in the shape parametrizations. If a requirement in terms of maximum displacement exists, an upper or lower bound can be suitably imposed on the maximum amplification of the shape parameter that causes it.

In order to demonstrate the methods and workflows detailed in the previous sections, a set of industrial applications concerning both CSM and CFD analyses have been selected and are hereinafter presented.

8.5 Structural Applications

The two structural studies described in the following sections pertain the use of an in-house structural continuum-discrete adjoint-variable solver. Such a structural adjoint solver, written in python programming language, is coupled with ANSYS Mechanical through the ACT technology and it is characterized by an high degree of flexibility, being able to be run with any CSM solver that allows the exchange between numerical outputs and boundary conditions (Groth et al. 2018).

The RBF-based mesh morphing is performed by using the RBF Morph ACT extension exploiting its ACT^2 functionality, whilst an unconstrained steepest descent method was employed to drive the shape optimization according to the adjoint sculpting approach.

8.5.1 Bracket

This section pertains the use of the adjoint sculpting technique to perform the structural shape optimization of a bracket, whose baseline geometry is shown in Fig. 8.10. The bracket is made of structural steel with a Young's modulus of 200 GPa and a Poisson's ratio of 0.3, it is constrained on one side by keeping fixed the hole and it is loaded on the opposite side with a 5000 N force directed along the x axis shown in Fig. 8.11.

The target of the optimization is to reduce the displacement of the free end along the direction of the load adding mass to the mechanical component. By using the adjoint sculpting workflow and adopting a gradient-based (deepest-descent) technique, the automatic optimization reaches the local minimum guaranteeing that the addition of mass is reduced at the minimum value.

On the left side of Fig. 8.11 the sensitivity magnitude map of the loaded end displacement with respect to each mesh node absolute translation is shown. The

Fig. 8.10 Bracket geometry in the baseline configuration

morphing set-up is conceived to maintain the hole and the free end undeformed, modifying only the areas near the bending. The remaining portions of the geometry are left free to be deformed acting as buffer zones. In Fig. 8.11 (right) the fixed source points are shown in red, whilst the set of moving nodes is visualised in green. By taking into account the sensitivity map (left) and the moving source points (right), it can be understood how the sculpting action suggested by the adjoint solver can be suitably filtered in order to modify only the wanted areas.

The step length is varied during the optimization computing at each cycle, in view of guaranteeing a maximum displacement of 1 mm. Nine optimization cycles are carried out in order to obtain a displacement reduction of more than 22% with respect to the baseline geometry. In Fig. 8.12 the baseline (left) and the optimized bracket meshes (right) are compared.

In Fig. 8.13 (left) the evolution of the free end displacement is plotted for each optimization cycle. The same increased stiffness can be achieved by widening uniformly the thickness of the whole component. In Fig. 8.13 (right) the maximum displacement with respect to the thickness, obtained by means of the morpher tool using surface offset modifier (see Sect. 6.3.4), is shown.

It is worth to observe that adopting such a traditional approach the same performance is achieved increasing the mass of 9%, whilst employing the adjoint approach the same result is met with a 6% extra mass only. This difference becomes more evident if the structural geometry is more complicated.

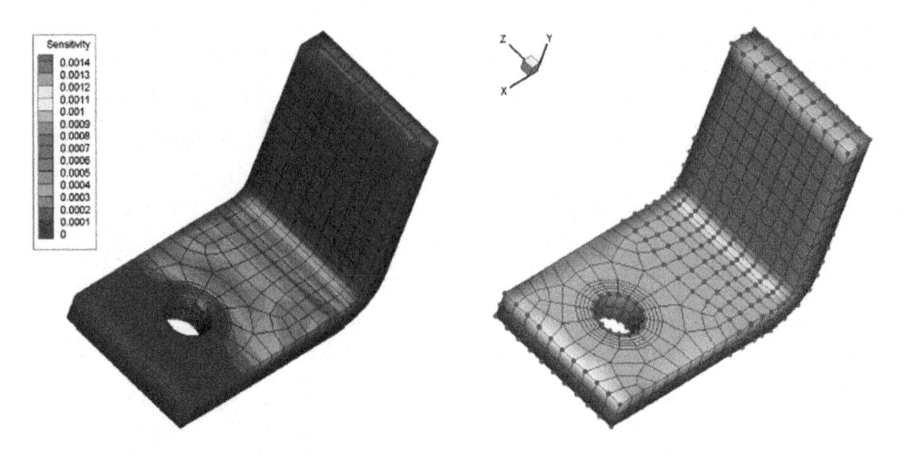

Fig. 8.11 Free end displacement sensitivity map with respect to mesh movement (left) and RBF set-up for the adjoint sculpting workflow (right)

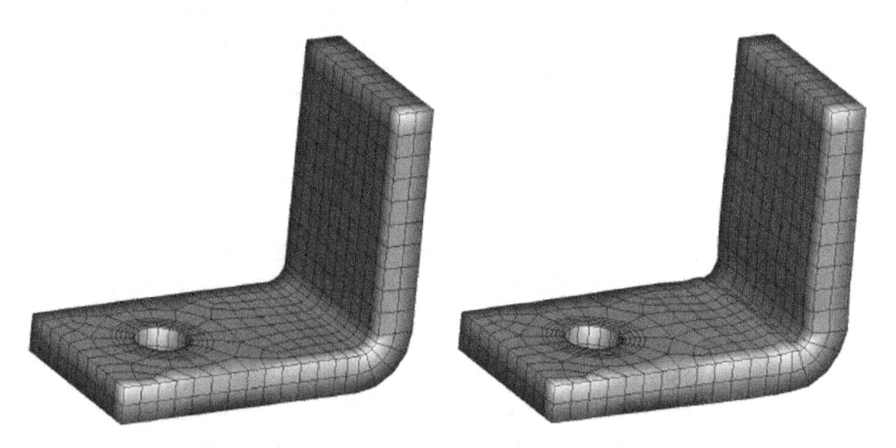

Fig. 8.12 Baseline (left) and optimized (right) bracket mesh

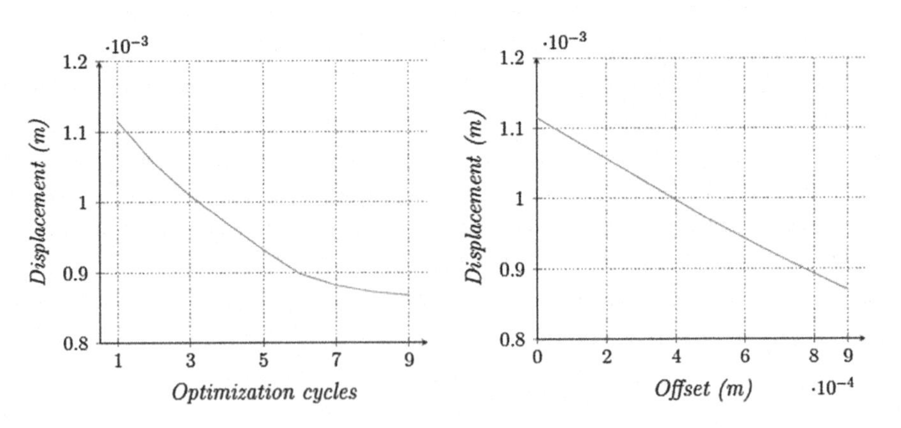

Fig. 8.13 Bracket displacement reduction for each sculpting cycle (left) and in function of the thickness increase (right)

8.5.2 T-Beam

This section details of a cantilever T-beam shape optimization performed with the objective to reduce the loaded tip displacement following the same workflow and methods adopted for the bracket.

In Fig. 8.14 (left) the baseline geometry of the beam is shown. A 10000 N load is applied to the surfaces highlighted in green (left side) along the y positive direction, while maintaining the red surface (right side) fixed. The material used is a steel with a Young's modulus of 200 GPa and a Poisson's ratio of 0.3. The RBF set-up, shown in Fig. 8.14 (right), is conceived to maintain undeformed the loaded end applying fixed source points, visualized in red, while smoothly deforming the remaining surfaces of the structure through moving points, shown in green. Also in this case a small buffer zone is left in proximity of the fixed surface in order to accommodate mesh displacements.

The sensitivity map of the free end translation with respect to nodal displacement is depicted in Fig. 8.15.

The shape variation suggested by the adjoint solver foresees the enlargement of the root section where the bending moment reaches the highest value. The target of a 25% reduction of the displacement at the free end is achieved after twenty-four adjoint sculpting optimization cycles as shown in Fig. 8.16.

The baseline and the optimized geometries of the T-beam mesh are shown in Fig. 8.17 respectively on the left and right side.

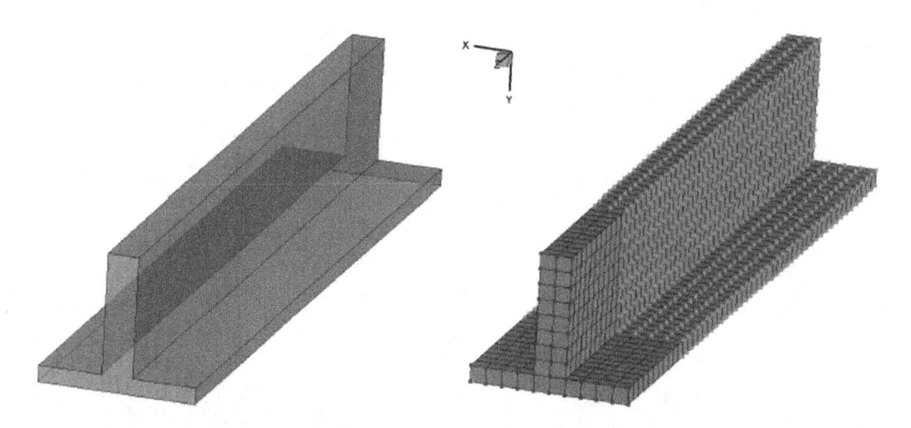

Fig. 8.14 T-beam geometry (left) and RBF set-up for adjoint sculpting (right)

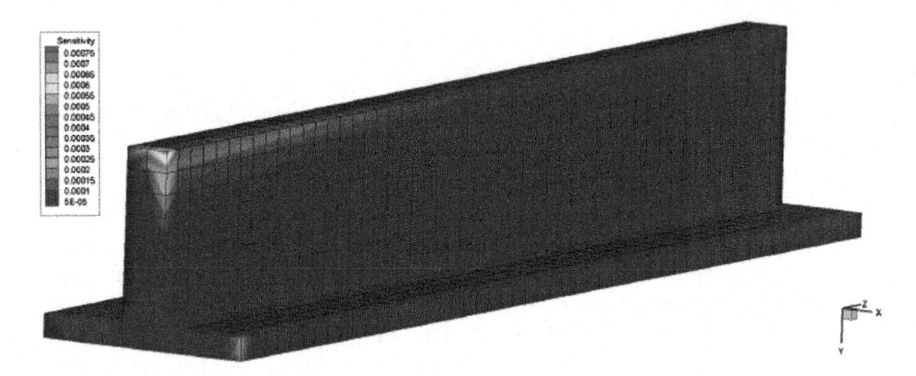

Fig. 8.15 T-beam sensitivity map of tip translation with respect to nodal displacement

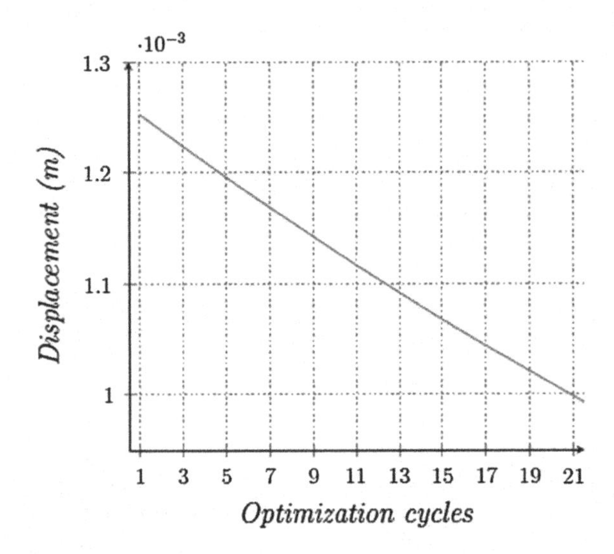

Fig. 8.16 T-beam displacement reduction for each sculpting cycle

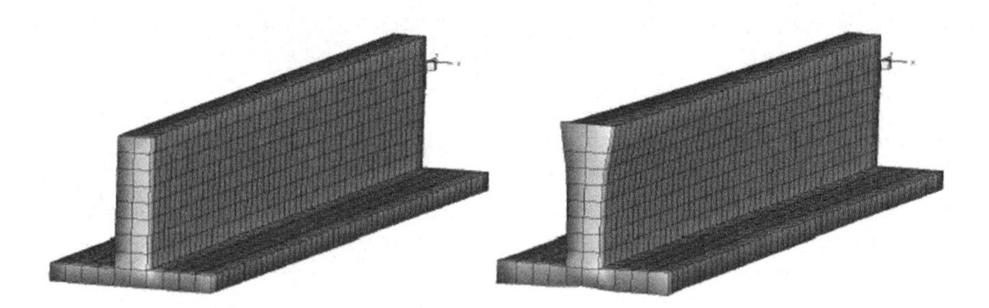

Fig. 8.17 T-beam baseline (left) and optimized (right) geometries

8.6 Fluid-Dynamic Applications

The four fluid-dynamic optimizations described in the following sections are carried out in two working environments. In particular, the first two applications, namely the Taurus glider and the drivAer car, are performed through the RBF4AERO platform implemented during the 7th Framework Programme project RBF4AERO, a European project aiming at developing the RBF4AERO Benchmark Technology, an integrated numerical platform and methodology to efficiently face the most demanding challenges of aircraft design and optimization (RBF4AERO, 2013). Such a platform makes use of the stand-alone version of the RBF Morph technology to apply morphing and of two optimizers, a stochastic one and a gradient based tool. The stochastic optimizer consists of an in-house evolutionary algorithm based (EA-based) optimizer trained by off-line surrogate evaluation models (metamodels), whilst the gradient based optimizer makes use of an in-house continuous adjoint solver working with the OpenFOAM suite (OpenFOAM, 2017). This latter, in particular, enables the adjoint preview as well as the adjoint sculpting that is accomplished adopting the steepest descent approach. For all the reported RBF4AERO-based applications, the required noise filtering is handled by employing the RBF least squares smoothing technique, adopting a smoothing radius variable in function of the nodal density of mesh surfaces (see Sect. 8.3.3).

With regard to the shape optimization of the remaining applications, that is the airbox and 90° curved duct, sensitivities were achieved using the discrete adjoint solver implemented in the finite volume commercial CFD solver ANSYS Fluent. In these cases, the add-on version of the RBF Morph technology is used instead.

8.6.1 Taurus Glider

This section showcases the application of both the adjoint preview and adjoint sculpting to handle a typical aeronautical design problem concerning the improvement of the aerodynamic performance of a glider while manoeuvring at high angles of attack (AoAs). This problem, tackled in the RBF4AERO project, was already studied in a series of papers using zero order methods (Costa et al. 2014a, b; Biancolini et al. 2016a) and a gradient-based one (Kapsoulis et al. 2016).

The Taurus glider, shown in Fig. 8.18, designed and manufactured by Pipistrel d.o.o. Ajdovscina, is a side-by-side two-seat self-launching ultra-light glider made out of composite materials. Its wing is located in a vertical central position with respect to the fuselage, whilst its longitudinal position is aft the maximum section, behind the cockpit, in a positive pressure gradient region.

This configuration implies that, in the fuselage junction region, the wing suffers an extra increase in flow velocity close to the leading edge and a significant adverse pressure gradient in the trailing edge region that cause a leading edge separation at

Fig. 8.18 Taurus glider

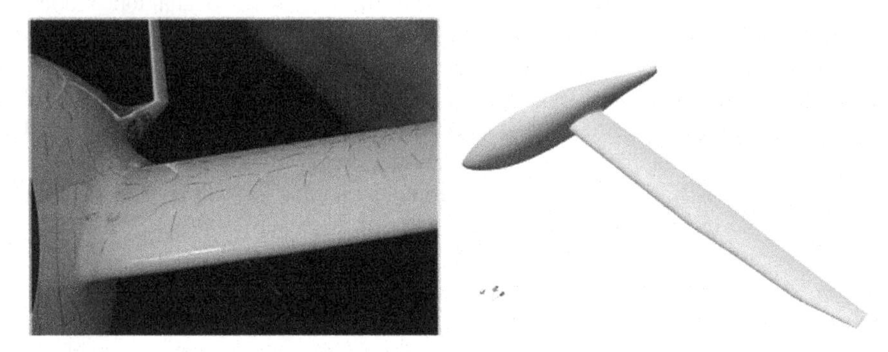

Fig. 8.19 Flow detachment altitudes (left) and baseline glider geometry (right) employed for numerical calculations

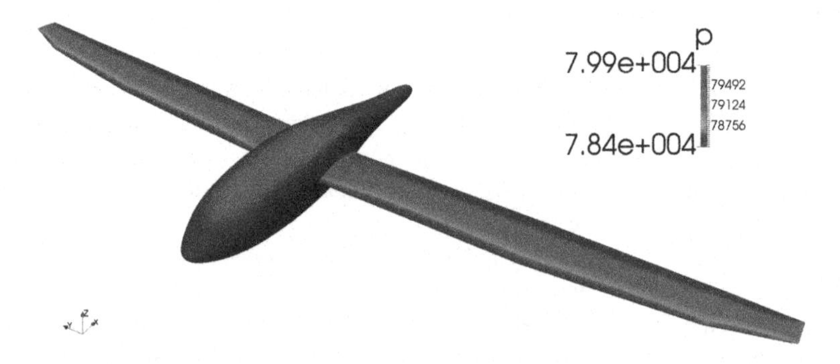

Fig. 8.20 Pressure contours for the baseline Taurus model

high AoAs. The occurrence of such flow conditions was demonstrated experimentally by performing flight tests as shown in Fig. 8.19 (left).

The target of this study is to perform shape optimization in order to diminish the separation while enhancing the aerodynamic efficiency of the aircraft whose baseline geometry is shown in Fig. 8.19 (right).

The flow conditions considered are Mach = 0.08, Re = 10^6 calculated with respect to the wing chord (0.8 m), altitude = 6561.68 ft (2000 m) and 10° for the AoA with respect to aircraft waterline. Numerical computations are performed using an incompressible OpenFOAM solver taking into account turbulence using the single equation Spalart-Allmaras model. In Fig. 8.20 the pressure contours computed for the baseline configuration are shown.

In Fig. 8.21 the wall shear stress contours (left) are depicted for the baseline configuration. As expected, flow detachment is clearly visible on the wing, fuselage and their junction area. Velocity streamlines in Fig. 8.21 (right) highlight the flow behaviour over the wing surface. The physical phenomenon registered during the experimental flight is numerically reproduced and then confirmed. With the applied boundary conditions, drag and lift coefficients calculated for the baseline configuration were Cd = 0.09652 and Cl = 1.25462, resulting in an aerodynamic efficiency of 12.99.

As already outlined, the objective of this case is to improve the aerodynamic efficiency by reducing drag, in large amount due to flow detachment. The following sections respectively detail how such a target is gained by means of the adjoint preview and the adjoint sculpting.

8.6.1.1 Glider Case: Adjoint Preview

For this specific application, ten shape parameters acting on the fuselage and on the wing junction without altering the wing shape are evaluated using the adjoint

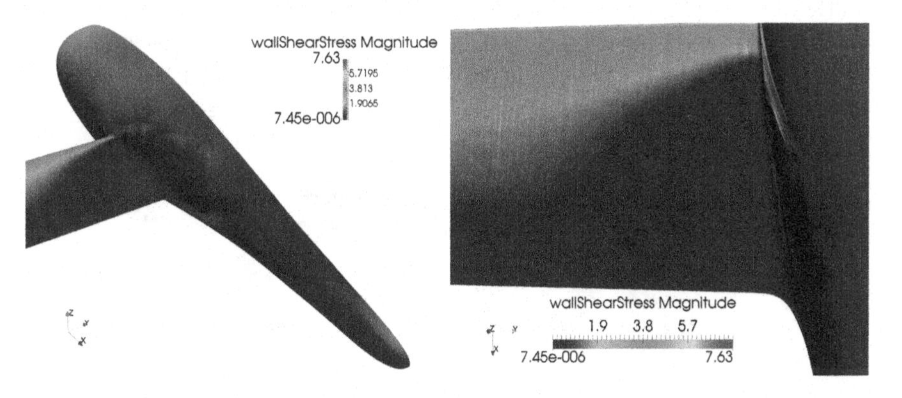

Fig. 8.21 Wall shear stress map (left) and velocity streamlines (right) in the flow detachment area for the baseline configuration

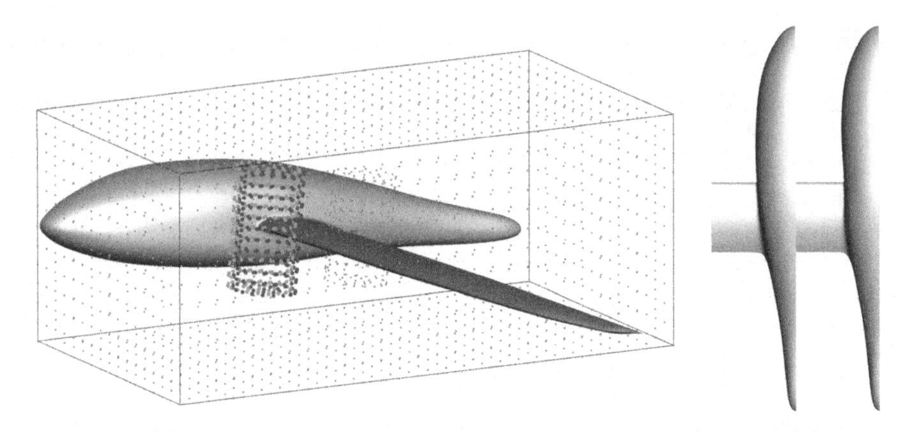

Fig. 8.22 RBF set-up for the shape modification acting on the fuselage near the leading edge (left) and its effect before and after its application (right)

preview method. The four most important shape parameters for the objective function (glider drag) are then employed in an EA-based optimization.

Being the approach employed to set up the RBF problem very similar for all shape variations, only the morphing set-up of the shape modification dealing with the modification of the fuselage and fairing area near the wing leading edge is detailed.

The objective of this shape parameter is to restrict the fuselage width before the wing in the attempt to influence flow velocity in the separation area. To obtain this modification, a set of moving source points belonging to a cylinder is defined near the fuselage in the area to be directly deformed (green points in Fig. 8.22). Moving these points towards the surface (final position is coloured in orange), while maintaining the fuselage near the trailing edge area fixed, the geometry is deformed as shown in the right image of Fig. 8.22. By defining a box of fixed source points wrapping the fuselage, it is possible not only to delimit the morphing action bringing deformations to zero on the box surface, but also to avoid displacements on the symmetry plane. In this way, the deformation of the fuselage in the unwanted area is avoided and a smooth deformation near the wing is reliably gained. One of the fundamental constraints that had to be respected by all shape parameters is to maintain the wing profile unaltered during deformations. Being the wing profile

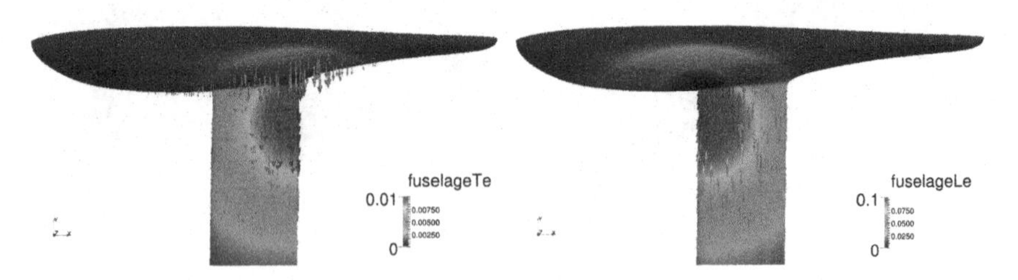

Fig. 8.23 Shape parameters acting on the fuselage and fairing near the wing (Color figure online)

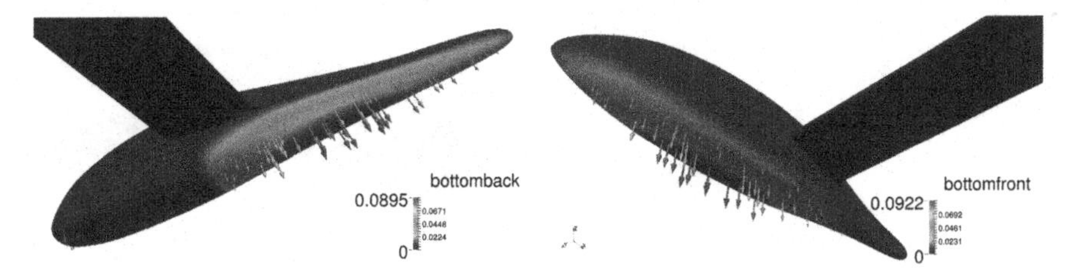

Fig. 8.24 Shape parameters acting on the lower part of the fuselage

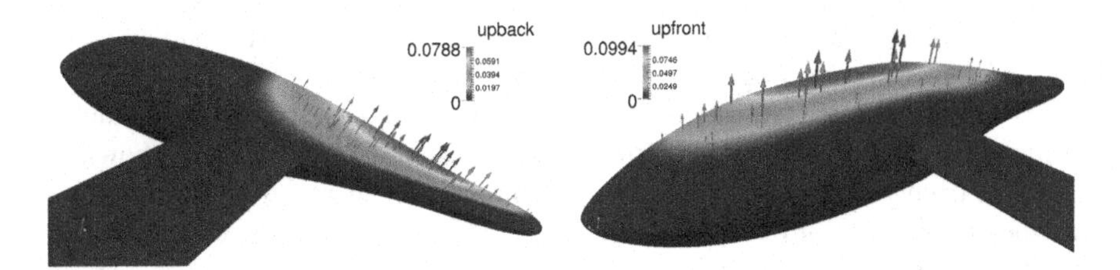

Fig. 8.25 Shape parameters acting on the upper part of the fuselage

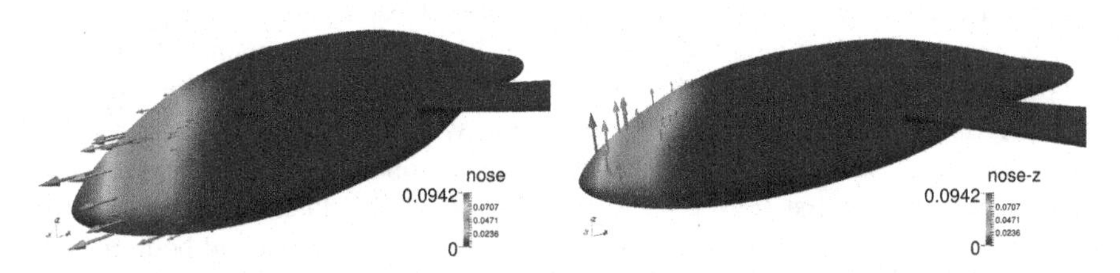

Fig. 8.26 Shape parameters acting on the glider nose

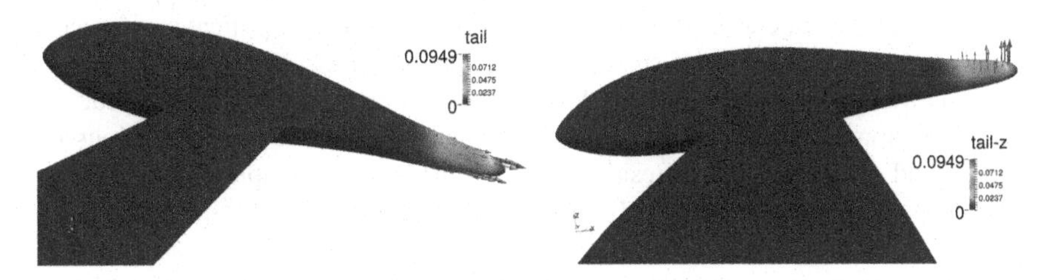

Fig. 8.27 Shape parameters acting on the glider tail section

constant near the wing-fuselage junction, this condition is automatically fulfilled by imposing the motion of source points in the wing direction.

The mesh velocities obtained for the ten shape parameters are shown from Figs. 8.23, 8.24, 8.25, 8.26 and 8.27, where the name of the corresponding RBF solution is also reported.

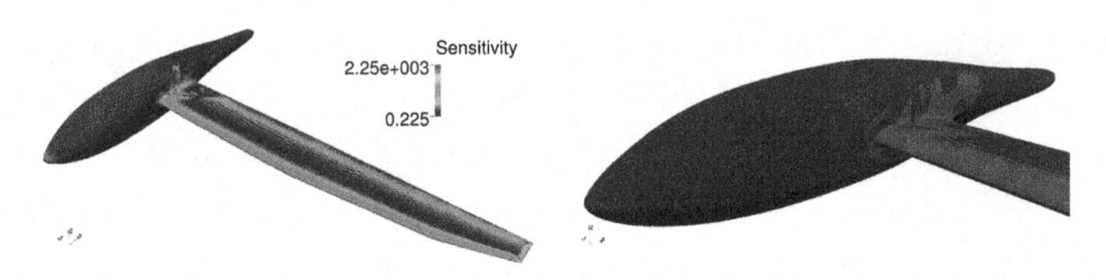

Fig. 8.28 Magnitude of drag sensitivity with respect to normal displacements (left) and direction (right)

The distribution of the sensitivity magnitude of drag with respect to surface mesh normal displacements, obtained using the adjoint solver, is shown in Fig. 8.28 (left). To have a clear view on how surfaces should be moved in order to maximize or minimize drag, its sensitivity with respect to normal displacements is shown in Fig. 8.28 (right) using just two colours. Relating to this latter distribution, the areas in blue are where an inward displacement would minimize the objective function, whereas the areas in red are where the displacement should be directed outward to get the same effect.

This information is used in the adjoint preview workflow, together with the ten mesh velocities maps, to calculate the influence of each geometrical parameter. Evaluating the absolute value of the shape sensitivities values obtained for the ten shape parameters, the modification acting on the wing-fuselage junction near the leading edge is judged to be the shape modifier with the greatest impact on drag. The nose shape parametrization is the second most important variation having half the influence of the first (FuselageLe). The four shape modifiers adopted to carry out the EA-based optimization are then FuselageLe, nose, upfront and bottomback.

The EA-based optimization employed to perform this process is applied to a metamodel trained using a full-factorial sampling for all shape parameters, obtaining a DOE table populated with 36 design points (DPs). To respect packaging and functional constraints, the upper and lower bound amplifications for each shape parameter are imposed calculating the maximum displacement they would cause on the glider shapes. The EA-based optimization set-up is tuned in order to calculate a maximum number of 40 CFD evaluations and 500 approximations by RSM-based evaluations. As a result of the optimization, the amplification value of each shape parameter is retrieved obtaining an efficiency improvement of 19.7%

Table 8.1 Aerodynamic coefficients for baseline and optimal configurations using the adjoint preview method

Parameter	Baseline	Optimum
Cd	0.09652	0.08333
Cl	1.25462	1.29640
Aerodynamic efficiency	12.99	15.55
Efficiency relative variation	–	19.7%

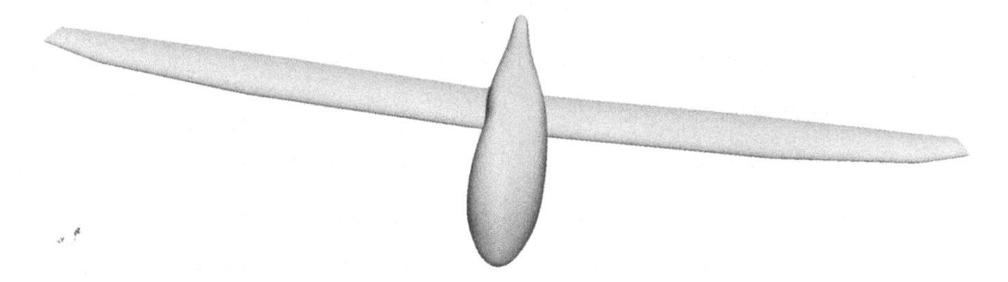

Fig. 8.29 Baseline (right) and optimized glider geometry (left)

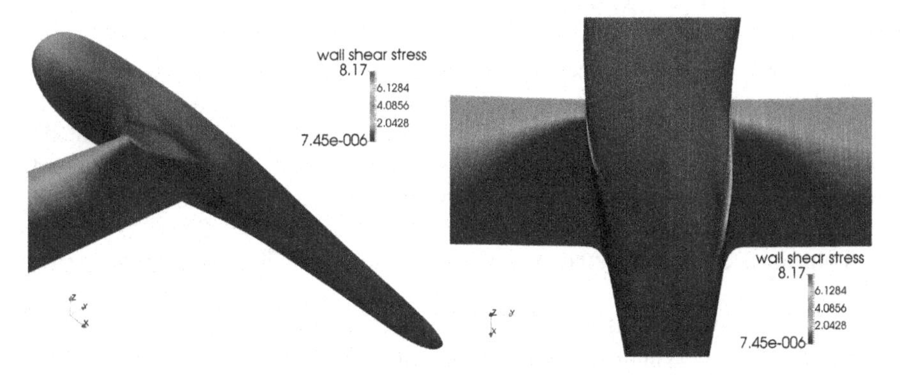

Fig. 8.30 Baseline (right) and optimized glider geometry (left)

with respect to the baseline configuration. The aerodynamic coefficients for the baseline and optimal configuration are shown in Table 8.1.

In Fig. 8.29 the baseline geometry (right) is compared to the optimized one using the adjoint preview workflow coupled with the RBF4AERO EA-based optimizer. As it can be seen, the shape parameter that was identified by the adjoint preview workflow as the most important, namely the one insisting on the fuselage near the wing leading edge area, is the one producing the largest shape variation in the optimized geometry.

The wall shear stress distribution on the optimized glider is shown in Fig. 8.30 (left). Comparing the baseline and the optimized geometries, the reduction of the area interested by flow separation can be appreciated. In Fig. 8.30 (right) the streamlines of both geometries, respectively the original (on the left) and the modified one (on the right), are compared.

In the specific flow conditions considered for this test case, the reduction of the fuselage section positively affects the flow investing the wing fuselage junction by decreasing separation and preventing the leading edge stall.

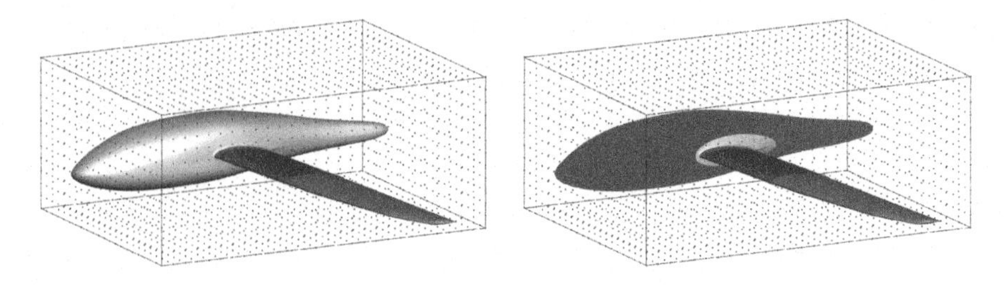

Fig. 8.31 Adjoint sculpting fixed (left) and complete (right) RBF set-up for the glider case

8.6.1.2 Glider Case: Adjoint Sculpting

As previously described in Sect. 8.3, the adjoint sculpting approach is characterized by the flow directly sculpting the surfaces of interest. The sensitivity of the drag with respect to normal displacements calculated in baseline configuration of the glider model by the adjoint solver, are shown in Fig. 8.28. Being an evolutionary method that depends on the evolving shape of the glider, the sensitivity maps will change at each optimization cycle according to the mutating flow. At the first optimization step, however, the expected shape must be similar to the one suggested in Fig. 8.28 (right).

As shape modifications are required to act on the fuselage and wing junction only, a fixed set is generated to constrain the motion of the wing and to reduce the morphing action within an area near the glider. The fixed RBF set (red points on the wing) is shown in Fig. 8.31 (left). Zero motion source points are also defined on the surfaces of a box to control volume smoothing similarly to what done for the shape modifications used with the adjoint preview workflow.

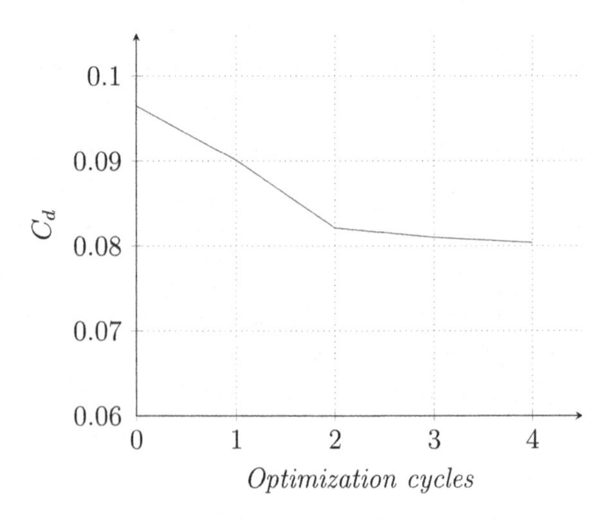

Fig. 8.32 Drag coefficient reduction for each sculpting cycle

Table 8.2 Aerodynamic coefficients for baseline and optimal configurations using the adjoint sculpting approach

Parameter	Baseline	Optimum
Cd	0.09652	0.08038
Cl	1.25462	1.21245
Aerodynamic efficiency	12.99	15.08
Efficiency relative variation	–	16.1%

This time, however, the box is enlarged in order to wrap all the fuselage so as to allow deformations in the symmetry plane in view of obtaining shape variations on the fuselage centre line as well. The wing is fixed to maintain the original profile. At each adjoint sculpting optimization cycle the displacements suggested by the sensitivities are added as moving source points to the set-up, leaving a deformable buffer zone on the wing-fuselage junction area. The complete RBF set-up for each optimization cycle is shown in Fig. 8.31 (left).

To avoid ill-posed problems, a smoothing filter (see Sect. 8.3.3) using the least squares method was employed imposing a sub-sampling radius of 10 mm. The optimization cycles are guided using a steepest descent approach with bisection search starting with a maximum displacement of 10 mm.

Four optimization cycles are accomplished to reach a final drag decrease of 16.7% with respect to the baseline configuration as depicted in Fig. 8.32.

In Table 8.2 the aerodynamic coefficients of the baseline and optimal model are compared.

The final improvement of the aerodynamic efficiency is 16.1%. It should be noticed that a single objective function, namely the drag force, is used to drive the optimization obtaining a good improvement also in terms of aerodynamic efficiency.

In Fig. 8.33 the baseline glider geometry (blue) is superposed to the optimized one (red). It can be seen how the outward displacements with respect to the fuselage are consistent to the drag sensitivities shown in Fig. 8.28 (right). Final wing surfaces are perfectly coincident to the original ones.

Fig. 8.33 Baseline (blue) and modified glider (red) superposed after the adjoint sculpting optimization

8.6.2 DrivAer Car Optimization

To tackle a typical automotive application and demonstrate the adjoint preview workflow effectiveness on a meaningful industrial case, the optimization of the DrivAer car model is accomplished. The DrivAer car geometry is a test-case developed by the Technical University of Munich (Heft et al. 2012) to close the gap between the over simplified models employed for external automotive aerodynamic validations, such as the Ahmed (Ahmed 1984) and SAE body, and the complex geometries of specific vehicles used for industrial research and development. While the former are often too generic and too different from actual production cars, making difficult to transfer tools and methodologies working on such geometries to complex real life applications, the latter are generally unavailable to the public making them a not viable solution for academic research. The DrivAer model was produced to overcome these problems, basing its geometry on the combination of two production vehicles, an Audi A4 and a BMW 3 series, and thus offering a meaningful geometry representing a challenging industrial application. Experimental campaigns were performed in wind tunnels to offer a large validation database. Both the model and experimental data are freely provided for academic research. Three main geometries —Fastback, Estate Back, Notchback—are available each with different variations to the model in terms of details. The model employed in this study is shown in Fig. 8.34.

The configuration chosen is the fastback one with mirrors, wheels and a smooth underbody. The goal of this study was to demonstrate the shape optimization using the adjoint sculpting workflow to minimize the DrivAer model drag force (Papoutsis et al. 2015). Six typical automotive shape parameters were employed to perform the optimization. The resulting deformation velocities are shown in Figs. 8.35, 8.36 and 8.37.

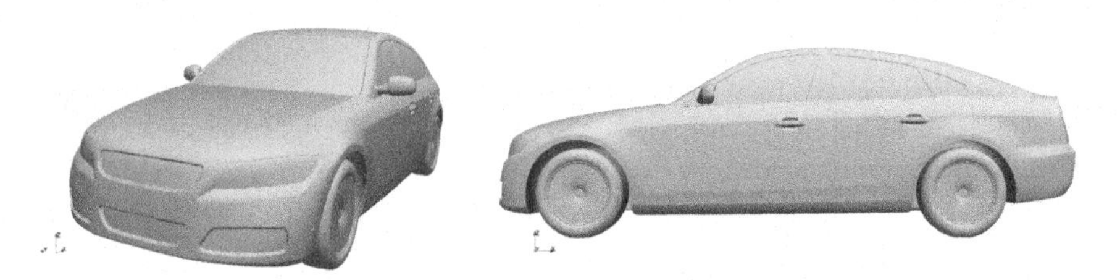

Fig. 8.34 DrivAer car geometry employed for adjoint sculpting optimization

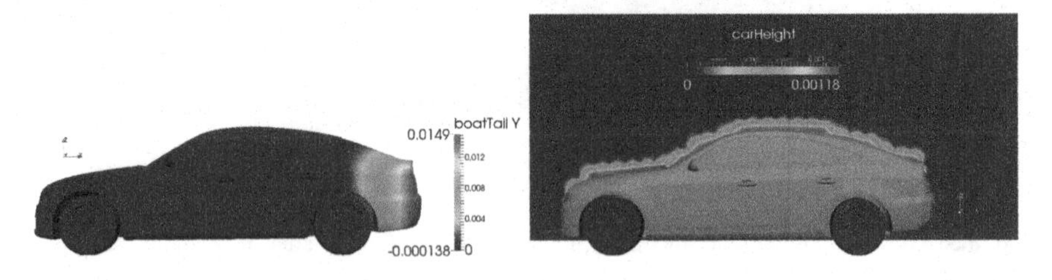

Fig. 8.35 Boat tail (left) and car height (right) shape modifiers deformation velocities

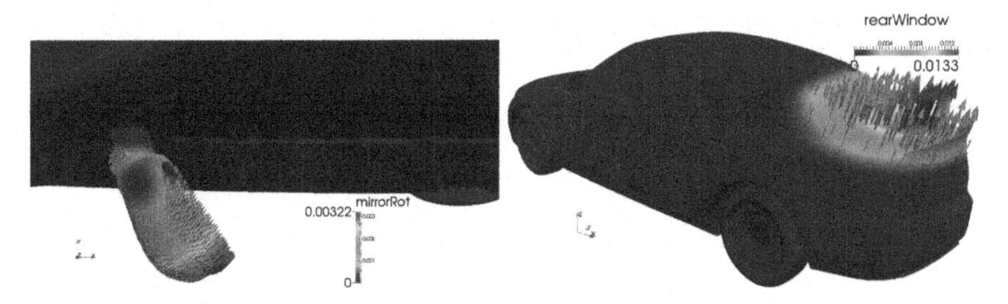

Fig. 8.36 Mirror rotation (left) and rear window angle (right) shape modifiers with deformation velocities

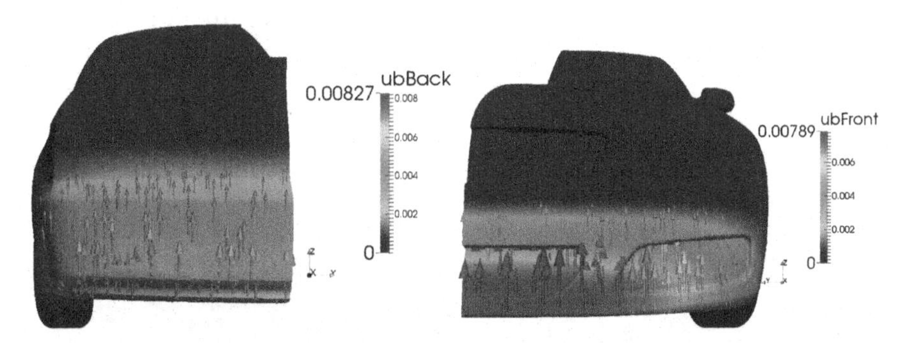

Fig. 8.37 Back (left) and front (right) underbody shape modifications and their mesh deformation velocities

A 3.8 million cells mesh was employed and CFD calculations are carried out using OpenFOAM coupled with the National Technical University of Athens (NTUA) continuous adjoint solver available through the RBF4AERO platform. The turbulence was modelled using the single equation Spalart-Allmaras model with wall functions. While the flow around the body is not steady, at each optimization cycle the flow and adjoint solutions are computed at steady state proceeding backwards in time. This strategy is applied to avoid the difficulties introduced by the unsteady adjoint solution. As a result, the calculated objective function varies around a mean value during calculation. A deepest descent method with fixed step length is employed for all parameters. After 15 optimization cycles the DrivAer mean drag was reduced by 7% as shown by the plot in Fig. 8.38 reporting the ratio between the drag of the morphed configuration and the drag on the baseline versus the optimisation cycles.

The obtained optimal geometry is shown in Fig. 8.39 where the applied displacement map and vectors are visualised. As expected, the most important shape modifications are focused on the rear car area where the wake causes most of the drag.

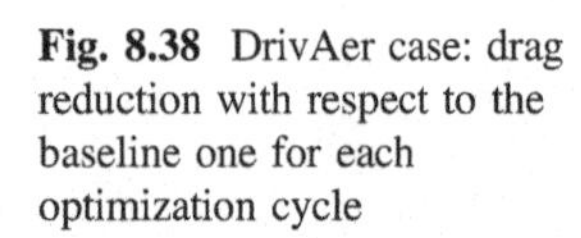

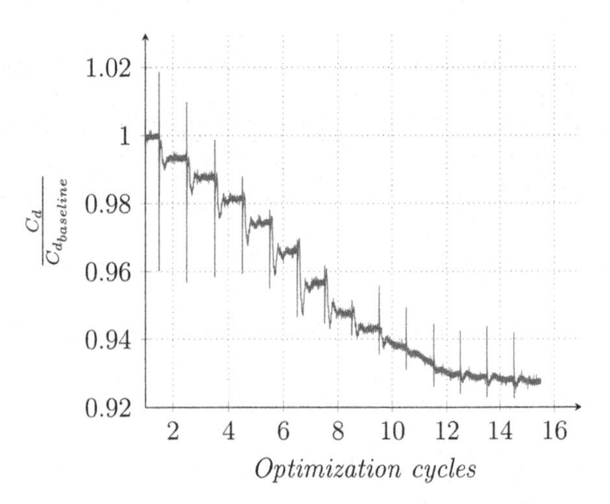

Fig. 8.38 DrivAer case: drag reduction with respect to the baseline one for each optimization cycle

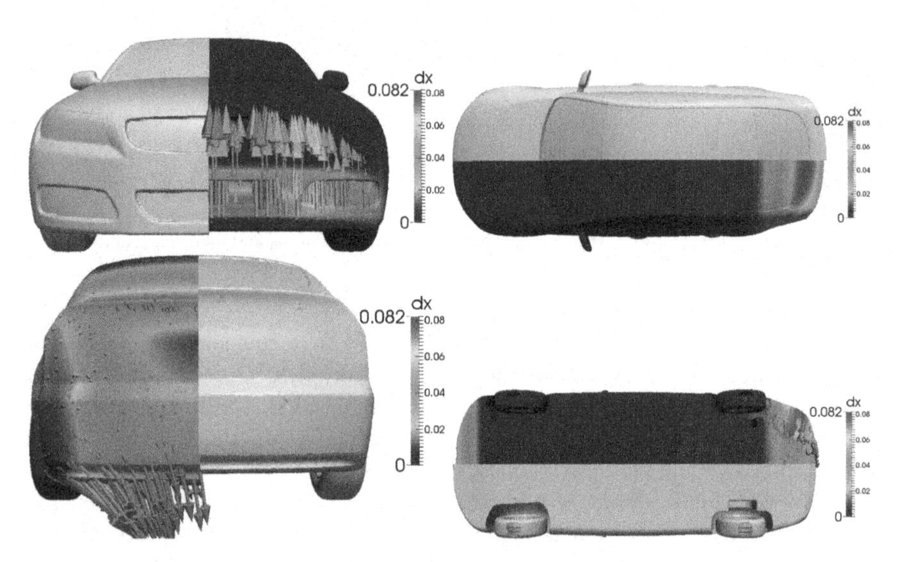

Fig. 8.39 Front, top, back and underbody shape modifications and absolute displacement vectors

The trunk is lowered by amplifying the rear window angle as shown in detail in Fig. 8.40, where the baseline (right) and optimized (left) geometries are compared. Another important geometrical variation is caused by the boat tail modifier reducing the trunk section of the car. On the front, the bumper is slightly raised to better penetrate the air.

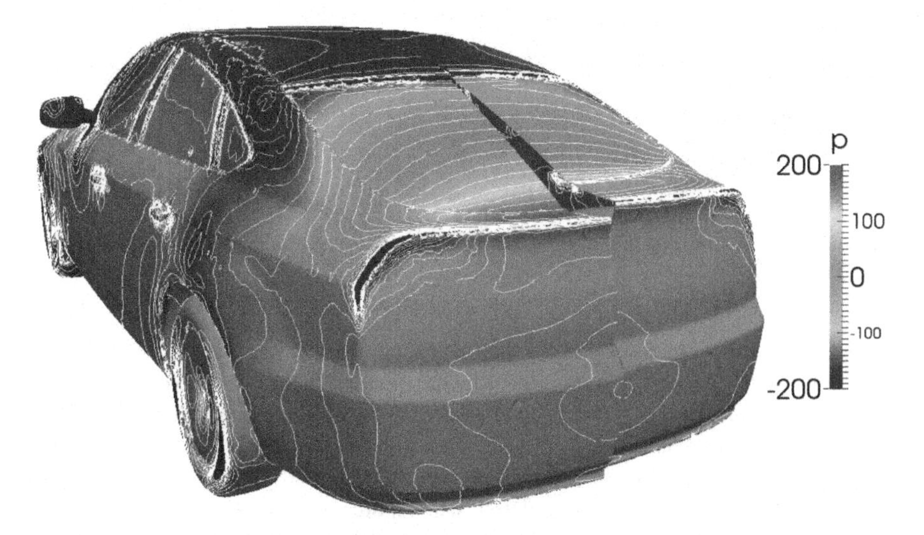

Fig. 8.40 Pressure coefficient distribution on the optimized shape (left side of the car) and on the original one (right side of the car)

8.6.3 Airbox Runners Flow Balance Optimization

In order to demonstrate the adjoint preview workflow for an internal flow case, a shape optimization of a three-cylinder automotive airbox is performed (Biancolini 2014). The geometry taken into account is shown in Fig. 8.41. As it can be seen, the three runners draw a circumference arc to connect the plenum to cylinders leaving plenty of space for shape morphing and, consequently, for the optimization.

The steady-state pressure contours for the baseline configuration are shown in Fig. 8.42 opening one runner at a time.

In this case the optimisation goal is to reduce both the pressure drop and the unbalance among runners. To achieve these results, three analyses are carried out for each configuration, evaluating the influence of the shape parameters when each runner was opened. The fluid dynamic and the adjoint solver of the ANSYS Fluent solver are used for accomplishing computing.

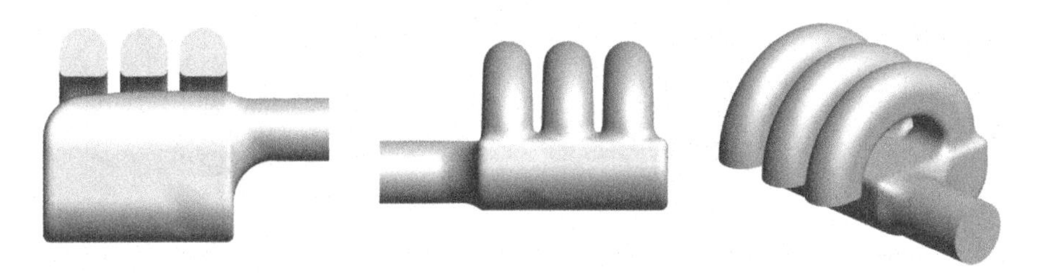

Fig. 8.41 Airbox case: geometry employed for adjoint preview optimization

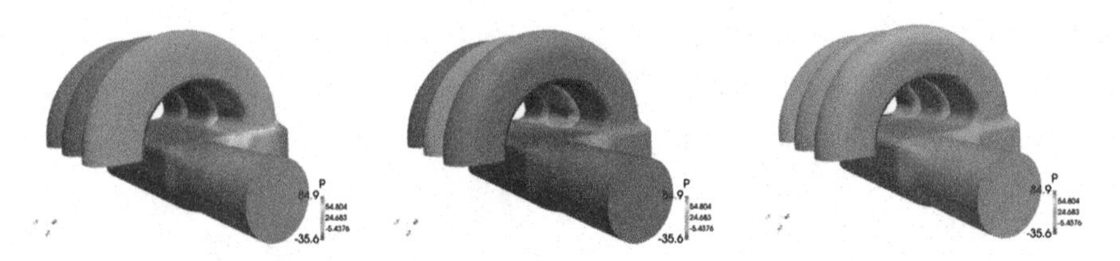

Fig. 8.42 Airbox case: pressure contours for the baseline configuration opening one runner at a time

For what previously outlined, such a shape parametrization is finalised by deforming each runner radially towards the plenum so as to exploit the large clearance available. To respect both packaging and functional constraints, the external portion of each runner and the inlet duct are kept fixed. A total of thirty shape parameters, ten for each runner, are generated by defining along each runner ten evenly spaced cylinders populated with source points pushing towards the surfaces. Being the approach used to set-up the RBF problem the same for each shape variation, only one set-up is hereinafter detailed.

In Fig. 8.43 the set-up for a shape parameter acting on the runner-plenum junction area is shown. Boundary conditions for the deformation are composed by a first set of fixed source points that keep unchanged the external section of the runners, the inlet duct and a portion of the plenum (red points in Fig. 8.43), and by a second set of moving points defined on a cylinder moving towards the runner along the radial direction (the position of the source points before and after apply morphing are coloured in green and orange respectively).

Optimization is guided by a steepest descent logic, reducing at each optimization cycle the pressure loss for each runner and tuning the so obtained amplifications in order to decrease the unbalance among runners. The optimization history for drop

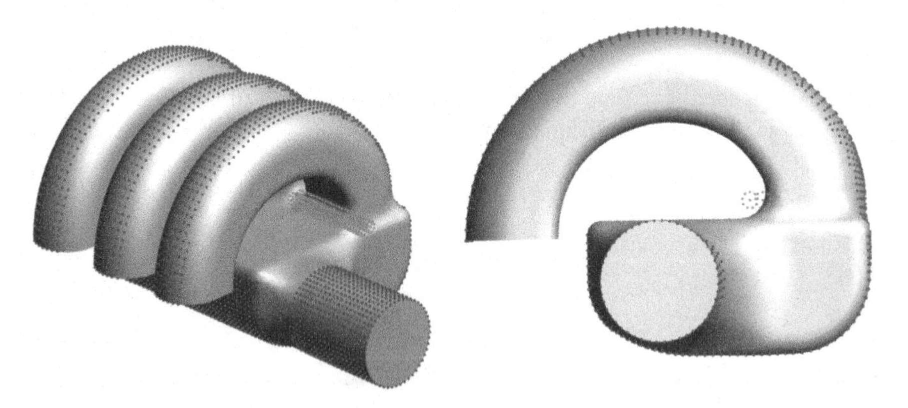

Fig. 8.43 Airbox case: RBF setup for a shape modification acting on a single runner (Color figure online)

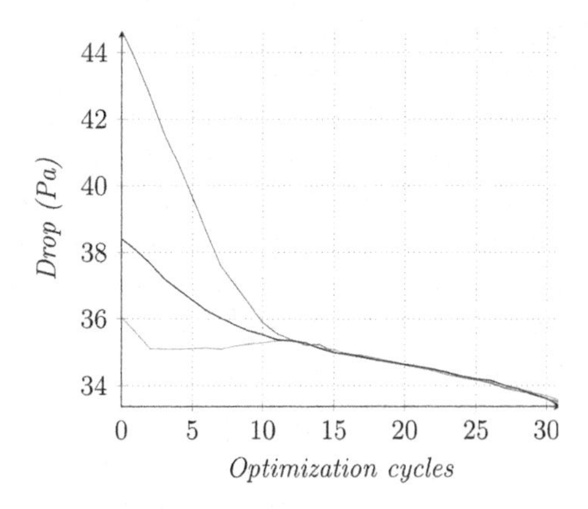

Fig. 8.44 Airbox case: Pressure drop reduction for the three runners at each optimization cycle

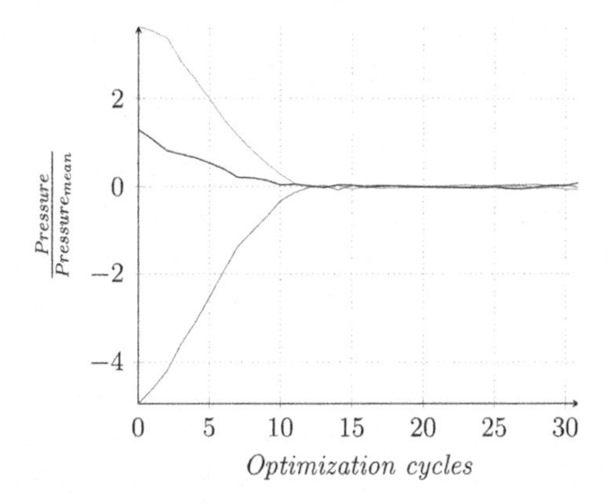

Fig. 8.45 Pressure unbalance for the three runners at each optimization cycle

and unbalance reduction is shown in Figs. 8.44 and 8.45 respectively. Twelve optimization cycles are required to make the pressure unbalance between the runners almost null. A total of thirty cycles are carried out obtaining a 15.28% reduction in the overall pressure loss.

In Fig. 8.46 the pressure contours for the optimized geometry in steady flow conditions are shown for each runner opening.

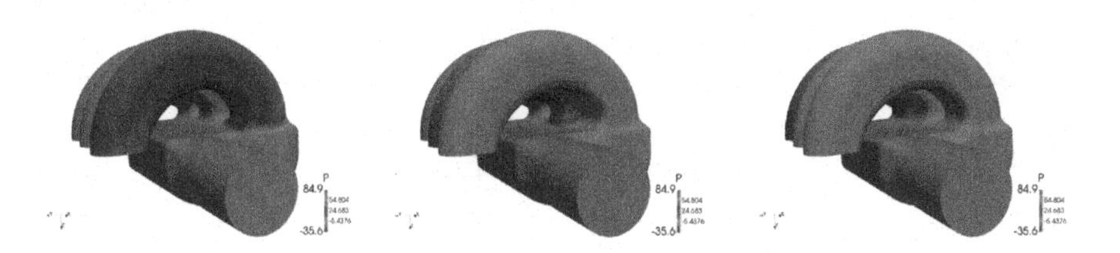

Fig. 8.46 Airbox case: pressure contours for the optimized configuration opening

8.6.4 90° Curved Duct

The aim of this test case previously introduced in Fig. 6.20 of Chap. 6, is to decrease the pressure loss between the inlet and outlet of the 90° duct whose CAD model in the baseline configuration is shown Figs. 8.42 and 8.47.

The minimization of the pressure loss is gained adopting the adjoint sculpting approach and performing the optimization through the coupling between the RBF Morph Fluent Add On, the ANSYS Fluent CFD solver, its adjoint solver and the morpher embedded into ANSYS Fluent whose functioning is based on the FFD technique (Petrone et al. 2014).

Running the CFD analysis on the baseline shape numerical model, whose hexahedral mesh is composed by about 60,000 cells, a pressure jump of 33 Pa between inlet and outlet is calculated.

The optimization is carried out employing the steepest descent method. After performing 4 optimization cycles the gained optimal discretised shape is depicted in Fig. 8.48 together with the baseline one. As visible, the sensitivities suggested by the adjoint solution cause a morphing action that produces a sort of inflation of the external surface of the curved portion of the duct.

Figure 8.49 shows the pressure distribution over the duct and the streamlines within it for the baseline (bottom right) and the optimal configuration (top left) gained through adjoint sculpting. The pressure drop for the optimized shape is 22 Pa (33% of relative reduction).

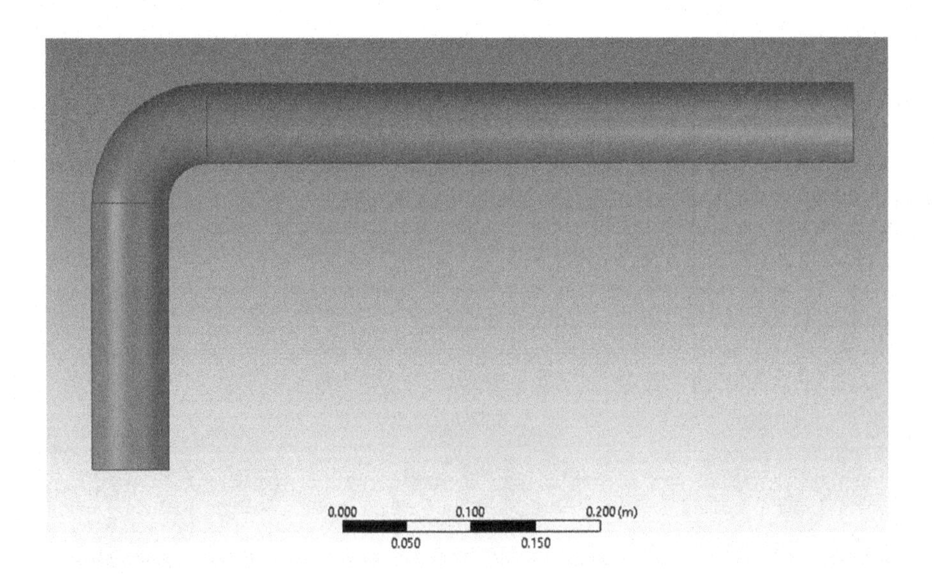

Fig. 8.47 CAD model of the baseline 90° curved duct

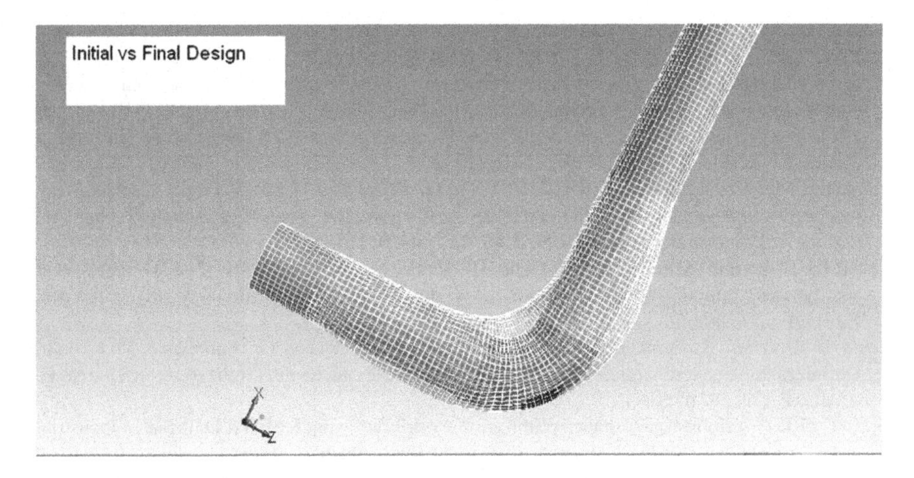

Fig. 8.48 Optimal shape of 90° duct model

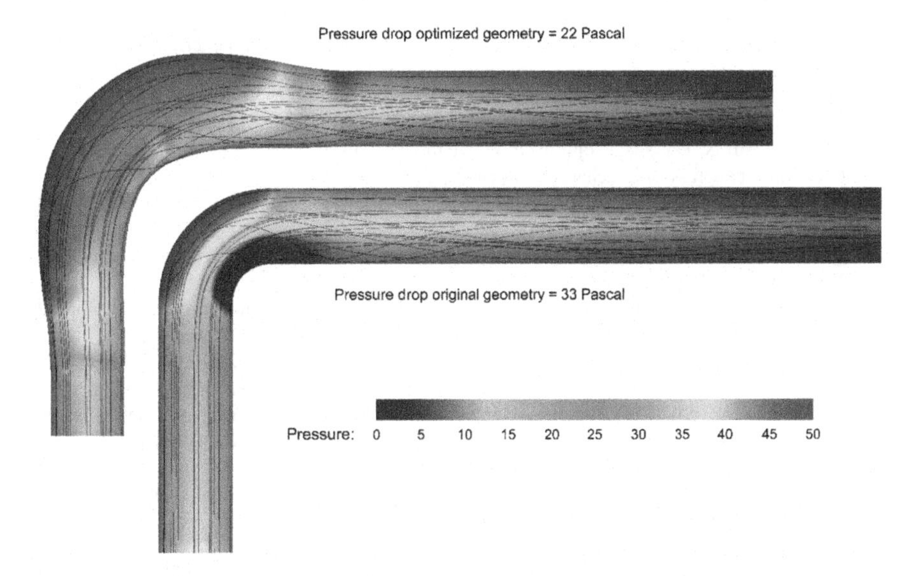

Fig. 8.49 Pressure distribution and streamlines of the baseline and optimised duct

References

Ahmed SR, Ramm G (1984) Some salient features of the time-averaged ground vehicle wake. SAE Technical Paper 840300

Arora JS (2016) Introduction to optimum design. Academic Press. ISBN: 978-0-12-800806-5. https://www.elsevier.com/books/introduction-to-optimum-design/arora/978-0-12-800806-5

Ashlock D (2005) Evolutionary computation for modeling and optimization. Springer-Verlag New York. ISBN: 978-0-387-31909-4. http://www.springer.com/gp/book/9780387221960
Biancolini ME (2014) How to boost fluent adjoint using RBF morph. International Conference on Automotive and Electronics Technologies 2014. Proceedings of a meeting held 9-10 October 2014, Tokyo, Japan. Automotive Simulation World Congress 2014 and ANSYS Electronics Simulation EXPO 2014. ISBN: 9781510841031
Biancolini ME, Costa E, Cella U, Groth C, Veble G, Andrejašič M (2016) Glider fuselage-wing junction optimization using CFD and RBF mesh morphing. Aircr Eng Aerosp Technol 88 (6):740–752. https://doi.org/10.1108/AEAT-12-2014-0211
Costa E, Biancolini ME, Groth C, Cella U, Veble G, Andrejasic M (2014a) RBF-based aerodynamic optimization of an industrial glider. In: 30th international CAE conference, Pacengo del Garda, Italy
Costa E, Porziani S, Biancolini ME, Groth C, Cella U, Veble G, Andrejasic M (2014b) Ottimizzazione aerodinamica di un aliante industriale mediante l'utilizzo di RBF, A&C. Analisi E Calcolo 64:31–48
Groth C (2017) Adjoint-based shape optimization workflows using RBF. Ph.D. thesis in industrial engineering, Cycle XXIX, University of Rome "Tor Vergata"
Groth C, Chiappa A, Biancolini ME (2018) Shape optimization using structural adjoint and RBF mesh morphing. Procedia Structural Integrity 8:379–389
Haftka RT, Gürdal Z (1992) Elements of structural optimization, ser. Solid mechanics and its applications. Dordrecht: Springer Netherlands, vol. 11, no. March
Heft AI, Indinger T, Adams NA (2012) Introduction of a new realistic generic car model for aerodynamic investigations. SAE International, warrendale
Kapsoulis DH, Asouti VG, Papoutsis-Kiachagias E, Giannakoglou K, Porziani S, Costa E, Groth C, Cella U, Biancolini ME, Andrejasic M, Erzen D, Bernaschi M, Sabellico A, Urso G (2016) Aircraft & car shape optimization on the RBF4AERO platform. In: 11th HSTAM international congress on mechanics, Athens
Mohammadi B, Pironneau O (2009) Applied shape optimization for fluids, vol 53, no 9. Oxford University Press, Oxford, p 9
Montgomery DC (2012) Design and analysis of experiments, vol 2, John Wiley & Sons, Inc. ISBN: 978-1-118-14692-7
Othmer C, Papoutsis-Kiachagias EM, Haliskos K (2011) CFD optimization via sensitivity-based shape morphing. In: Proceedings of 4th ANSA & μETA international conference
Papoutsis-Kiachagias EM, Porziani S, Groth C, Biancolini ME, Costa E, Giannakoglou KC (2015) Aerodynamic optimization of car shapes using the continuous adjoint method and an RBF morpher. EUROGEN 2015, 11th International Conference on Evolutionary and Deterministic Methods for Design, Optimization and Control with Applications to Industrial and Societal Problems, Glasgow, UK, September 14-16, 2015.
Papoutsis-Kiachagias EM, Giannakoglou KC, Porziani S, Groth C, Biancolini ME, Costa E, Andrejasic (2018) Combining an OpenFOAM-based adjoint solver with RBF morphing for shape optimization problems on the RBF4AERO platform. In: OpenFOAM® selected papers of the 11th workshop, Springer International Publishing, ISBN: 978-3-319-60846-4 (in press)
Petrone G, Hill C, Biancolini ME (2014) Track by track robust optimization of a F1 front wing using adjoint solutions and radial basis functions. In: 32nd AIAA Applied Aerodynamics Conference, AIAA AVIATION Forum (AIAA 2014-3174). https://doi.org/10.2514/6.2014-3174
Vanderplaats GN (2005) Numerical optimization techniques for engineering design. Vanderplaats Research & Development, Incorporated. ISBN: 9780944956021

Chapter 9
Advanced RBF Mesh Morphing for Multi-physics Applications with Evolutionary Shapes

Abstract In many multi-physics fields the evolution of model shapes can be predicted in advance or determined in an evolutionarily manner during computing. For such applications, the analyst can use RBF mesh morphing to reliably handle geometrical variations and build-up automatic and efficient workflows. Depending on the extent of the modification between one shape configuration and the successive one, either a single or sequential multi-stage morphing approach (see Sect. 6.1.3) can be employed. A complete description of mesh morphing based on Fast RBF is provided in Chap. 6. In the following sections the basic principles of the application of mesh morphing for the simulation of evolutionary shapes are provided. The method is first explained step by step with a practical example about snow accretion and then advanced usage examples of ice accretion are provided. Successively, the chapter includes examples of the application of mesh morphing in FEA for Fracture Mechanics demonstrating how crack update and propagation can be consistently addressed. Shape optimisations of structural parts using the biological growth method (BGM) are finally presented.

9.1 Mesh Morphing of Evolutionary Shapes

The ability to morph a mesh o a computational model in order to generate a new shape has been already presented in Sect. 6.5 mesh update requirements are in this case driven by an evolutionary physics that can be computed by the CAE solver and that provides input at nodes of the computational grid. Usually the complexity of the evolutionary models has to be tackled with a specific customization of the CAE solver so as to properly address the evolution (particle deposition, local driving force to move nodes on a crack front), including the morphing task that takes, as input, the deformations to be applied typically available as a field at either mesh nodes or elements centroids locations.

It is worth to notice that the Fast RBF mesh morphing workflow to be implemented for such evolutionary cases may be conducted according to what previously

M. E. Biancolini, *Fast Radial Basis Functions for Engineering Applications*,
https://doi.org/10.1007/978-3-319-75011-8_9

introduced for the adjoint sculpting described in Sect. 8.3. If the local surface movements suggested by the adjoint solver were considered as a possible evolutionary physics, that topic would have been included in the present Chapter but it was instead included in Chap. 8 because it is correlated with the gradient based shape optimisation driven by adjoint sensitivity. Although the BGM, covered in Sect. 9.7, is conceived for shape optimisation, it mimics a biological evolution and, as such, it well fits the contents of the present Chapter. Whatever the single choice that drives the placement among the topics of the book, it is important to say that the generic evolutionary shape concept herein presented has specific details in common with the adjoint sculpting. In fact, in both scenarios the amount of morphing perturbation applied to the current shape has to be metered in view of avoiding too much mesh distortion and of keeping valid the physical analysis that computed the field (this is a first order approximation and the CAE solution of the perturbed mesh will change the amount of deformation). Furthermore, the noisy behaviour that might arise with the adjoint field can be noticed in workflows where the deformation is computed by a multi-physics solver meaning that the noise filtering method, presented in Sect. 8.3.3, could be properly tuned and applied to the available data before updating the mesh.

When evolutionary data are not directly available on the numerical grid, the field to be applied may be computed adopting the implicit surface approach method described in Chap. 5. In the context of the present chapter we assume that the evolution input is computed directly on the computational mesh.

The mesh morphing problem can be faced using Fast RBF and it usually requires not only the input at the controlled boundaries (surfaces in 3D cases, curves for 2D cases), but also the information necessary to decide how much propagating the deformation field (up to a far boundary or limiting it into a specific domain) and preserving other parts of the mesh close to the one to be updated and belonging to the deformation domain.

To gain a better insight on the use of the evolutionary mesh morphing, the method is showcased in the next section through a practical detailed example concerning snow accretion.

9.2 Snow Accretion

In this section, a technique to simulate snow accretion on a three-dimensional body through RBF mesh morphing is described. The feasibility of the proposed numerical practice is tested adopting an interactive process that concerns the coupled use of three tools: the ANSYS Fluent CFD solver, the add-on version of the RBF Morph software and Mathcad. Some of the tasks of the workflow are conducted using the standard text user interface (TUI) commands of ANSYS Fluent (ANSYS 2017) and RBF Morph, whilst the Mathcad tool is specifically employed to accomplish the operations in which raw data need to be suitably processed to enable their usage in the suggested process.

The just introduced technique foresees that two steps of operations are performed in sequence. Such phases are respectively detailed in the following sections. The effectiveness of the concept is then demonstrated in Sect. 9.2.5 where the results by Allain et al. (2014) are summarized.

9.2.1 Computation of Snow Growth Using CFD

The first step of the method concerns the CFD simulation and snow data generation. Using the flow field results characterizing the fluid-dynamic solution in steady-state conditions, the snow accretion tool, embedded into the CFD solver, calculates the snow data over the surfaces of interest: the nodal snow thickness and nodal normal (including area). Such an accretion tool consists of a simplified scheme with particles' collection enabled at front surfaces only. The same technique was adopted by Allain et al. (2014) that implemented a model working in the respect of the approach earlier proposed by Trenker and Payer (2006) for the prediction of snow deposition on high-speed trains. Some outputs gained by means of the aforementioned tool are shown in Fig. 9.1 where the streamlines of the flow transporting the snow (left) as well as the snow accretion speed map on the Morel body (right) are reported.

In specific, the snow thickness information is stored adopting the user defined memory (UDM) feature of the CFD tool (ANSYS 2017). Finally, the snow data are automatically gathered and exported in an ASCII format file whose structure is reported in Table 9.1.

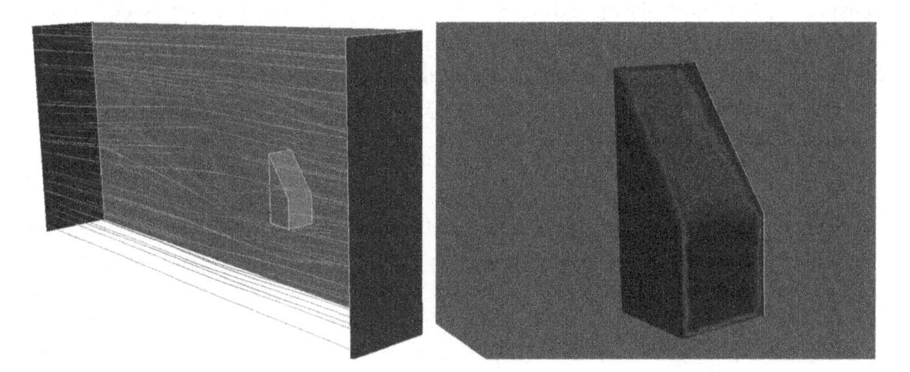

Fig. 9.1 CFD model used to predict snow particle deposition on the Morel body: streamlines of the flow transporting the snow (left) and snow accretion speed map (right)

Table 9.1 Snow accretion case: data structure exported by ANSYS Fluent solver containing the thickness information at nodes

Node number	x-coordinate	y-coordinate	z-coordinate	Thickness	x-face-area	y-face-area	z-face-area
1	0.5	0.25	−0.01	3.37e−5	2.413e−5	2.058e−5	0
2	0.5	0.26	0	2.708e−5	2.65e−5	0	2.832e−5
3	0.5	0.25	0	6.94e−6	8.773e−6	1.046e−5	3.337e−5
4	0.5	0.35	−0.09	6.27e−5	4.595e−5	−1.411e−5	−7.373e−12
…	…	…	…	…	…	…	…
852	0.512	0.367	−0.062	5.412e−5	4.2e−5	−3.052e−5	3.741e−12

For each node the index, the x, y, z coordinates, the snow thickness and the x, y, z components of the oriented area are stored

9.2.2 *Pre Processing of Snow Data Using Mathcad*

In the second step snow data are imported in the Mathcad tool and processed to generate the cloud pf RBF centres. The parsing task is specified in Eq. (9.1): the ASCII file content is stored in the **DATA** matrix using the **READPRN**(·) function, **submatrix**(·) function and the column extraction operator $^{<\cdot>}$ allows to break the data in original position $\mathbf{POS_0}$, oriented normal $\mathbf{N_{dir}}$ and scalar thickness **thick**.

$$\begin{gathered} \text{DATA} := \text{READPRN}(\text{“P1 – accr.dat”}) \\ \text{POS}_0 := \text{submatrix}(\text{DATA},1,\text{rows}(\text{DATA}),2,4) \\ \text{N}_{\text{dir}} := \text{submatrix}(\text{DATA},1,\text{rows}(\text{DATA},6,8)) \\ \text{thick}: = \text{DATA}^{\langle 5 \rangle} \end{gathered} \tag{9.1}$$

New accreted positions at surface nodes after a certain growth time are computed multiplying the nodal unit vector and the corresponding nodal thickness. Such processing is detailed in (9.2) where $\mathbf{POS_1}$ are computed visiting all the points running the **i** variable with a **for** cycle for all the **rows**(·) of the dataset **DATA**. The coordinate of the individual point **P** and the oriented area $\mathbf{A_n}$ are extracted using the transpose$^{\mathbf{T}}$ and the column extractor operator $^{<\cdot>}$. The unit vector is computed dividing the oriented area by its modulus, the thickness is computed multiplying the **thick** variable by the amplification λ, whilst the new position $\mathbf{P_n}$ is calculated moving the point in the normal direction by the amplified local amount and then assigned to the $\mathbf{POS_1}$ matrix.

$$
POS_1 := \left| \begin{array}{l} \text{for } i \in 1..\text{rows}(DATA) \\ \quad \left| \begin{array}{l} P \leftarrow \left(POS_0{}^T\right)^{\langle i \rangle} \\ A_n \leftarrow \left(N_{dir}{}^T\right)^{\langle i \rangle} \\ n \leftarrow -\dfrac{A_n}{\left|A_n\right|} \\ t \leftarrow thick_i \cdot \lambda \\ P_n \leftarrow P + n \cdot t \\ \left(POS_1\right)^{\langle i \rangle} \leftarrow P_n \end{array} \right. \\ out \leftarrow POS_1{}^T \end{array} \right. \tag{9.2}
$$

The position of the surface mesh nodes at original and growth configurations can be visualised in Mathcad as shown in Fig. 9.2. The amplification can be tweaked to get a good graphical representation or normalised, during calculation stages, to prescribe the maximum thickness growth to be used in a single calculation step.

9.2.3 Exploring Accretion Data Using RBF

The applications presented in this chapter are based on the RBF industrial implementation of the software RBF Morph. This section provides the details of the RBF processing to be used for the completion of the workflow and is intended as a guide to put the reader in condition to write her/his own implementation from scratch.

The functions provided in Sect. 4.1 of Chap. 4 for fitting and evaluating RBF are here adopted. Original node positions are first arranged as a nested vector containing the position of the nodes as components. The resulting cloud of source points $\mathbf{x}_k$ construction is detailed in (9.3).

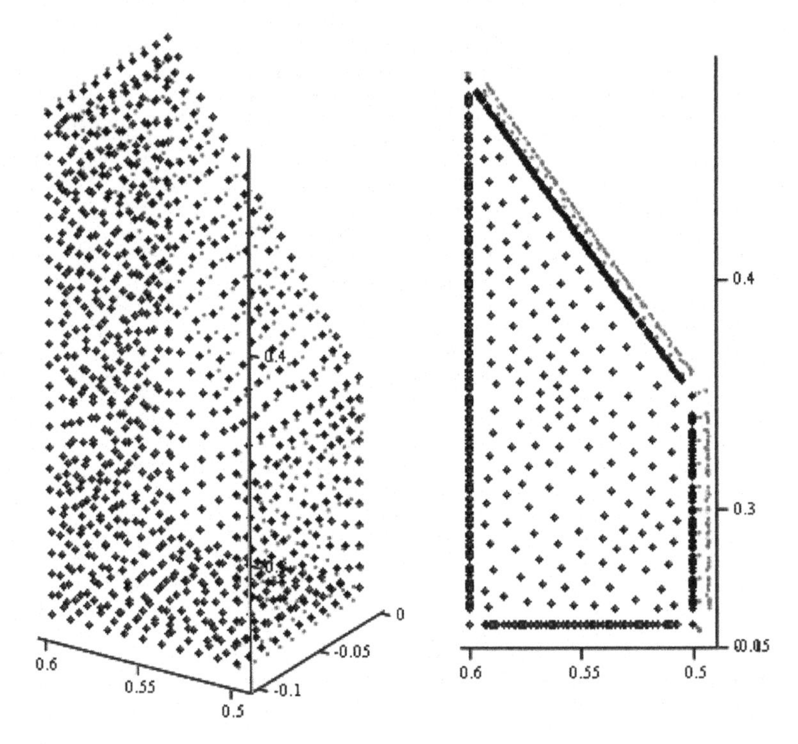

Fig. 9.2 Cloud of snow points at original (black larger) and final (red smaller) configurations as visualised using Mathcad software in a perspective view (left) and in a side view (right). The original growth is magnified 100 times

$$x_k := \begin{array}{l} \text{for } i \in 1 \ldots \text{rows}(POS_0) \\ \quad x_{k_i} \leftarrow \begin{pmatrix} POS_{0_{i,1}} \\ POS_{0_{i,2}} \\ POS_{0_{i,3}} \end{pmatrix} \end{array} \tag{9.3}$$

Input $\mathbf{g_k}$ for the field at source points is given by the three components of the displacement field computed subtracting the initial positions $\mathbf{POS_0}$ to the final positions $\mathbf{POS_1}$ (9.4).

$$g_k := POS_1 - POS_0 \tag{9.4}$$

The coefficient of the RBF are then computed using the **RBF_fit(·)** function (9.5) and the local growth $\mathbf{x_{RBF}}(\cdot)$ at a generic point is accessed by the **RBF_eval(·)** function (9.6).

$$\begin{pmatrix} \gamma \\ \beta \end{pmatrix} := \text{RBF\_fit}(\mathrm{x_k}, \mathrm{g_k}) \tag{9.5}$$

$$\mathrm{x_{RBF}}(\mathrm{x}) := \text{RBF\_eval}(\gamma, \beta, \mathrm{x_k}, \mathrm{g_k}, \mathrm{x}) \tag{9.6}$$

It is important to notice that such RBF can be used to propagate the field in the volume mesh if a compact supported RBF is adopted as the C2 of Table 2.1 of Chap. 2. If a full supported RBF is chosen, additional RBF sources are required to define a zero field either at the domain boundary or at an auxiliary box that can be generated according to (4.11) of Sect. 4.6.1 as done in the industrial completion of this study given in Sect. 9.2.4.

The effect of RBF interpolation is demonstrated in Fig. 9.3 where regular paths on the surface are explored to plot the thickness of the snow at five parallel locations.

The deformation field can be computed in the volume mesh using the same RBF by visiting all the nodes of the volume grid so that the deformation effect at the surface can be propagated to suitably adapt the full computational domain.

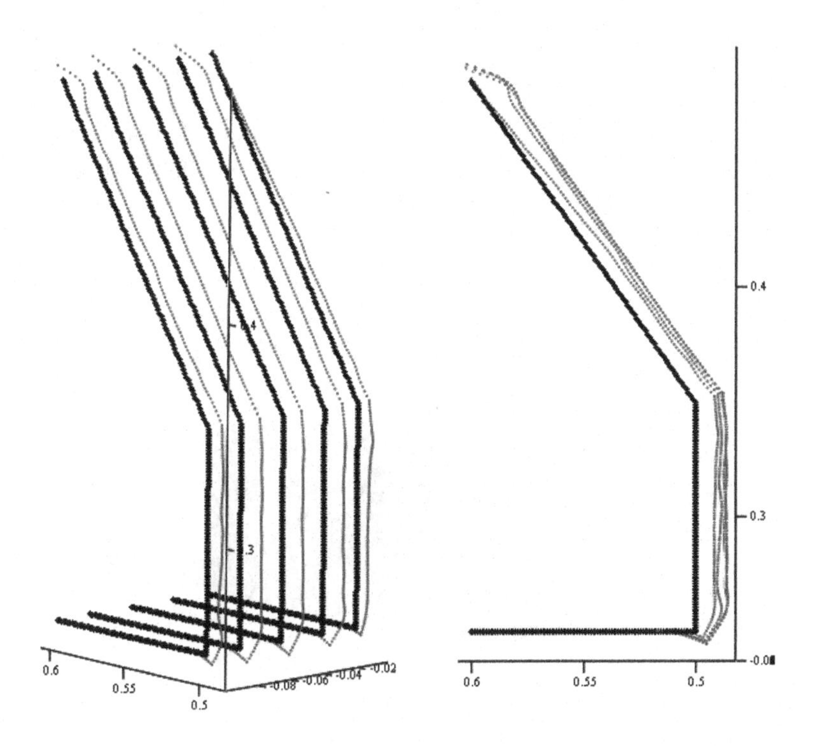

Fig. 9.3 The RBF is explored on the points of five paths (black larger) along the surface, moved in growth position (red smaller) by applying the computed local field in a perspective view (left) and in a side view (right). The original growth is magnified 100 times

9.2.4 Mesh Update in the CFD Code

Using Mathcad the initial coordinates of the surface nodes together with the computed displacements are exported in a file according to a specific form such to enable the import in the RBF mesh morphing software in order to make these data drive surface morphing. In final automated workflow of the following example, the processing developed in Mathcad is replaced with a specific implementation carried out through the UDF feature of the ANSYS Fluent CFD solver.

The subsequent step pertains the RBF field set-up which comprises two main settings. The first is the generation of the source points steering morphing which are carried out just loading the file already available at the end of the previous step, whilst the second is the assignment of a volume defining the target for the mesh morphing action. Figure 9.4 shows the preview of the position of the source points created at the end of the RBF set-up and the position they have after applying morphing. In particular, the box-shaped domain assigned to delimit the morphing action is shown together with its source points laying over the box surfaces. The resolution of these source points, visualized in red colour, is defined in view of prescribing the zero displacement condition at such boundaries. In the same picture, the preview of the position of source points imposing the accreted surface displacements is also depicted. In detail, the red points denote their position before morphing, whereas the green points visualise the position they have after the application of the RBF field.

Figure 9.5 compares the original mesh (left) and the morphed mesh (right) of the model subjected to a maximum growth equal to 7 mm; a longitudinal cut of the volume mesh allows to understand that the prisms layers at surface are updated without substantially affecting their spacing.

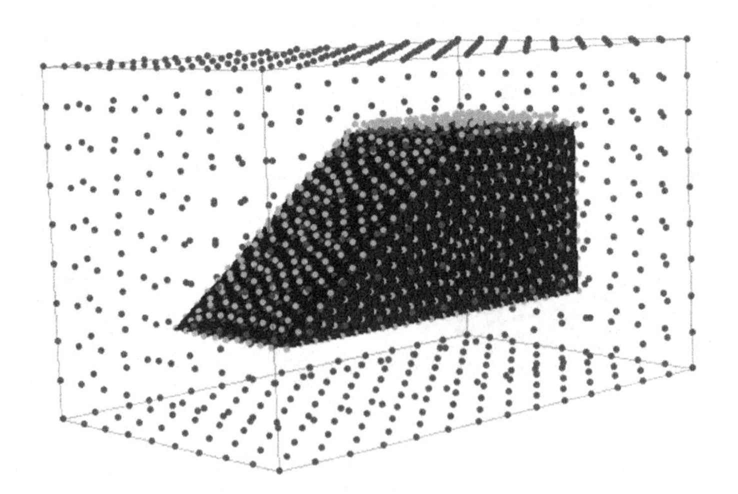

Fig. 9.4 Snow accretions case: source nodes of the RBF set-up

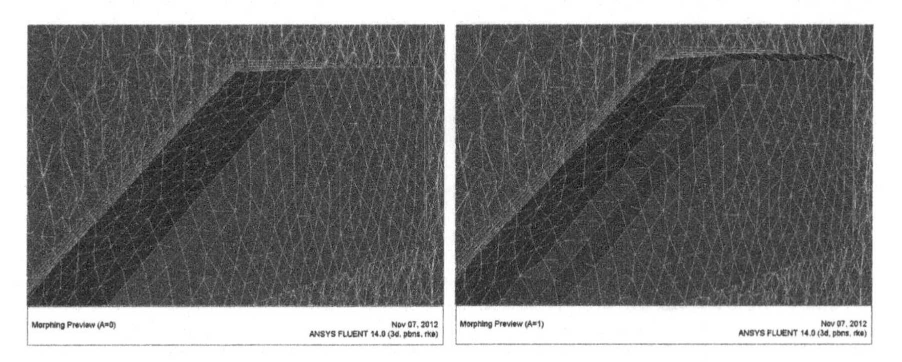

Fig. 9.5 Snow accretion case: preview of the morphing action on a cut of the volume mesh. The undeformed mesh (left) is compared with the deformed one (right) for a maximum snow growth equal to 7 mm

Since the amount of both cell (150,000 cells) and RBF source points (1200 centres) is quite small, the RBF computing takes just 2 s to finalise the RBF fit and 3 s to apply morphing. After this latter operation the mesh quality, considered in terms of maximum cell skewness, remains quite good (0.84) because it shows a slight worsening with respect to the value characterizing the original mesh (0.79).

As previously stated, the explorative procedure described in this section envisages the manual intervention of the analyst. However, the same workflow can be straightforwardly automated using both the UDM and UDF feature of ANSYS Fluent instead of Mathcad. Moreover, the automated process may be designed to work cyclically in view of performing an evolutionary simulation of snow accretion in function of time.

9.2.5 Validation of Snow Growth Predictions

The full implementation of the aforementioned workflow was presented by Allain et al. (2014) which study is summarized in the present section. In order to minimize the impacts of extreme winter conditions on rolling stock, the French National Railway Company SNCF undertook a research aimed at exploring solutions to reduce the snow accumulation on high-speed trains. The first part of the study focused on the collection of data to identify the mechanism of snow accumulation. Specific tests for generic simple-shaped prisms and a railway shape geometry at 1/2 reduced scale were used in a climatic wind tunnel. Figure 9.6 shows on the left how different simple-shaped prims are arranged in the rig, details of typical end test conditions are reproduced on the right. The 1/2 reduced scale model of TGV is shown on the left of Fig. 9.7 where the device for putting in rotation the axles is detailed on the right.

Fig. 9.6 Location of the simple-shaped prisms under the supporting planes (left), snow accumulation on the large prism faces at the end of a 10 min test period (right)

Fig. 9.7 TGV model set-up in the climaticwind tunnel (left), detail of the drive belt system of the wheel axles without protecting shields (right)

Snow and ice growths and localisations were acquired at different values of the wind speed and air flow temperature for snow particles at various angles of impact. The second part of the study consisted in reproducing these mechanisms by numerical simulations. A very good correlation between computational results and measurements was registered for both the simple shapes, as demonstrated in Fig. 9.8, and for both the 1/2 scale model of the TGV as demonstrated in Fig. 9.9 where a zoom on front bogie shaft and rear shaft is represented.

9.3 Icing Simulation

Flying in icing atmospheric conditions can be a serious safety problem. Ice build-up generally occurs when supercooled droplets impinge on aircraft surfaces, most commonly the engine and the windward face of the wings, causing a variation in the overall vehicle fluid dynamics and manoeuvrability that may lead stall to suddenly occur at angle of attacks (AoAs) lower than the design ones. Ice

Fig. 9.8 Snow accumulation on prisms: comparison between numerical simulations (top) and wind tunnel observations (bottom)

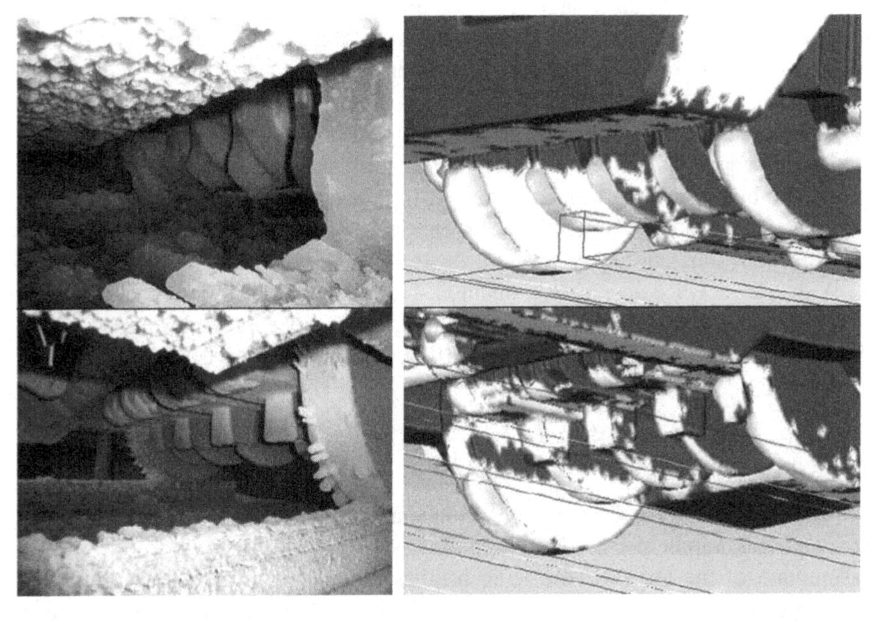

Fig. 9.9 Snow accumulation on 1/2 reduced scale TGV model: comparison between numerical wind tunnel observations (left) and simulations (right)—Zoom on front bogie shaft and rear shaft

accretions typically increase the drag force and decreases the lifting characteristics of the aerofoil, so more power and a greater AoA is required to maintain flight conditions.

Wing icing causes not only stall to occur at lower AoAs, but an uneven ice distribution can also diminish dangerously the vehicle manoeuvrability. Although icing accidents are just a small percentage of the total aviation accidents, in the USA only aircraft icing was responsible, or a concurrent factor, in 803 aviation accidents between 1975 and 1988 (Cole and Sand 1991), in 583 between 1982 and 2000 (Petty and Floyd 2004) and in 228 from 2006 to 2010 (Appiah-Kubi 2011). Understanding and studying the icing problem is then of capital importance, and an ice accretion analysis has thus become a must during the aircrafts design process. Ice build-up can be roughly investigated by means of flight tests, wind tunnels or numerical simulations. Flight tests are the most realistic and the most expensive too, so they are just used in certain conditions or in the final stage of the design. Wind tunnels can recreate exact ice shapes, but the control over the dimensionless parameters can be very hard. On the other hand, CFD is widely used for simulating icing because it is a low cost alternative to experimental campaigns and it can reliably catch many important features of the icing process, allowing to manage a large number of correlated parameters.

A typical icing accretion model consists of two main modules that work seamlessly together. The first one deals with the droplets trajectory and determines the collection efficiency distribution over the whole body, while the second one computes a thermodynamic analysis to establish, given the collection efficiency, the ice thickness at any given point. The results from the accretion model can be thus used to model a new shape in the analysis environment used for the CFD simulation. Therefore, a direct connection between the ice accretion model and the CFD solver is needed to be able to cyclically compute back the flow field for the collection efficiency determination of the new ice accreted configuration. Being able to generate a high quality grid in the shortest time possible is the major challenge in icing simulation as it is a potential bottleneck of the whole ice accretion analysis, in terms of both time and accuracy.

With regard to ice accretion simulation, two methodological approaches can be considered when using RBF mesh morphing. Such procedures, developed within the RBF4AERO Project (RBF4AERO 2013) platform, are respectively referred to as follows:

1. frozen (or constrained);
2. on-the-fly (or evolutionary).

According to the frozen approach, icing simulations are carried out by imposing, at specific iterations of the CFD computing, steady or unsteady depending on the assumption of the analysis, the icing profiles previously calculated at predefined instants of time by means of an icing accretion tool (icing data). Using these data and acting on the source points extracted from surface mesh, the geometry actually covered by ice can be suitably moulded thanks to the exact local control capability of RBF, and the cells' nodes of the target volume are accordingly adjusted by volume mesh smoothing. The workflow of such an approach is shown on the left side of Fig. 9.10 through a block diagram.

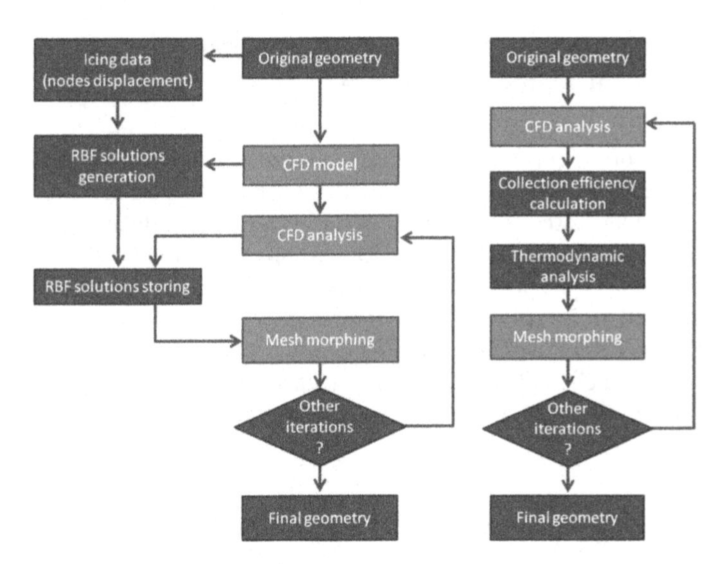

Fig. 9.10 Frozen (left) and on-the-fly (right) approach workflow for icing simulation based on RBF mesh morphing

The second icing approach pertains the use of an accretion code that, in conjunction with a CFD solver, dynamically modifies the numerical grid according to the calculated ice accretion. The workflow of the on-the-fly approach is illustrated in Fig. 9.10 (right), with the same logic already used for the frozen approach.

As visualised in Fig. 9.10, the frozen approach foresees the accomplishment of two subsequent stages. In the first one, 2D or 3D ice profiles are evaluated for each icing step through an ice accretion tool and, then, exported according to a predefined format in which the coordinates and displacement components for each point of the accreted surface are specified (.pts file format in the case that RBF Morph tools are used). These data, referred to as icing data, constitute the input for the generation of RBF solutions that are finally saved for successive use. The second stage deals with the icing simulation in which a certain number of CFD iterations are performed and the stored RBF solutions are properly applied to update the grid of the numerical model during computation.

Test cases respectively dealing with the application of the two aforementioned approaches are described in Sects. 9.4 and 9.5. It is worth to comment that there are different scenarios in which the shape of fluid wetted surface are changed by particle depositions or erosion. In Sect. 9.2 the case of snow deposition, that is not substantially different from ice growth here presented, has been tackled. In Riccio et al. (2017) the problem of formation of deposits on superheaters and reheaters in biomass and on coal-fired combustion systems is addressed using predictive models for deposition on such heat exchanger tubes using CFD.

9.4 Icing Simulation Tests Cases According to the Constrained Approach

The application of the constrained approach to simulate ice accretion on 2D and 3D models concerning the aviation field is described in the following sections. The first of the described test cases deals with the application of well-validated and complex icing profiles that are applied on two 2D aerofoils (Biancolini 2014), whilst the second one pertains the icing simulation on the 3D model called HIRENASD (Costa et al. 2014). In particular, the former involves the NACA0012 aerofoil, which has been widely used in icing tests over the years, and the GLC305 profile which belongs to a class of typical business jet aerofoils. The development and the results gained on these aerofoils are detailed in the next sections respectively.

9.4.1 2D Icing Case: NACA0012 and GLC305 Aerofoils

To verify the consistency of the constrained approach in simulating ice accretion, the ice accreted profiles of the LEWICE 2.0 validation manual (Wright 2002) are considered as a reference. LEWICE 2.0 is an extensively tested ice accretion computer code developed by NASA that includes an analytical thermodynamic model for the freezing evaluation of the supercooled droplets. The final ice shape can be gained varying the atmospheric and meteorological parameters such as velocity, pressure, temperature, liquid water content, relative humidity and droplet diameter. As the calculated iced shapes are the result of specific atmospheric conditions and flight parameters, they are representative of a wide spectrum of real cases.

To run CFD simulations on iced shapes through the ANSYS Fluent solver with the purpose to investigate the effects introduced by the change of the aerofoil profile, six among the most challenging ice shapes are chosen in the LEWICE 2.0 validation manual. Considering the categorization adopted in this manual, the identification numbers (IDs) of the Runs which such profiles refer to are 401, 072503, 072501, 072504, 072704 and 073105, that are respectively shown in Fig. 9.11 from top left to bottom right. In order to enable the import of icing data in the RBF morpher (add-on version of RBF Morph software), the selected profiles are suitably processed so that the starting absolute coordinate and displacements of a set of source points from the surface mesh nodes belonging to the portion of the aerofoil involved by ice accretion are collected in an ASCII file according to a specific format required by the morpher.

The structured mesh used for both the analysed aerofoils is composed by 9800 quadrilateral cells discretizing the C-shape simulation domain. A schematic of the computational grid, reported in Figs. 9.12 and 9.13 for the NACA0012 and GL305 profile respectively, depicts how a fine density mesh close to the aerofoil is created to properly catch the viscous effects.

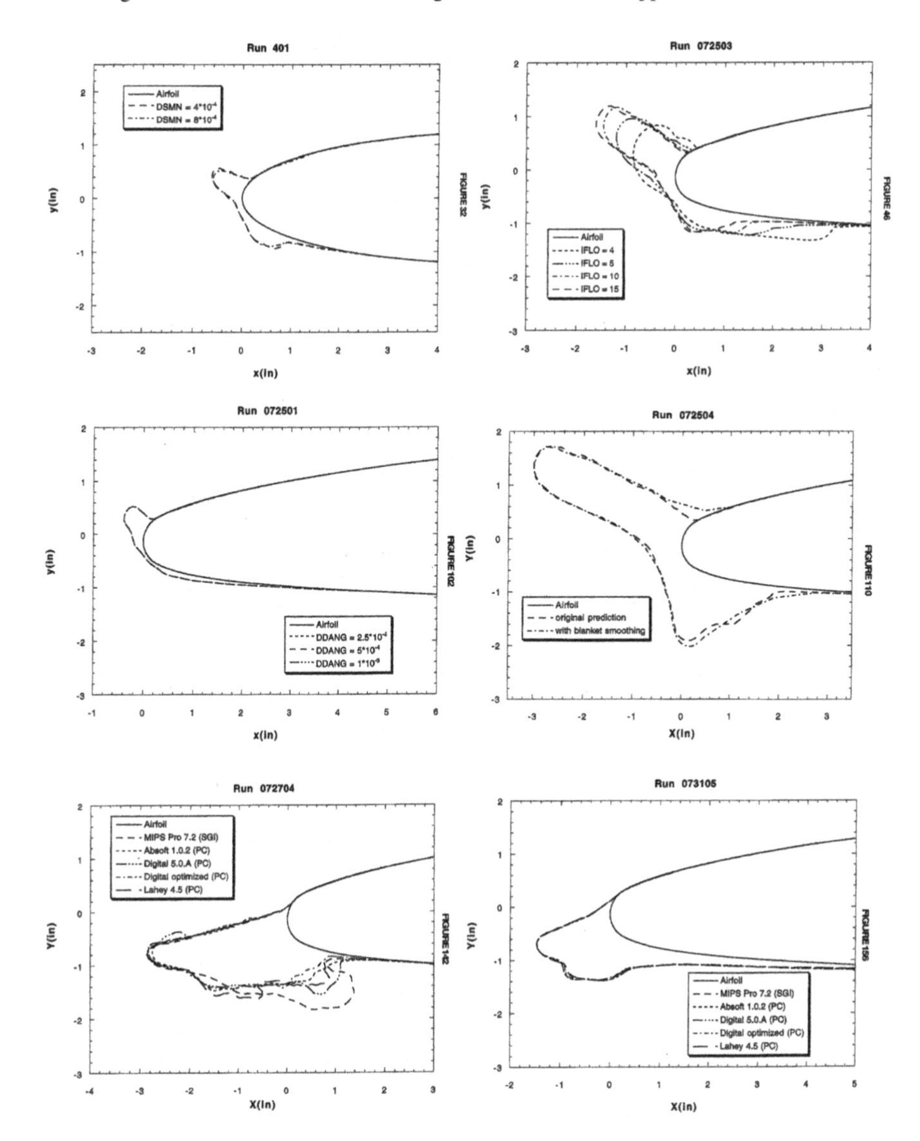

Fig. 9.11 2D icing case: Ice profiles chosen to run icing simulation

The same strategy is adopted for performing the simulation of all the considered iced profiles, using RBF solutions that share the same rationale. Relating to the morphing strategy, an incremental multi-stage approach is adopted. In fact, since the modification of the aerofoil due to the presence of ice is characterised by a large extent, it is too sever, or even impossible in some cases (e.g. concave accreted areas), to obtain the fully iced profiles just with a single morphing step. Given that, one or more intermediate stages are imposed between the initial and the final one,

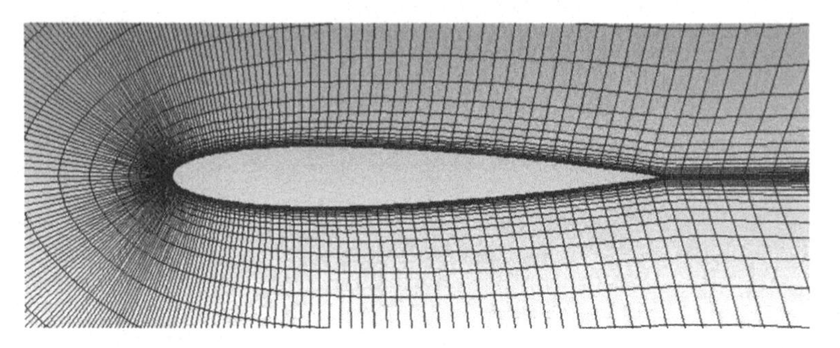

Fig. 9.12 2D icing case: NACA0012 aerofoil mesh

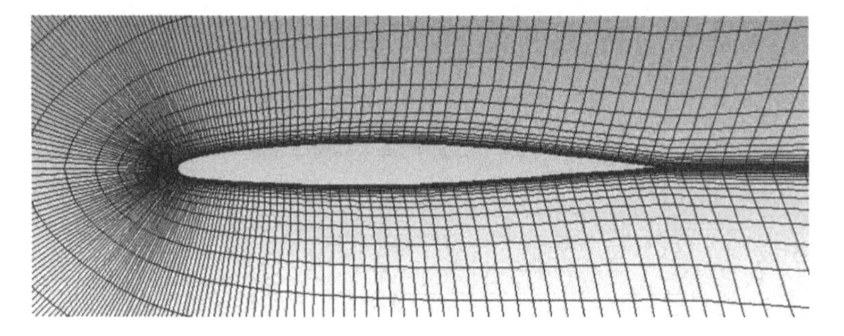

Fig. 9.13 2D icing case: GL305 aerofoil mesh

assuming a linear growing for ice. With regard to the set-up of the RBF field, the ice accretion is imposed exploiting the data of the accreted surface generated as already described, while the portion of the aerofoil requested to maintain its shape (zero ice thickness layer) is kept unchanged assigning a null motion to the source points extracted from the surface mesh nodes. Moreover, this latter condition is also imposed to the external boundaries (inlet and outlet surfaces) to prevent their deformation when the morphing action is applied. For each icing Run, Table 9.2 summarizes the ID, the type of used aerofoil and the number of morphing steps needed to properly simulate the final shape of the accreted profiles.

Table 9.2 2D icing case: Run ID morphing steps and used aerofoil

Run ID	Aerofoil	Number of morphing steps
401	NACA0012	4
072503	GL305	8
072501	GL305	4
072504	GL305	8
072704	GL305	12
073195	GL305	12

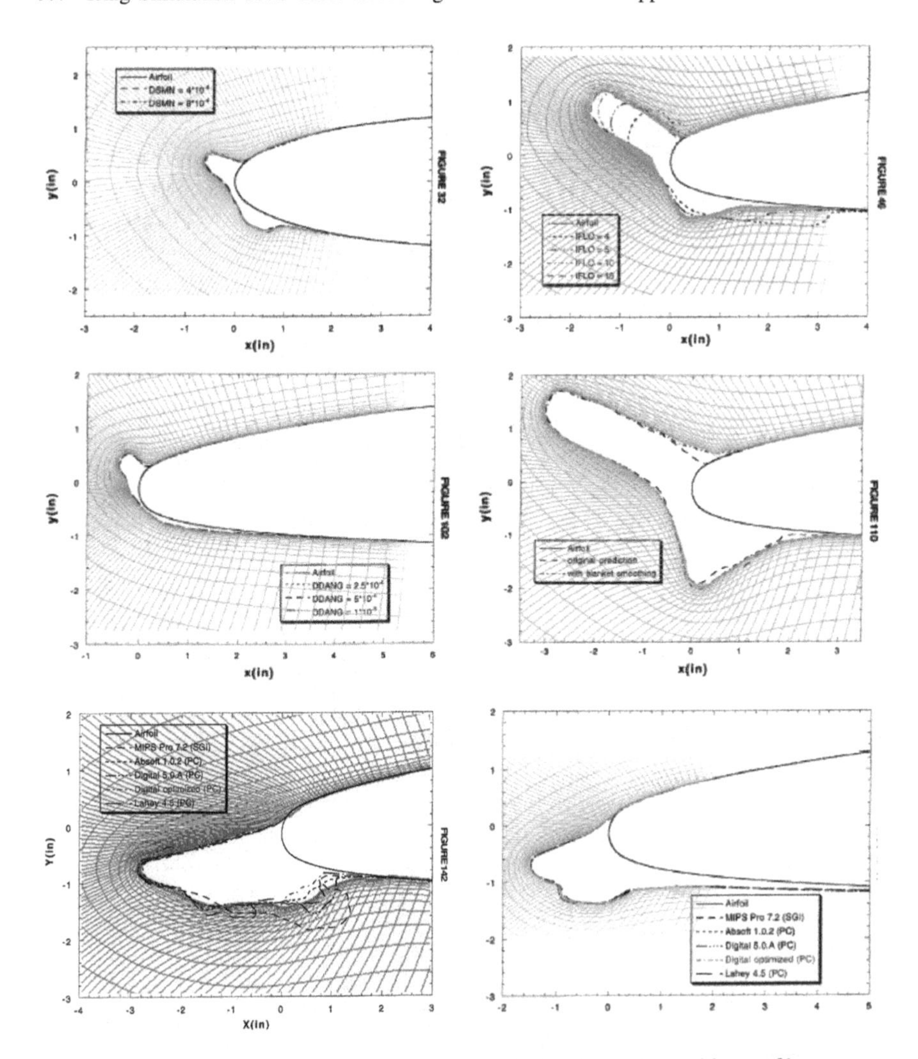

Fig. 9.14 2D icing case: comparison between numerical and experimental ice profiles

Maintaining the same order of the Fig. 9.11, the comparison between numerical profiles and experimental iced profiles is depicted in Fig. 9.14. In these images the completely morphed mesh is superposed to the ice accretion shapes taken from the LEWICE 2.0 validation manual. As visible, the action of morphing is able to accurately capture the complexity of the final geometries of the ice.

Figure 9.15 shows the morphing in the case of the most challenging of the considered runs, namely that with ID 072704. In this icing scenario, to reach the final shape by taking into account the dimension of the ice addendum, twelve morphing steps are adopted. To better appreciate the mesh movement during morphing, the evolution from the original to the final stage of the accretion is

Fig. 9.15 2D icing case: morphed grid for Run 072704

displayed from top left to bottom right. The big horn shape is very difficult to reproduce without a brand new mesh. In fact, there is a local stretch in the grid that causes a reduced nodal density due to the fact that the same amount of nodes on the boundary is maintained. To avoid a poor quality of the mesh after morphing and then to get better results, a refined mesh is advisable to be employed when dealing or expecting large ice accretions.

A CFD analysis is carried out for the just detailed Run to quantify the ice accretion influence on the aerodynamic efficiency of the aerofoil. Referring to the CFD solution set-up, a pressure based double precision solver with the $k - \varepsilon$ (two equations) turbulence model is used. Air is supposed to behave as an ideal gas and to respect the Sutherland viscosity model. As far as boundary conditions are concerned, a pressure far field condition is imposed at both inlet and outlet, whereas a standard slip wall is used for the wing profile. The SIMPLE scheme is set to couple pressure and velocity, whilst second order upwind spatial discretization scheme is used for density, momentum, turbulent kinetic energy, turbulent dissipation rate and the energy equations (ANSYS 2017).

The numerical outputs obtained at Mach 0.6 are visualized in Figs. 9.16, 9.17 and 9.18. In particular, for the clean, intermediate (sixth morphing step) and complete iced configuration in a range of value for the AoA between 0° and 14°, the first two figures respectively report the lift and drag curves (Fig. 9.16), and the full polar (Fig. 9.17). As expected, ice accretions are responsible for an increase of the drag force and a decrease of the lift coefficient, due to the deteriorated geometry that causes a premature flow separation.

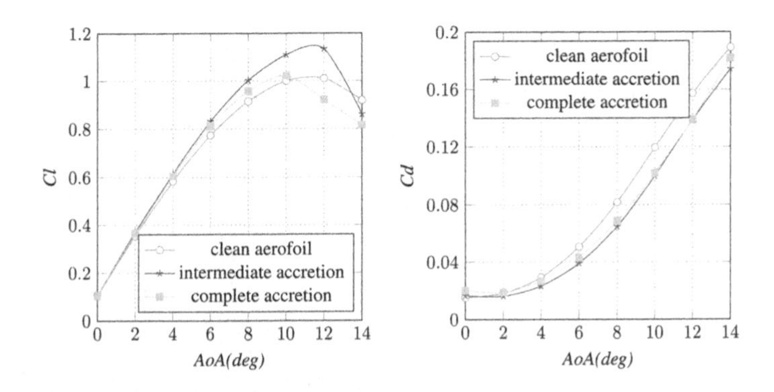

Fig. 9.16 2D icing case: drag and lift profiles for Run 072704

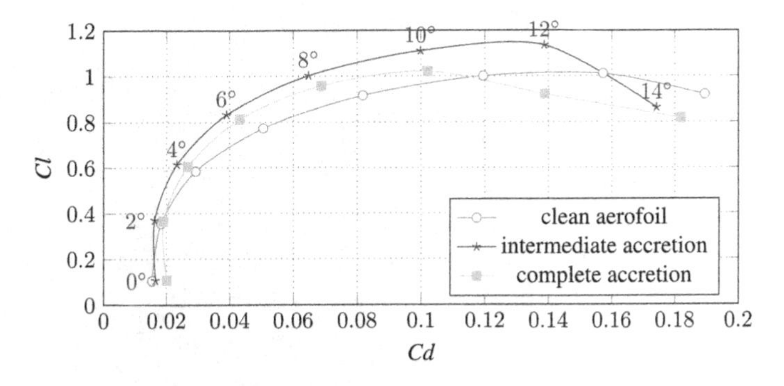

Fig. 9.17 2D icing case: polar plots for Run 072704

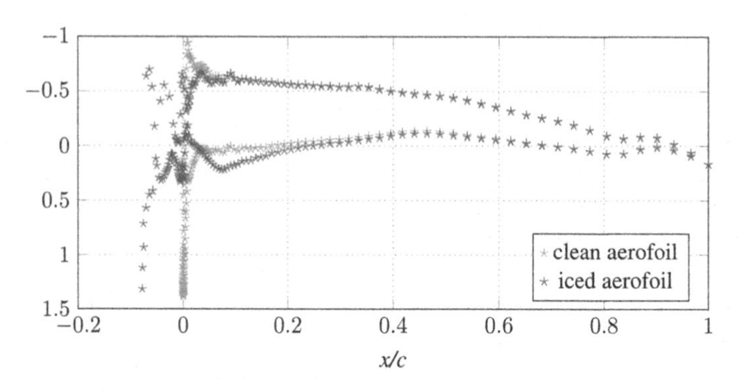

Fig. 9.18 2D icing case: Cp profiles for Run 072704

Figure 9.18 depicts the pressure coefficient along the aerofoil in the clean and fully iced configuration.

9.4.2 3D Icing Case: HIRENASD

This paragraph illustrates the numerical activities concerning ice accretion simulations on the 3D model referred to as HIgh REynolds Number Aero-Structural Dynamics (HIRENASD) (NASA 2012a), that is shown in Fig. 9.19 (Chwalowski

Fig. 9.19 HIRENASD wind tunnel model

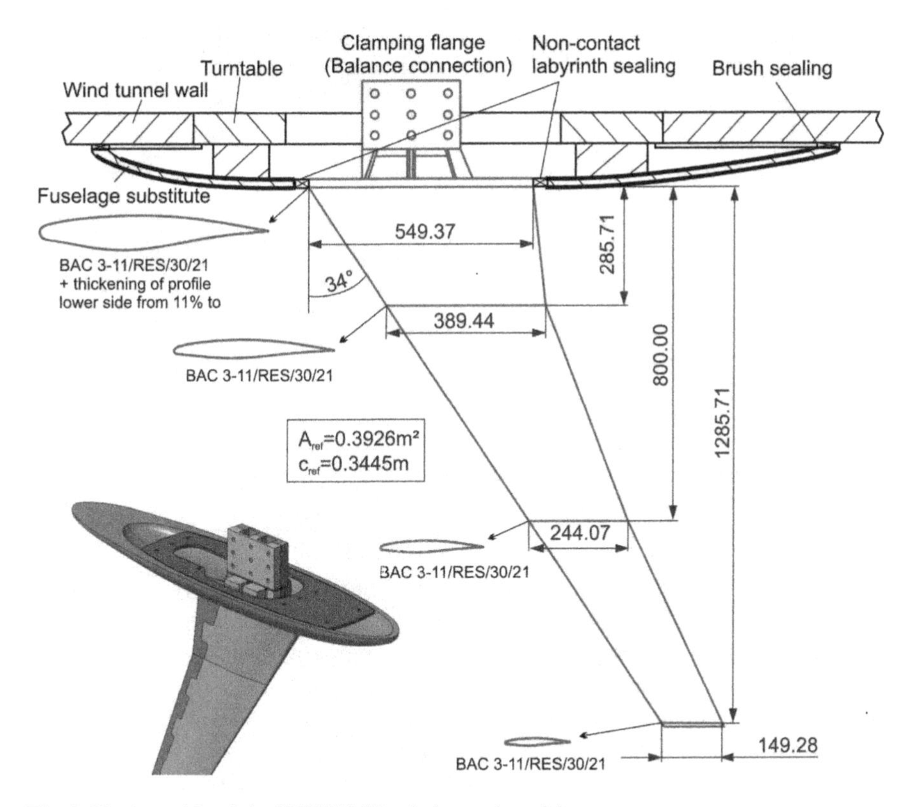

Fig. 9.20 Assembly of the HIRENASD wind tunnel model

et al. 2011). This model is one of the configurations supplied to participate to the 1st Aeroelastic Prediction Workshop (AePW) launched by NASA with the major purpose to assess the capability of the most advanced numerical methods in predicting static and dynamic aeroelastic phenomena and responses (NASA 2012b).

Such a model, tested in the Cologne European Transonic Wind tunnel (ETW), consists of a tapered 34° aft-swept wing with a BAC3-11/RES/30/21 supercritical aerofoil profile having a mean chord length (reference chord) of 0.3445 m and a wing span of 1.28571 m. The assembly of the HIRENASD wind tunnel model is depicted in Fig. 9.20 (Chwalowski et al. 2011).

The detailed tests case have been developed in the framework of the RBF4AERO Project (RBF4AERO 2013) using the icing tool employed by Piaggio Aero Space aerodynamic team that is linked to the CFD++ solver and the SU2 solver to solve the fluid-dynamic field. The flow conditions are imposed considering Mach = 0.5,

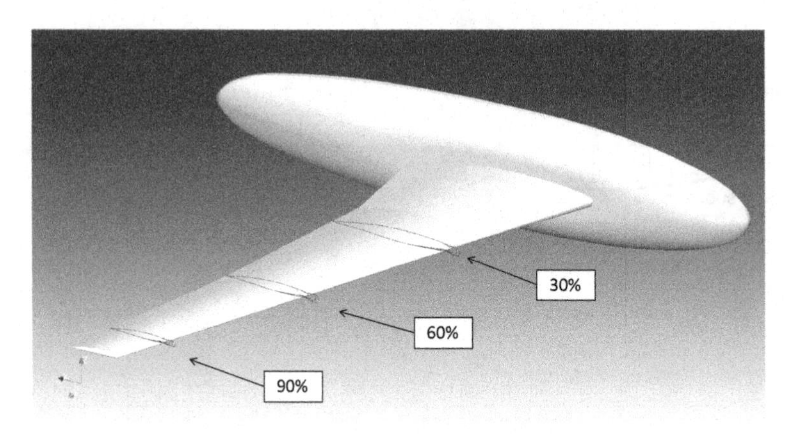

Fig. 9.21 3D icing case: iced sections location on HIRENASD wing

Re = 11.5×10^6, AoA = 1.5° and altitude = 4650 ft (1427.32 m). By means of the ice accretion tool the icing profiles at 7, 14 and 21 min for three wing sections were computed adopting the Langmuir-D distribution, imposing Delta ISA = −25° and the median volume diameter (MVD) equal to 20 μm. Such sections, shown in Fig. 9.21, are positioned at 30, 60 and 90% of the wing span, and were used to create the 3D iced surfaces through a blending surface.

The volume mesh and a detail of the surface mesh of the HIRENASD model are shown in Fig. 9.22 on the left and right side respectively. Exploiting the symmetry of the flow field, only half of the model has been meshed. The volume mesh is a multi-block structured hexahedral grid extended about 40 wing chords upstream the

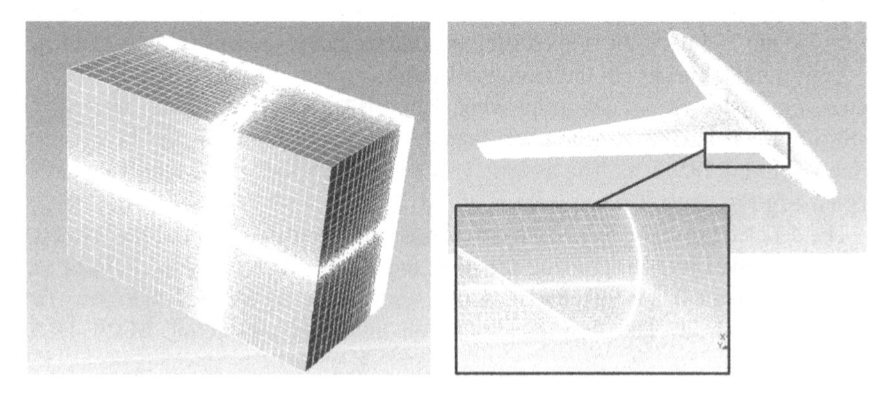

Fig. 9.22 3D icing case: mesh of the HIRENASD case

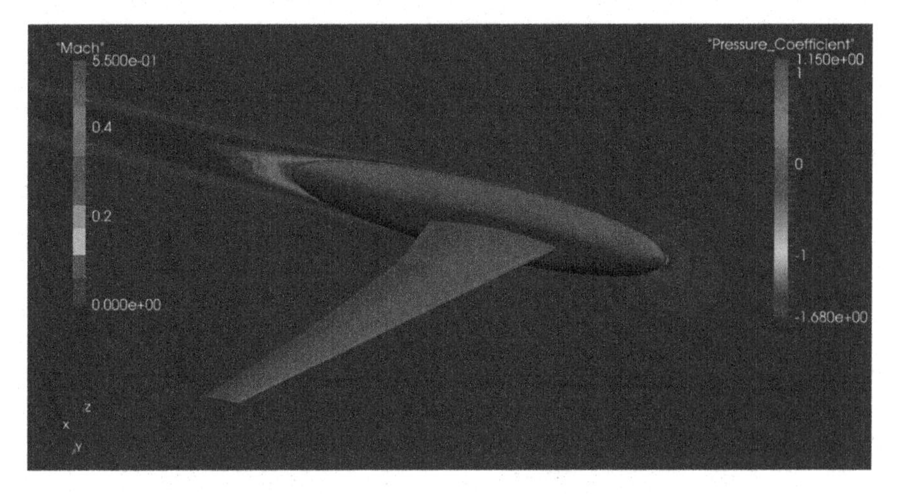

Fig. 9.23 3D icing case: Cp and Mach distribution in the baseline configuration

model, 42 downstream and 43 on the side with wall cells clustering aimed to not solve the boundary layer up to the wall.

The main parameters and models used for the SU2 simulations include the use of the Spalart-Allmaras turbulence model, the maximum number for CFL equal to 4, the Green-Gauss method for spatial gradients, the JST for convective method, the first order for the spatial integration of flow and turbulence, and the first order for time discretization (flow and turbulence) Euler implicit. As far as boundary conditions are concerned, a far field condition was set for all simulation volume boundaries apart from the symmetry condition that was applied to the symmetry surface. The wing and fuselage were set to no-slip wall.

The pressure coefficient (Cp) distribution on the fuselage and wing together with the contour of the Mach number on the symmetry plane are depicted in Fig. 9.23 for the ice growth configuration at 0 min (baseline).

Using the discretizations of those surfaces, the icing data required by the RBF morpher (stand-alone version of the RBF Morph software) to apply surface mesh smoothing, namely the starting position and the displacement of each node of the surface mesh at each icing step, have been generated.

Since the case is three-dimensional, the source points extracted from the surface mesh of the wing identify a cloud of points. Figure 9.24 reports, for example, the

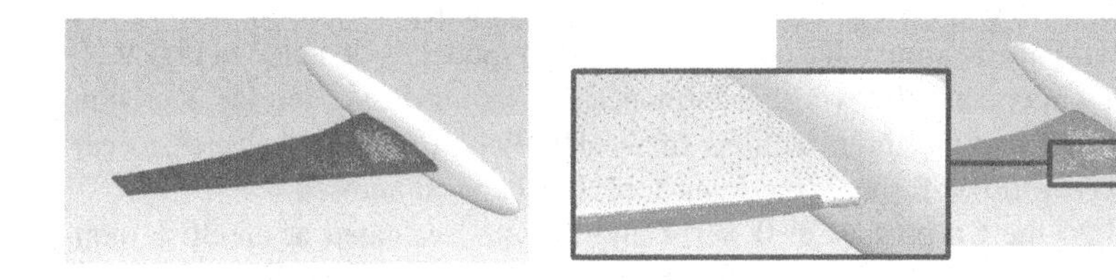

Fig. 9.24 Source nodes position before and after morphing for 7 min ice solution

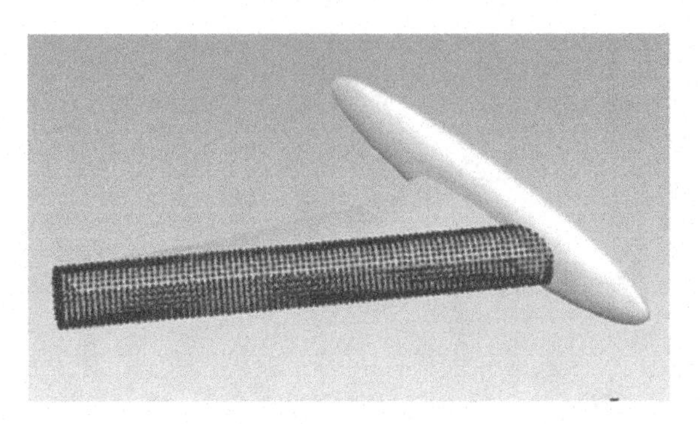

Fig. 9.25 3D icing case: source points preview of the cylinder-shaped delimiting domain for the HIRENASD model

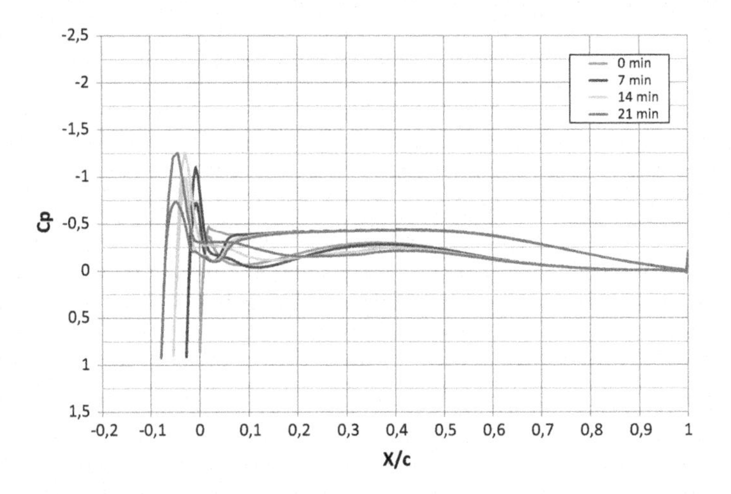

Fig. 9.26 Cp profiles at the 30% monitoring section

position of the source points before (left) and after morphing for the first-step solution corresponding to 7 min of ice accretion. Specifically, the nodes of the wing in the area that are not covered by ice have a null displacement, whereas for the remaining ones the displacement is dictated by icing data.

To delimit the action of morphing nearby the leading edge where the icing is expected to grow, a cylinder-shaped delimiting domain is generated. The source points distribution on such an entity of the morphing tool, used for the second-step solution to wrap the leading edge of the HIRENASD model, is depicted in Fig. 9.25.

To better understand the wing performances variation during the ice accretion, the Cp profile along the three sections of the HIRENASD model selected to calculate the iced profiles through the icing tool, has been monitored.

In Fig. 9.26 the Cp profiles at 0, 7, 14 and 21 min evaluated at the 30% monitoring section are depicted. The Cp is plotted as a function of the X/c coordinate,

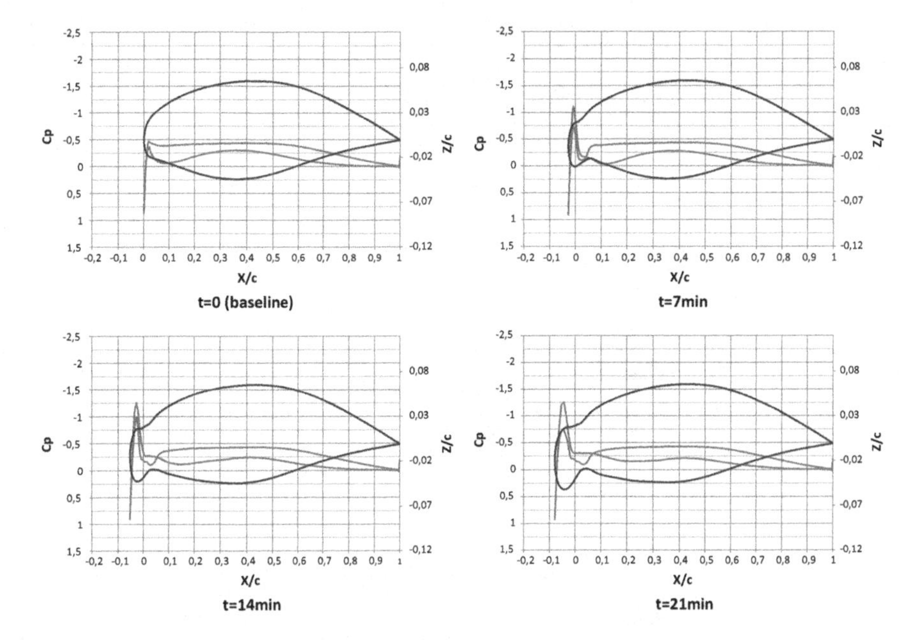

Fig. 9.27 Cp and wing section profiles at the 30% monitoring section

where X is the abscissa coordinate and c is the chord length of the current section in the baseline configuration. For values of X/c greater than 0.4.

Figure 9.27 shows the Cp and wing section profiles computed at the 30% monitoring section for the baseline configuration and for the configuration at the icing steps, respectively refer to 7, 14 and 21 min. The ice growth as well as how it influences the Cp are well visible.

The profile extent (about the 8% of the current section chord) reached in the last icing step evidences the great robustness on the used mesh morphing technique in handling challenging and complex profiles also for 3D cases.

9.5 Icing Simulation Tests Cases According to the On-The-Fly Approach

In this section the implementation and the verification of the second approach to simulate icing through the use of RBF mesh morphing and termed on-the-fly are described (European Commission 2013).

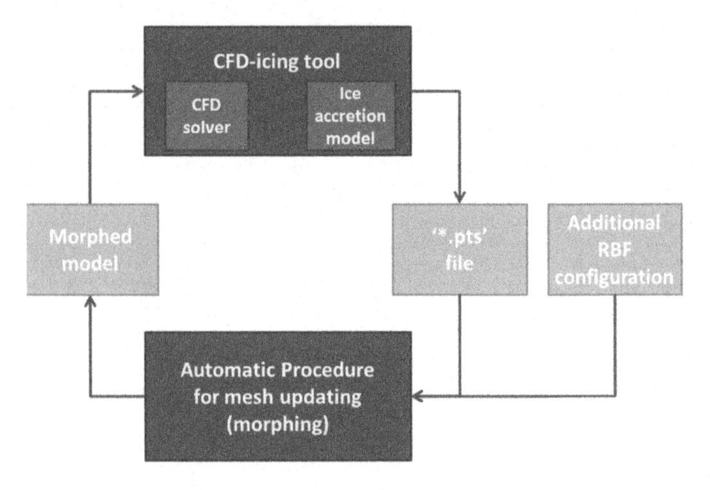

Fig. 9.28 On-the-fly approach framework

9.5.1 *Adopted Strategy for the On-The-Fly Icing Implementation*

The framework of the on-the-fly icing already summarised in Fig. 9.10 (right), has been conceived taking into account its characteristic workflow as well as the fact that it is required to be general and ready to be linked to different tools capable to run CFD-icing coupled computing. Given that, it has been designed according to what is illustrated in Fig. 9.28.

As shown in this latter diagram, the on-the-fly icing framework basically envisages the presence of the specific CFD-icing tool component that exchanges data with an automatic procedure which generates the RBF solution(s) and applies morphing to the computational case of the CFD solver employed by the CFD-icing tool. This latter component is mainly made up of the CFD solver and ice accretion model and, as it is required to be internally driven, it acts as a sort of black box which takes in input the morphed models and provides as output the .pts file for each icing step. The level of independence of the CFD-icing tool makes such a component replaceable and gives a general significance to the whole icing simulation procedure.

To ease the comprehension of the functioning of the icing approach under discussion by means of an explicatory example, the CFD-icing tool used for the verification study is detailed in the next paragraph.

9.5.2 *On-The-Fly Icing Test Case CFD-Icing Tool*

With regard to the on-the-fly verification test case, the CFD-icing tool was primarily composed of the ANSYS Fluent solver and an in-house developed icing accretion model. Such a model has been implemented through the UDF feature. Specifically, the subcomponents identifying the entire CFD-icing tool were the following:

- computational case (.cas file);
- icing model (ANSI-C-based routines working according to UDF);
- driver of the CFD-icing tool overall computing (journal file);
- driver of the internal CFD-icing tool computing (scheme and UDF file).

Referring to the ice accretion model, its implementation exploited the ANSYS Fluent wall film functionalities and its working was driven using customized UDF and scheme file. The obtained iced thickness along the aerofoil profile was directly converted inside ANSYS Fluent to a .pts file identifying the nodal displacements to be imposed along the surface normal. Since icing boundaries conditions were defined in the ANSYS Fluent case, the .pts file can be considered the only output of the ice accretion model. The simulated time was intrinsically accounted by the icing model and that information was recorded in a monitoring file generated during computing.

Several assumptions that do not change from one accretion step to the successive, are made for the test case. Among those are the icing boundary conditions (liquid water content, droplet size, etc.) and the fact that each CFD simulation, including the wall film, is started from scratch so that only rime ice is evaluated. Moreover, it is worth to mention that the specific accretion model is suitable only for 2.5D geometries for an intrinsic limitation of the ANSYS Fluent wall film feature.

9.5.3 *On-The-Fly Icing Verification*

The model used for carrying out the verification of the methodology developed for handling on-the-fly icing was the NACA 0012.

The computational model of NACA 0012 was the one already described in Sect. 9.4.1. In order to make the icing model operative, the Eulerian multiphase model has been used with air as primary phase (supposed to behave as an ideal gas) and water as secondary phase. As far as boundary conditions are concerned, the aerofoil surface has been imposed as wall and a pressure gauge condition is assigned to the inlet and outlet boundaries of the simulation volume to set a Mach number equal to 0.45 with an AoA = 3.5°. Relating to turbulence, the k-ε model with standard wall functions has been adopted.

To apply the aerofoil nodes smoothing dictated by the icing model, the Dynamic Mesh feature of the CFD solver has been enabled linking the NACA0012 surface to UDF. According to the working hypotheses of the (wall film) icing model, a series of steady state calculations identifying the simulation of each icing step have been run adopting a coupled scheme for pressure-velocity coupling, the Green-Gauss Node Based scheme for Gradient and the First Order Upwind for all solution parameters.

With regard to the RBF solutions set-up, the morpher managed two types of contribution to generate the RBF solution that drives morphing in view of updating the aerofoil profile at each icing step. The first one, that typically remains

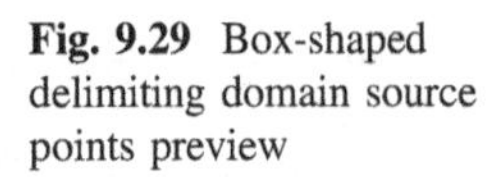

Fig. 9.29 Box-shaped delimiting domain source points preview

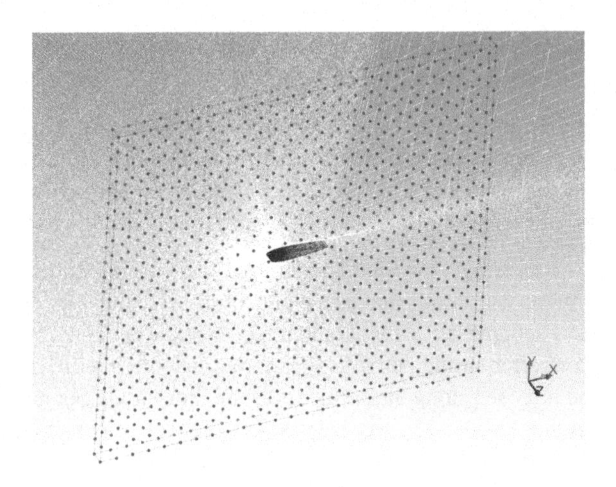

unchanged for all steps of icing, was constituted by the RBF solution limiting the morphing action. That solution was indicated as "Additional RBF configuration" in Fig. 9.28 and consists of a box-shaped domain used to delimit the volume for morphing application. The preview of the source points belonging to this latter RBF morpher entity is depicted in Fig. 9.29.

The second contribution consisted of the .pts file that specifies, for a particular icing step, the nodes' displacement to be applied to the current model that are specified with respect to the one of the previous icing step. The baseline aerofoil profile and the preview of source point for the application of the last icing step (t = 57.87 s) are illustrated in Fig. 9.30.

For testing purposes the CFD-icing tool based on ANSYS Fluent has performed 36 ice accretion steps. To show the gained results, four icing steps have been

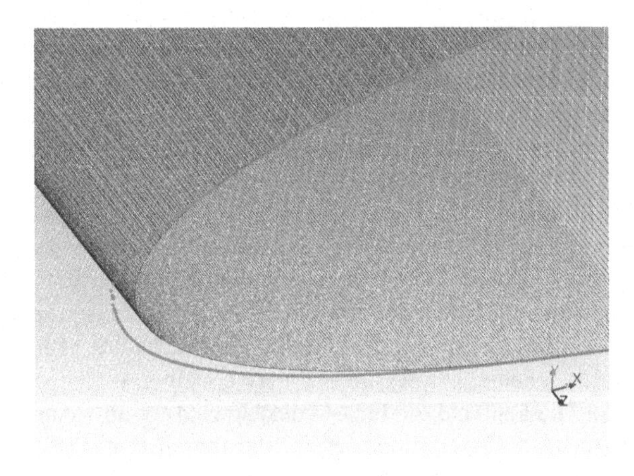

Fig. 9.30 Preview of the source points position after applying morphing (last icing step)

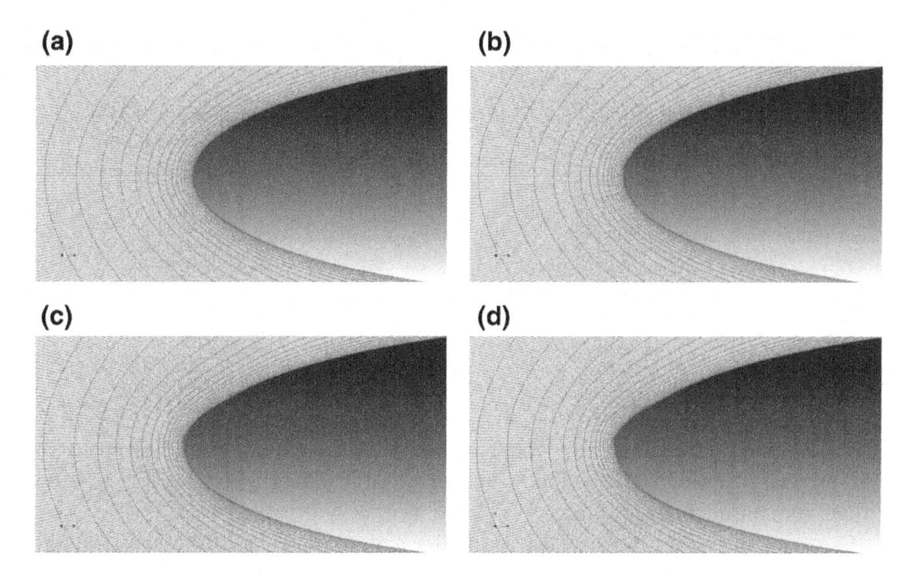

Fig. 9.31 Mesh for the icing representative steps

selected among those computed. The icing step number (ISN) of these stages, hereinafter termed icing representative steps, is 0, 12, 24 and 36 and the simulation time it mimics respectively is 0, 21.65, 38.47 and 57.87 s.

The mesh of the case for the icing representative steps is depicted in Fig. 9.31 from a lateral orthographic view. In particular, Fig. 9.31a–d respectively depict the baseline mesh (ISN-0), the mesh after 21.65 s (ISN-12), the mesh after 38.47 s (ISN-24) and the final mesh at 57.87 s (ISN-36).

In Fig. 9.32 the surface mesh of NACA 0012 for the icing representative steps are depicted in a unique visualization from an oblique perspective. By this overlapped representation it is possible to better appreciate the area of the aerofoil

Fig. 9.32 NACA0012 surface mesh at icing representative steps

Fig. 9.33 Morph accuracy check for ISN-12

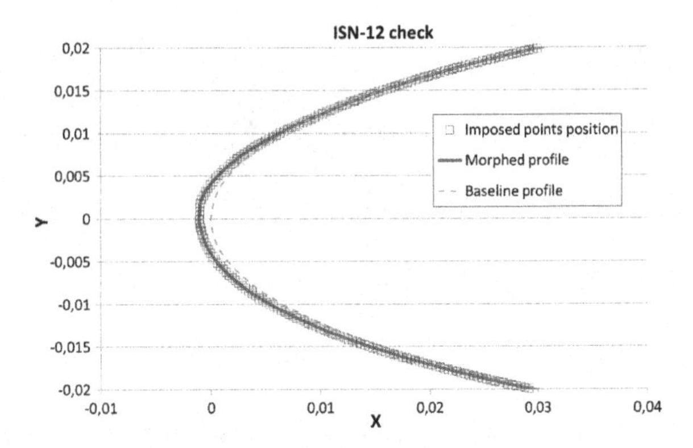

affected by the presence of ice. The asymmetry of ice growth is due to the AoA characterising the undisturbed flow, whereas the outstanding regions of the aerofoil surface remain unchanged.

To verify the correctness of the morphing action application, the shape modifications applied at ISN-12 and at ISN-36 have been checked. To this end, the comparison between the aerofoil nominal profile obtained through mesh nodes' displacements, which are stored in the Points file and processed by the RBF morpher, and the effective aerofoil profile (mesh nodes' location) after morphing, has been performed. The result of that comparison is depicted in Figs. 9.33 and 9.34 respectively for ISN-12 and ISN-36. The good alignment of profiles is highlighted by the correspondence between position of mesh nodes assigned by the icing tool (red squares) and the one of the morphed mesh (continuous line), and evidences the local control accuracy of morphing. For the sake of completion, the baseline profile is also reported in both the images by means of a dotted line.

Figure 9.35 depicts the dynamic pressure distribution of the air phase around the leading edge of the aerofoil profile on the mid-cutting plane of the computational case respectively at ISN-0 (Fig. 9.35a), at ISN-12 (Fig. 9.35b), at ISN-24 (Fig. 9.35c) and

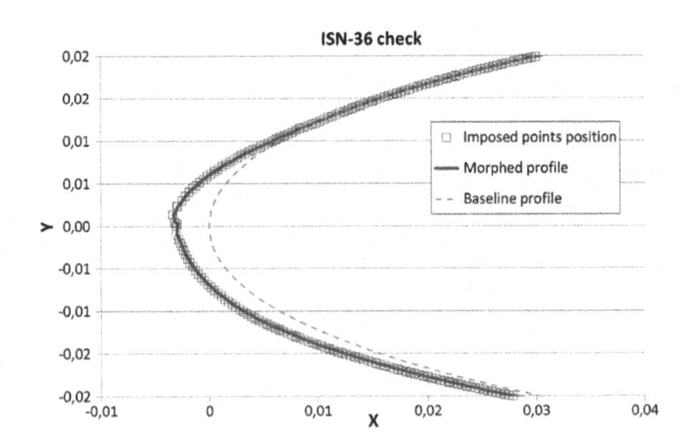

Fig. 9.34 Morph accuracy check for ISN-36

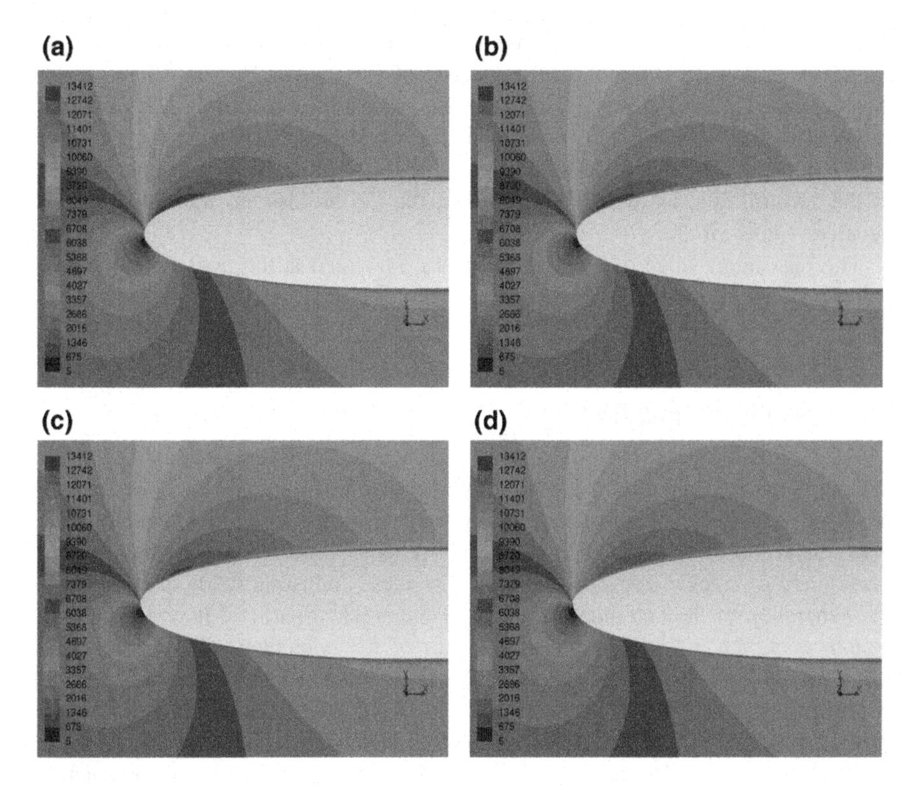

Fig. 9.35 Dynamic pressure map on the mid-cutting plane at icing representative steps

at ISN-36 (Fig. 9.35d), maintaining fixed the limits of the represented values. In this figure it is possible to notice the stagnation point and how the shape modification induced by icing influences the distribution around the leading edge.

Figure 9.36 shows a detailed view of the Cp profiles along the mid cross-section of the computational model for the four instants of simulation time already

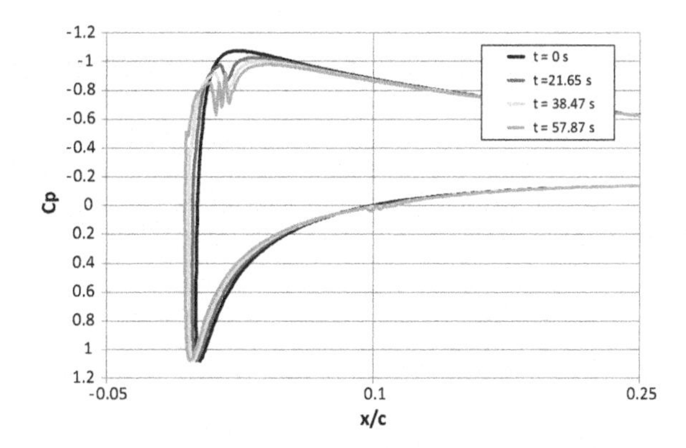

Fig. 9.36 Profile of Cp at icing representative steps (detailed view)

considered. The values of Cp are reported as function of the x/c, where c is the chord length of the aerofoil and x the abscissa of the aerofoil. For x/c values greater than 0.25 the Cp profiles are indistinguishable, whilst the effect of the ice growth in proximity of the leading edge is evident for x/c values lower than 0.1. The ice accretion causes a marked variation of pressure distribution along the upper surface of the aerofoil and, as the ice layer grows, the Cp profiles are translated toward negative values of x/c.

The maximum value of elongation for ice evaluated at the final time of simulation was 3% of the aerofoil chord.

9.6 Crack Propagation

Damage tolerant design requires the implementation of effective tools for fracture mechanics analysis suitable for complex shaped components. FEM methods are very well consolidated in this field and reliable procedures for the strength assessment of cracked parts are daily used in many industrial fields. Nevertheless, the generation of the computational grid of the cracked part and its update after a certain evolution are still a challenging part of the computational workflow and mesh morphing based on the use of RBF can be used to support this task.

As for ice growth workflow described in Sect. 9.3, the working scenarios can be substantially two. There are situations where a new shape of the crack is available as happens when a component in service is monitored using a non destructive testing (NDT) in a damage tolerant scenario: in this case the FEA model includes the crack at previous inspection and needs to be updated onto the new shape and size detected experimentally in order to assess crack stability and residual life. In other cases, such as when the residual life of the previous example is required to be evaluated, the evolution of crack shape needs to be predicted according to the local value of stress intensity factor (SIF) along the crack front, as computed in advanced fracture mechanics (FM) approaches such as K_I for linear elastic fracture mechanics or J integral for elastic plastic fracture mechanics: in this case the shape evolution is driven by the physics instead.

Mixed scenario may be also considered for situations in which the geometrical model of the crack is decided in advance (for instance elliptic with two parameters) and the original mesh has to be updated onto the target shape or evolved in the parametric shape according to FM calculations.

In the following sections two test cases dealing with crack propagation are shown: the semi-elliptical crack on a flat bar and the edge crack onto a notched round bar.

9.6.1 Semi-elliptical Crack on a Flat Bar

The example of morphing presented in this section consists in the evaluation of mesh morphing feasibility for addressing the case described in Galland et al. (2011) in which a mesh morphing approach based on the minimization of Dirichlet energy is proposed. The same mesh has been here used. The reference case adopted is summarized in Fig. 9.37 where the simple geometry, a square bar in tension with an edge crack (first image on the left), is shown together with the computational mesh (second image) and the FEA results at the beginning and at the end of crack propagation study (third and fourth images).

The study hereinafter described was conducted adopting the ANSYS Fluent working environment by means of the add-on version of the RBF Morph tool because at the time of the feasibility study the FEA implementation of the software was not available yet. The mesh was available in NASTRAN format (.nas), and crack propagation data were provided as the evolution of 88 nodes on the crack front laying over a horizontal plane.

Two approaches were investigated. The first approach uses, as input data, nodes on the crack front only. A poor accuracy results are gained because the mapped mesh area around the crack is not preserved. A refined approach that consists of the processing of the input data so that extra points at the boundary of the mapped region are used to control the morpher tool action, resolves the described unwanted effect. In fact, in this way the cross sections of the mapped mesh representing the volume surrounding the crack front are preserved. The latter approach is presented in detail in this section.

The basic assumption of this test case is that the crack front evolution is already known. In particular, the location of its mesh nodes for the two considered configurations are supposed to be supplied.

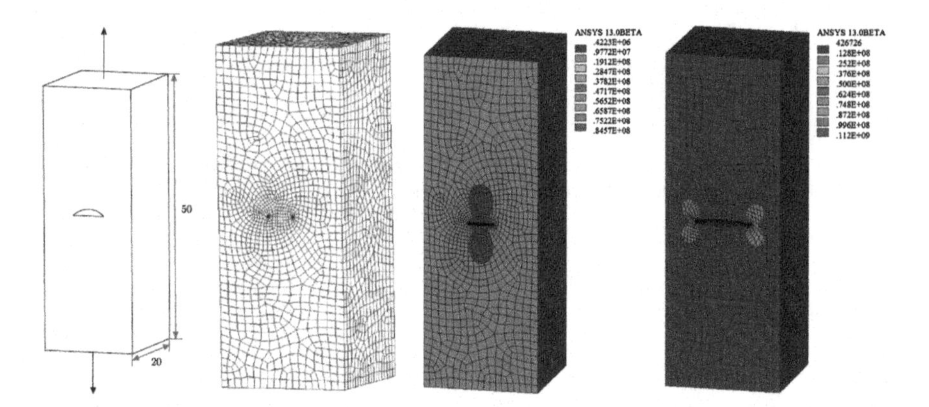

Fig. 9.37 Galland et al. (2011) study, from left to right. The tensile specimen geometry, a square bar with and edge crack, is sketched at left. FEA mesh with local refinement at crack front. FEA results for the original crack and the propagated crack

In the first stage of the procedure, the volume mesh is imported in the ANSYS Fluent environment. Successively, a two-step hierarchical morphing approach is adopted. According to that, the final volume mesh morphing is applied imposing the RBF solution computed in previous step as displacement field for model surface in order to precisely control different parts of the surface mesh of the case.

All the surface mesh nodes laying on the external surfaces and on the boundary of the crack front are used to control the volume morphing. These nodes are extracted and used as source points to finalise the RBF field set-up by means of the RBF mesh morphing software (see Fig. 9.38). Specifically, the red points are those kept fixed during morphing, the green points are forced to slide on the planar face in the respect of the crack front movement and the blue points have the displacement, that are controlled by the assigned crack front movement.

Similarly to what already described for icing simulation, the movement of crack front (blue source points) is assigned using the input file including the position and displacement of the 244 points (3.88) belonging to the crack (see Fig. 9.39). Such a file, identifying the original and final position of the front, is generated using the Mathcad tool.

The movement of the nodes of the front planar face of the model are controlled by fixing its boundaries through the red source points shown in Fig. 9.40, and

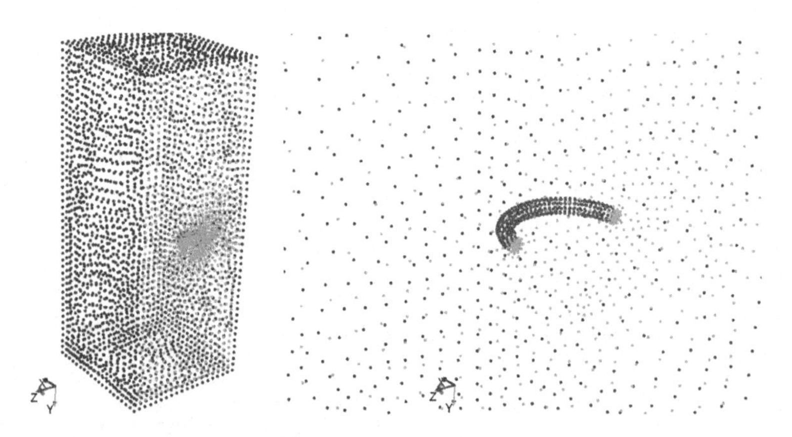

Fig. 9.38 Crack propagation case: RBF source points. The whole arrangement (left) and a detail around the crack (right) each colour represents a specific set of controlled points

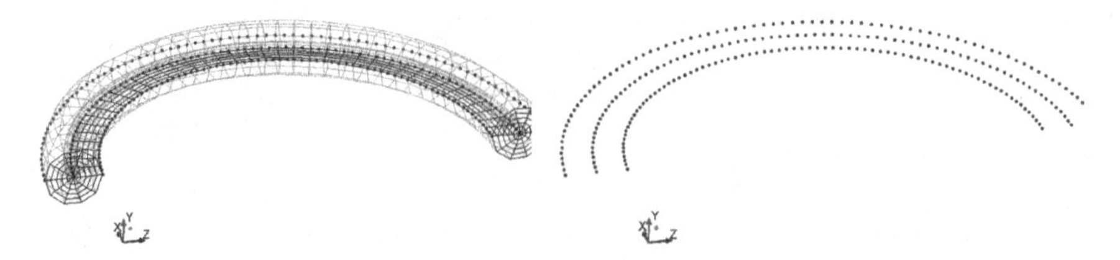

Fig. 9.39 Crack front source points

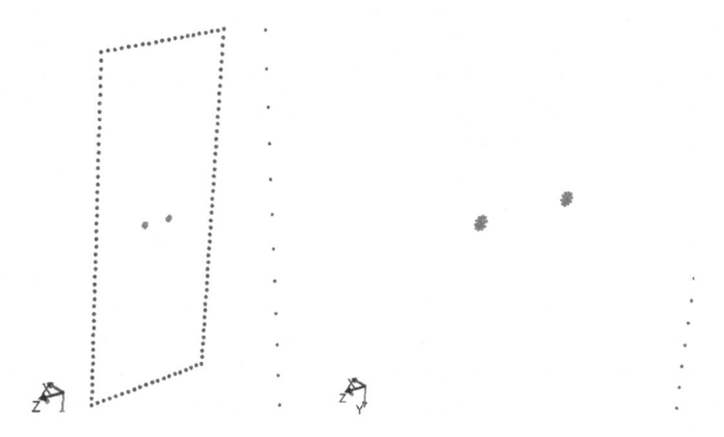

Fig. 9.40 Crack propagation case: fixed (left) and moving source points on the front face of the model (right)

prescribing rigid movements congruent with local crack front movement caused by the green source points on the extremal edges of the crack front (see Fig. 9.40).

The simulation of the crack propagation through morphing is previewed Fig. 9.41. In particular, starting from the baseline configuration (first image on the top left), the propagation is applied with an amplification value respectively of 0.33, 0.66 and 1. These latter configuration represents the fully propagated one.

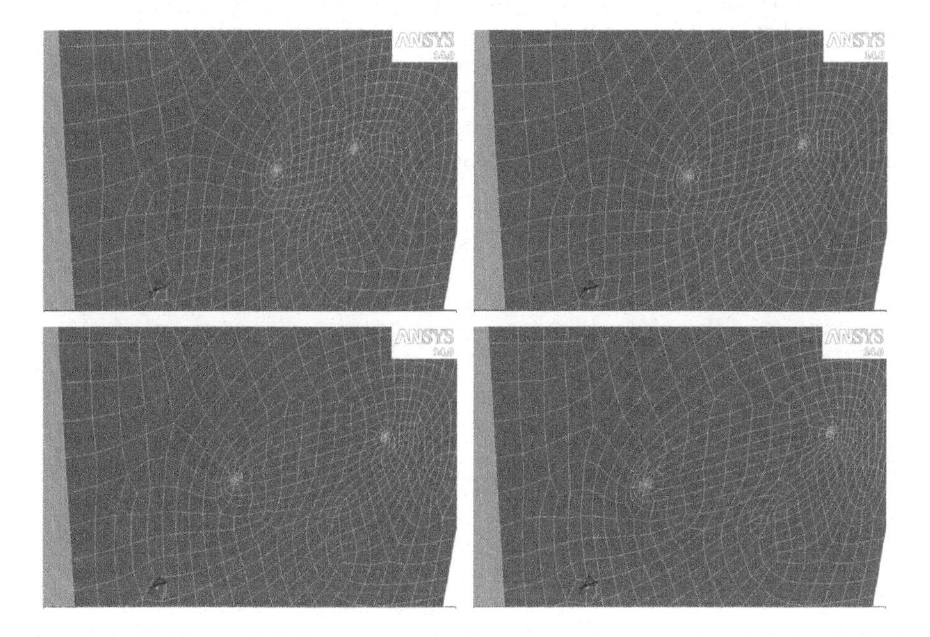

Fig. 9.41 Crack propagation case: morphing application preview

Table 9.3 Mesh quality evolution versus crack propagation

Morphing extent (%)	Max cell skew	Min orthogonal quality	Max aspect ratio
0	9.99960e−01	4.00935e−05	8.38617e+01
50	9.99929e−01	7.12791e−05	7.85819e+01
100	9.99995e−01	4.77582e−06	1.56665e+02

Original value, half way value and full

The quality of the mesh is summarized in Table 9.3 where quality of the volume mesh is summarized at three level of morphing amplifications: 0% (original mesh), 50% (half a way) and 100% (target final crack shape). Despite the overall bad quality of the starting mesh, it's worth to notice that the mesh validity is kept for the full range with a reasonable degradation of quality indexes.

9.6.2 *Edge Crack onto a Notched Round Bar*

The study here presented is derived from Biancolini et al. (2017) and is based on a workflow completely built in ANSYS Workbench. A round bar presents a circumferential notch with a constant curvature radius ρ, as depicted in Fig. 9.42. The diameter of the bar is equal to D in an un-notched cross-section and D_0 in the reduced cross-section. The circular section of the groove has its centre on the cylindrical surface of the bar so that $D_0 = D - 2\rho$. Total length of the bar is $4D$, supposed enough to extinguish any boundary effect before the flawed region. The bar is fixed at one edge and loaded by an axial force at the opposite.

The parameter set necessary to build and load the model is shown in Table 9.4. Crack modelling is embedded in ANSYS Mechanical by means of the Fracture Tool (FT). This tool allows inserting a fracture in a pre-existing mesh, with a local modification of the FE model. Input parameters define the geometry and the discretization of the flaw.

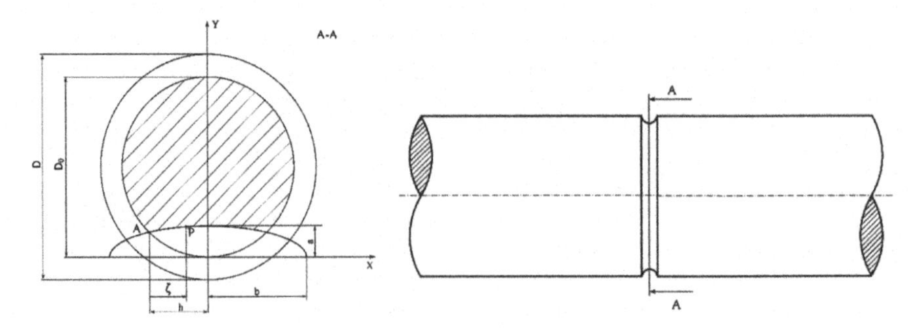

Fig. 9.42 Semi-elliptical crack front geometry (left); notched bar geometry (right)

Table 9.4 Main FE model parameters

Parameter	Value
D [mm]	20
D_0 [mm]	16
L [mm]	80
F [kN]	30

A three-dimensional FE model was adopted to determine the SIFs values along the crack front. Despite the symmetry of the problem, the whole bar is represented. In such a way the operations of the insertion and morph of the crack are straightforwardly accomplished thanks to the presence of buffer elements all around the affected zone. The mesh is composed by 10-node isoparametric tetrahedrons. A preliminary check of the model without the flaw is conducted in order to retrieve the stress concentration factor K_t in the reduced section. A convergence test is carried out increasing the number of elements to determine its asymptotic value. The match with the theoretical reference (Pilkey and Pilkey 2008) occurs for a mesh of 40 k elements ($K_{t,FEM} = 2.214$; $K_{t,theory} = 2.2$). However, acceptable correspondences take place for coarser meshes as well. A mesh of 29 k elements is chosen to be a satisfactory compromise between precision and numerical effort and is thus used for FM analyses.

As stated before, a baseline crack is inserted in the model by means of the automatic FT. It has the form of an elliptical-arc and it is assumed to exist at the notch root (Fig. 9.42 left). Quarter-point wedge elements are used around the crack front in order to model the stress field singularity. The baseline semi-elliptical crack, used for subsequent morphing actions, was set according to the following parameters: $a = 1.6\,\text{mm}$, $\alpha = 1$, $LCR = 0.3\,\text{mm}$, where a is the crack depth, $\alpha = a/b$ is the crack aspect ratio and LCR is the largest contour radius of wedge elements around the crack front.

The dimensionless quantities are introduced in order to generalize the investigation. Under a tensile load of 30 kN a static analysis is run where pointwise values of the SIFs are the desired output.

In the present case, the SIFs correspond to K_I mode I of loading (Anderson 2017). A dimensionless curvilinear abscissa and dimensionless SIF, normalised with respect to h (Fig. 9.42 left) and reference stress σ_F respectively are defined as follows:

$$\varsigma^* = \frac{\varsigma}{h} \tag{9.7}$$

$$K_I^* = \frac{K_I}{\sigma_F\sqrt{\pi a}} \tag{9.8}$$

where:

$$\sigma_F = \frac{2F}{\pi D_0^2} \tag{9.9}$$

The first check entails a comparison of dimensionless SIFs along ς^* for different values of the aspect ratio $\alpha = \frac{a}{b}$. A series of flaw geometries are built by means of the FT and, separately, obtained from the baseline configuration by a morphing action. Homologous SIFs distribution are compared pairwise with a perfect match (Fig. 9.43 left).

Figure 9.43 right shows the dimensionless SIFs for different aspect ratios. The obtained results are in good agreement with those presented by Carpinteri et al. (2003).

It is useful to provide detailed information on the way the morphing operation was conducted in view of avoiding an unacceptable degradation of mesh quality. As stated before, wedge elements constitute the appropriate mesh topology to catch the singularity of the stress field. They are arranged in circles with their tips converging on the line of the crack front. The intersection of the tube of elements along the flaw with the crack plane provides three curves: the central is the trace of the crack front, a tube radius shifts the inner and the outer offset lines from the centre (Fig. 9.44 left).

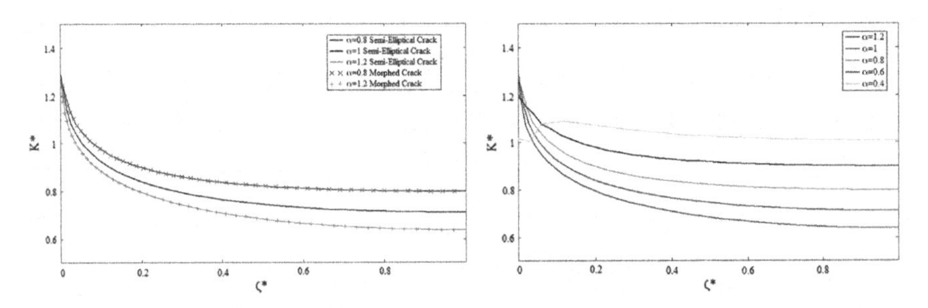

Fig. 9.43 Comparison of dimensionless SIFs obtained with FT and morphing the baseline configuration (left); dimensionless SIFs for Semi-elliptical cracks obtained with ANSYS FT (right)

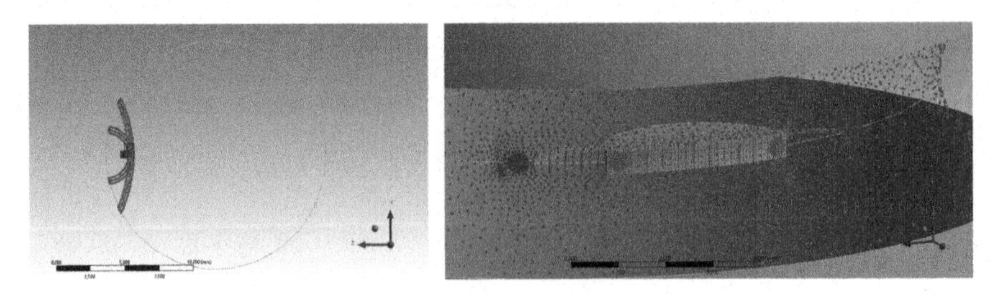

Fig. 9.44 Auxiliary circle geometries used as support for morphing actions (left); nodes preview of mesh morphing (right)

Table 9.5 Crack shape dimensions for parametric analysis

a [mm]	b [mm]	α [.]	SIF_{max} [MPa $\sqrt{mm}$]	SIF_{min} [MPa $\sqrt{mm}$]
1.30	1.30	1.00	400.24	227.95
1.30	1.95	0.66	340.16	262.24
1.30	3.90	0.33	340.93	203.33
1.30	13.00	0.10	356.86	144.48
1.40	1.40	1.00	396.69	237.33
1.40	2.10	0.66	348.92	271.29
1.40	4.20	0.33	350.55	211.02
1.40	14.00	0.10	365.17	156.13
1.60	1.60	1.00	433.75	238.47
1.60	2.40	0.66	368.85	275.16
1.60	4.81	0.33	368.49	238.08
1.60	16.00	0.10	381.13	178.96
1.70	1.70	1.00	394.92	263.64
1.70	2.55	0.66	375.78	296.97
1.70	5.11	0.33	379.40	235.43
1.70	17.00	0.10	390.08	188.27
1.80	1.80	1.00	452.18	245.86
1.80	2.70	0.66	389.13	284.09
1.80	5.40	0.33	386.92	261.38
1.80	18.00	0.10	398.12	201.58

The displacement field that turns the baseline configuration into a morphed one is assigned relying on these curves. Three new lines on the crack plane identify a new configuration of the flaw. The RBF field is the one that moves each baseline curve onto its new position.

A parametric study starting from a baseline configuration was performed in order to assess the developed approach. Crack shape was modified varying its aspect ratio and its dimensions as reported in Table 9.5. The aim is to push the morphing action to the limit identified by mesh quality after morphing.

Table 9.5 shows that, with a proper strategy, mesh morphing can impose large displacement to the mesh preserving numerical stability. The retrieved results are in line with the expected trends and all the analyses successfully ended up despite mesh deformation. In Fig. 9.45 it is possible to notice the baseline configuration and the mesh after the morphing action.

The crack growth follows Paris-Erdogan law reported in Eq. (9.10) where C and m are material dependent coefficients. In the present case study, the values of such constants were extracted from AFCEN (2007). In particular, $C = 7.5 \times 10^{-10}$ and $m = 4$. A zero-to-maximum load fluctuation (i.e. $\sigma_{min} = 0$) is assumed to compute

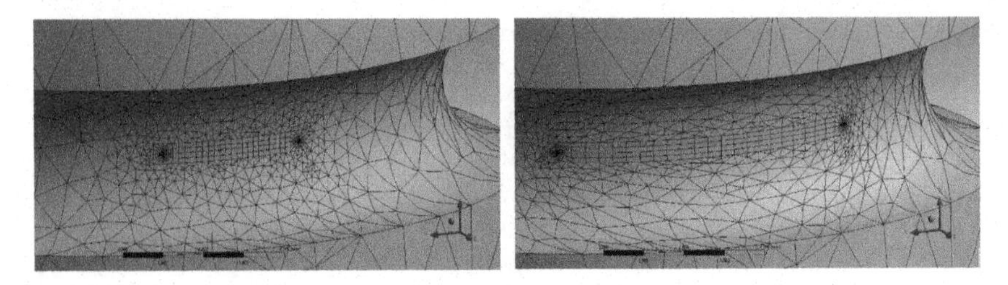

Fig. 9.45 Baseline crack front in the notched bar (right); deformed crack after morphing action (right)

fatigue parameters for the bar. In this case ΔK_{eff} is equal to the K_I for each node of the crack front.

$$\frac{da}{dN} = C\left(\Delta K_{eff}\right)^{m} \tag{9.10}$$

A direct approach to simulate flaw propagation is to update the crack front according to the distribution of the SIFs. The results previously reported showed that the maximum value of K_I is attained at a near-surface point, while its minimum value is encountered at the deepest point. This observation suggests a two-parameter model. As a circular arc approximates the crack profile (Fig. 9.46 left), its centre can move along the symmetry axis of the cross section in order to represent different aspect ratios. Three points define the circumference providing the arc: two lying on the perimeter of the reduced cross-section (points A and C) and a third point on the symmetry axis (point B).

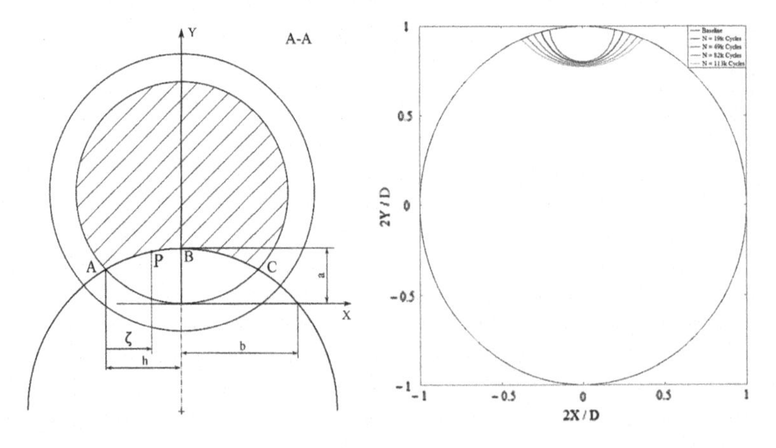

Fig. 9.46 Circular crack front geometry (left); crack front advancement (right)

Table 9.6 Crack front dimensions evolution and related number of cycles

a [mm]	B [mm]	α [.]	N_{cyc} [k cycles]	SIF_{max} [MPa $\sqrt{mm}$]	SIF_{min} [MPa $\sqrt{mm}$]
1.60	1.60	1.00	0	433.75	238.47
1.65	2.15	0.76	18.83	385.89	265.31
1.71	2.56	0.63	48.90	377.55	288.93
1.77	3.27	0.54	81.71	381.82	310.61
1.82	3.48	0.47	113.08		

A FEM analysis supplies the maximum and the minimum values of the SIFs distribution along the crack front. All data necessary to update the flaw profile are thus available. The two outer points of the arc move a given distance δ forward, along the perimeter of the notched section. The advancement of the symmetry point is:

$$\Delta \mathrm{a} = \delta \left(\frac{K_{min}}{K_{max}} \right)^{m} \tag{9.11}$$

The corresponding increment of the number of fatigue cycles is:

$$\Delta \mathrm{N} = \frac{\delta}{C(K_{max})^{m}} \tag{9.12}$$

The procedure recurs until a desired N is reached. Figure 9.46 right shows the crack advancement obtained with this method. Starting a circular 1.6 × 1.6 mm configuration the crack front moves into the notched section and after 113 k cycles assumes 1.82 × 3.87 mm dimensions. All the steps between baseline and ending crack shapes are collected in Table 9.6 through which is possible to notice how the crack aspect ratio decreases with the flaw growth.

9.7 Biological Growth Method

The biological growth method (BGM) is a shape optimisation method for the stress of structural parts based on the biological growth analogy proposed by Heywood (1969) and Mattheck and Burkhardt (1990), which consists in adding material where stresses are high and in removing it where they are low.

The rationale behind this heuristic (but effective) analogy is that biological structures as tree trunks and animal bones evolve by adding new layer of materials at surface with a stress promoted growth rate, i.e. the higher is the stress the more is the material added to withstand its effect. The concept can be extended including material removal on areas at low stress level; according to the last approach Heywood used a photoelastic technique to locally adjust the boundary of an unsatisfactory design to obtain a uniform stress along the boundary of a stress concentrator. Such an approach allowed reducing significantly the stress concentration factor K_t values for optimised fillet profiles in the Rolls-Royce “Merlin” aero engine. Mattheck and Burkhardt (1990) presented a 2D FEA study capable to predict the shape evolution observed in two framework tree structures naturally grown with self rounded notches and proposed this concept as a CAE based optimisation approach demonstrating the optimisation of a plate with a circular hole and of a chain link.

It is worth to notice that whilst loaded biological structures can grow only driven by local stress, when the analogy is exploited as a design tool, according to the equation (9.13), the method can act by adding or removing material. In the work of Mattheck and Burkhardt (1990) the local volumetric growth rate $\dot{\varepsilon}_v$ on a soft layer of element at the surface (i.e. a volume that mimics a tree trunk growth ring) is computed according to the von Mises stress (σ_{Mises}) and a threshold stress $\left(\sigma_{ref}\right)$ to be tuned on the basis of allowable stress for the specific design.

$$\dot{\varepsilon}_v = k\left(\sigma_{Mises} - \sigma_{ref}\right) \tag{9.13}$$

A refined evolution of the method can be found in the study of Waldman and Heller (2015) where the full implementation of a BGM method suitable for shape optimisation of holes in airframe structures with multiple stress peak locations is given. The driving force formula (9.13) evolved in a more complex one that is ruled through the relationship (9.14):

$$\begin{gathered} d_i^j = \left(\frac{\sigma_i^j - \sigma_{th}^j}{\sigma_{th}^j}\right) s \cdot c, \quad \sigma_{th}^j = \max\left(\sigma_i^j\right) \ \ if \ \ \sigma_i^j > 0 \\ or \ \ \sigma_{th}^j = \min\left(\sigma_i^j\right) \ \ if \ \ \sigma_i^j < 0 \end{gathered} \tag{9.14}$$

The FEA implementation published by Waldman and Heller is based on a custom-written FORTRAN 90 program: during each iteration the stresses are computed at all the nodes around the hole boundary. The i-th boundary node, which is located in the j-th stress subregion, is then moved in the direction of the local outward normal to the boundary by a distance d_i^j. This distance is computed using (9.14) where σ_i^j is the tangential stress at node i in the j-th stress subregion, σ_{th}^j is the stress threshold corresponding to the peak positive or negative stress occurring in the j-th stress subregion, c is an arbitrary characteristic length, and s is a step size scaling factor. Due to the selection of the stress threshold, in all cases there is a material removal, so the optimal shape will encompass the initial starting shape, which is advantageous in the context of shape reworking of an existing component.

One of the challenge posed by the implementation of BGM is the update of the mesh onto the new shape predicted at each iteration. Mattheck and Burkhardt (1990) proposed a FEA solver based mesh morphing implementation that accommodates the shape of the soft layer using thermal expansion, whereas Waldman and Heller (2015) method is based on mapped remeshing.

The examples presented in this section are based on the evolutionary mesh morphing principles already described in this chapter implemented by coupling the ACT tool of ANSYS Mechanical solver, for the computation of displacement field to be applied at the surface of the part, with the RBF Morph ACT Extension, for the updating of the volume mesh.

9.7.1 Stress Reduction on a Cantilever Beam

The BGM effectiveness concept is here showcased by a simple example: a cantilever beam loaded with a lumped load at the free edge. The geometry and the results in terms of von Mises stresses are shown in Fig. 9.47 where the solid model

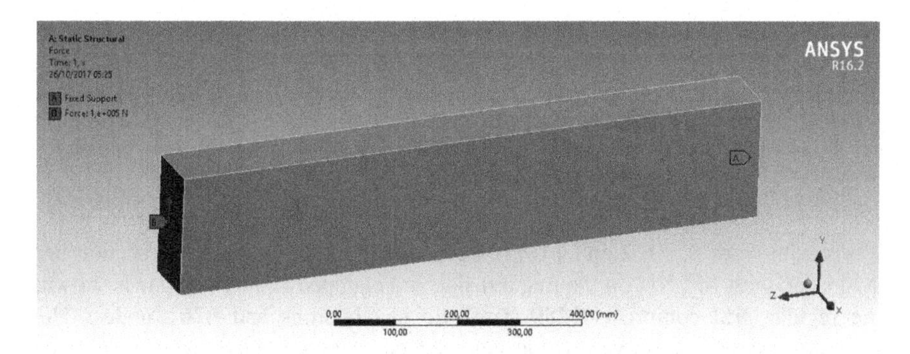

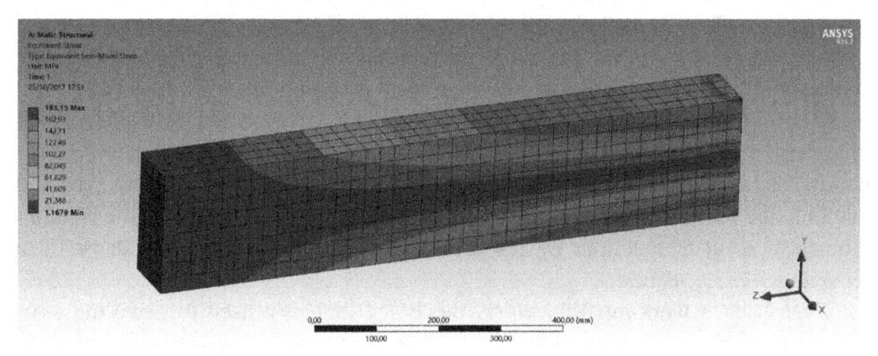

Fig. 9.47 FEA model of the cantilever beam. A box with sides 0.1, 0.2, 1.0 m is used to prepare the mesh, the left face is loaded with 100 kN the right one is fixed. Von Mises stress highlights the unloaded regions coloured with blue (darker in B&W) shades

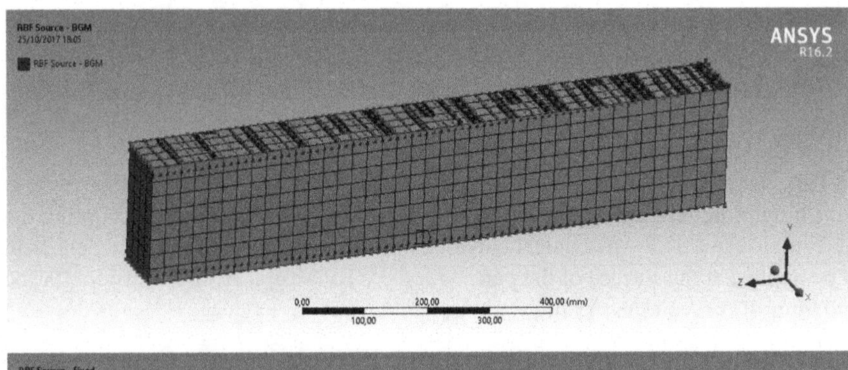

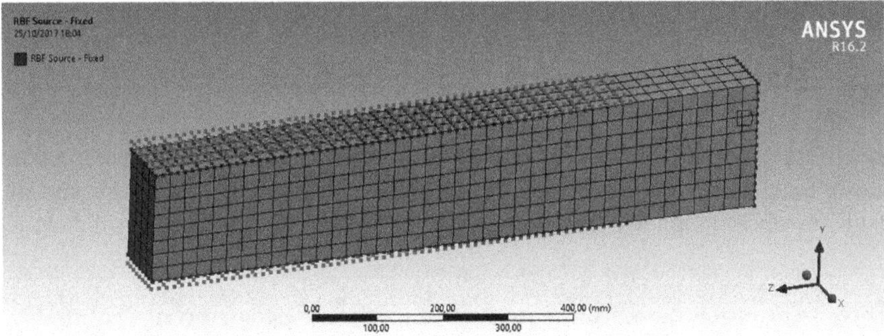

Fig. 9.48 BGM information are computed at top and bottom faces using a threshold stress of 140 MPa, the noise observed at the region close to the clamped edge has been excluded keeping only the nodes that are 0.2 m apart from the constraint. The BGM constant is tuned to limit the maximum deformation of a single optimisation step

(a box with sides 0.1, 0.2 and 1.0 m) representing a rectangular cross section bar and loaded with 100 KN on the free left side and clamped at the right one is shown. The regular grid comprises 1280 parabolic hexahedrons and 6761 nodes. The maximum stress is observed at the clamped edge, as expected, where the three dimensional effect of the solution is noticed.

It is worth to say that the representation of the clamped edge obtained by constraining all the nodes at the face is an ideal assumption and it introduces local peaks at constrained surface. The load at the edge, represented as an ideal load uniformly distributed on the free edge face, allows to have a regular stress distribution. Stress analysis clearly demonstrates that the neutral axis is stress free and that only the portion of the beam close to the clamped edge is loaded as the bending moment is a linear function. Basics stress theory predicts a parabolic shape for a rectangular cross section.

Given such a working framework, the BGM has been used to drive the optimisation. As it can be noticed in Fig. 9.48, using the BGM information the RBF

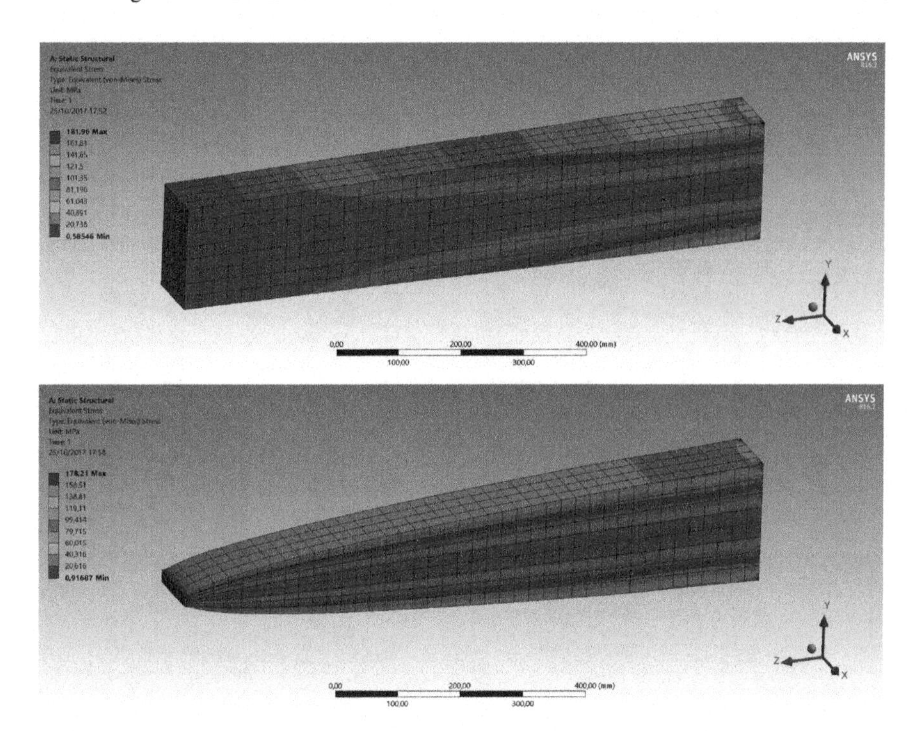

Fig. 9.49 The optimal shape is gained after 8 iterations. Peak stress is not changed and an almost uniform stress is gained on the top and bottom faces. A volume reduction equal to 32.6% is registered

set-up is not directly applied on the full face to avoid the noisy signal at nodes close to the surface, but an intermediate RBF is adopted to exclude the first 0.2 m of the beam, i.e. the height of the cross section. BGM parameters are defined using 140 MPa as a threshold stress that produces all the sculpting inward in the controlled area while allowing to have a small deformation at the end of the controlled surface. The morphing is applied using the linear RBF ($\varphi(r) = r$), and the surface not controlled is gently deformed up to the clamped edge that is fixed in the RBF set-up.

Shape optimisation results are presented in Fig. 9.49. In particular, the comparison between the original one (top) and the optimised model resulting after 8 BGM iterations is depicted. As visible, stress distribution has been regularized and the top stress at the faces reach the target value of 140 MPa. As expected the optimal beam is lighter and it is characterised about by 33% of volume saving.

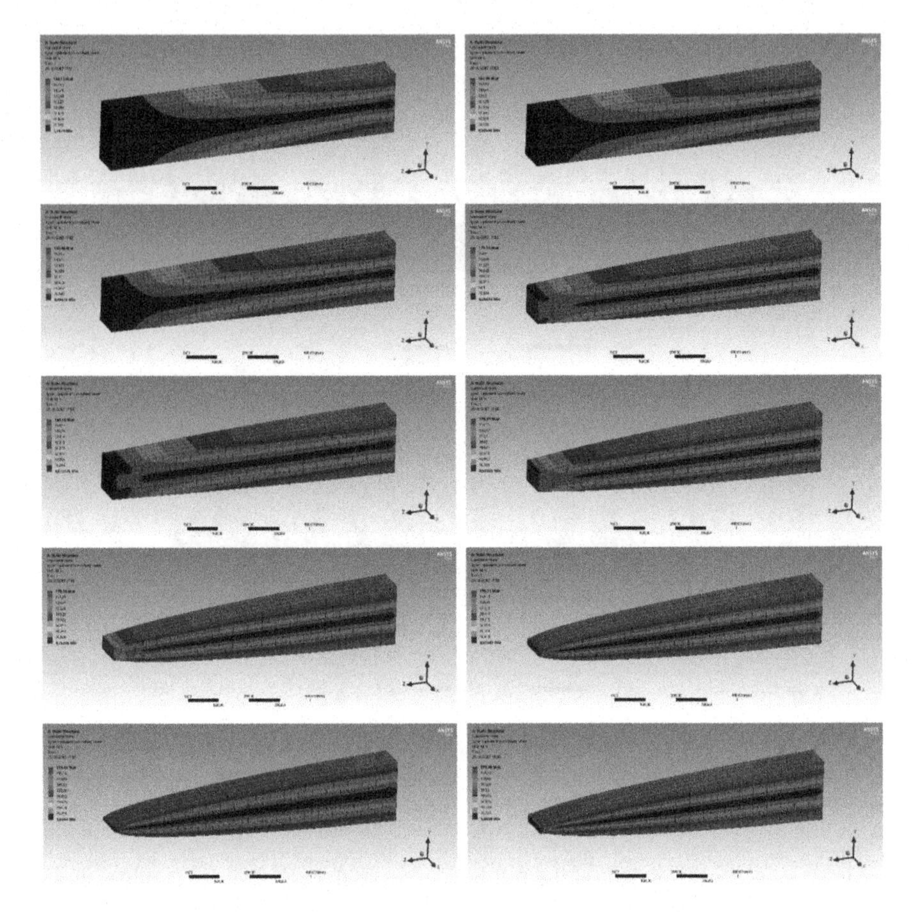

Fig. 9.50 Full BGM evolution. At each iteration the mesh resulting from the previous one is used for calculating the stress solution and for updating the BGM field. Step 8 is still valid and adopted as the final result. At Step 9 an un-physical stress at the free edge is produced because of mesh collapsing. At Step 10 the shape is recovered

A summary of shape evolution across all the iterations is given in Fig. 9.50. The maximum displacement at each step is constant and at the ninth step a bad mesh is generated. The mesh is automatically recovered at step 10 but this is not a rule. As for other evolutionary sculpting methods (see for instance adjoint automatic sculpting of Sect. 8.3), it is really important to properly set-up the amount of variation at each optimisation cycle and the stop criteria.

9.7.2 *Stress Reduction at a Turbine Blade Fillet*

A more representative design scenario is represented in Fig. 9.51, where simplified boundary conditions are applied to a turbine blade. The mesh morphing set-up for the volume mesh is explained in Fig. 9.52.

The morphing domain is limited to the volume mesh surrounding the fillet; fixed nodes are added to set a null morphing at the boundary of the morphing domain.

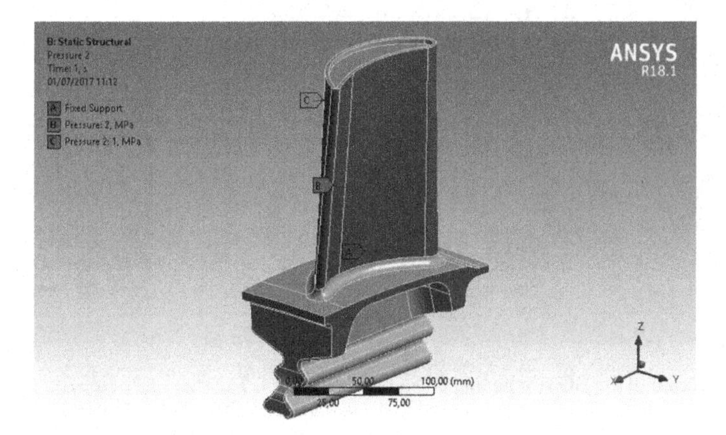

Fig. 9.51 Blade model geometry with boundary conditions highlighted

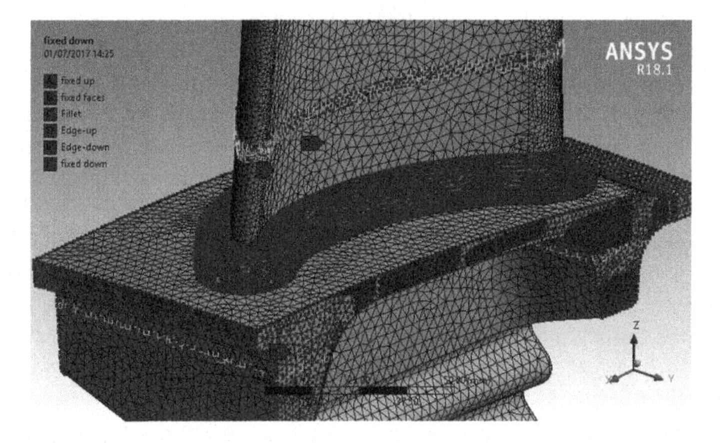

Fig. 9.52 Mesh morphing set-up for the volume mesh

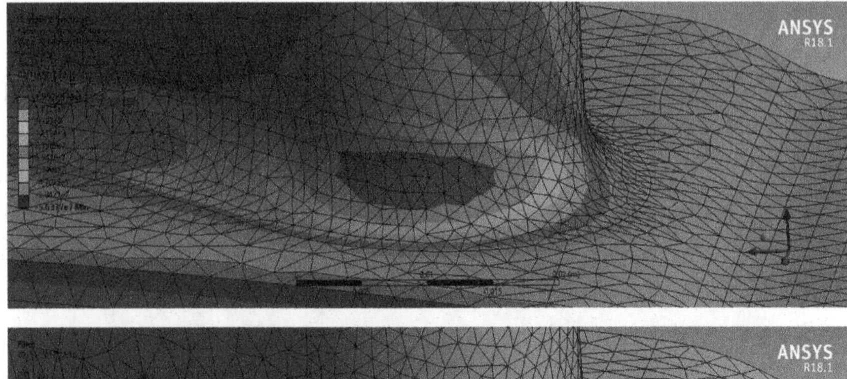

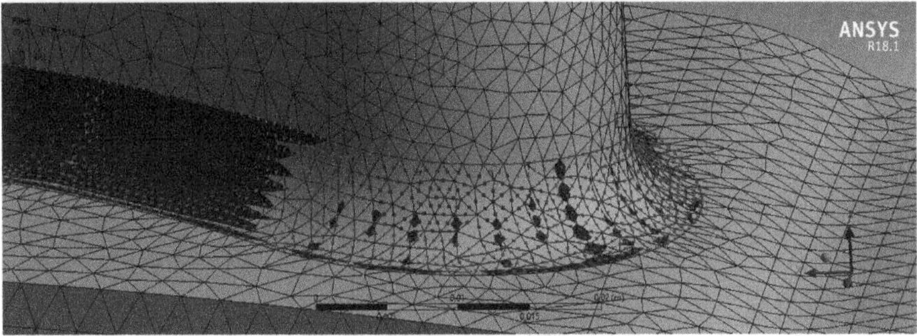

Fig. 9.53 Detail of the BGM sculpting effect on the hot spot. The stress map (top) on the surface is used to compute the displacement normal to the surface with a threshold equal to 150 MPa. The preview (bottom) allows to identify the growth region where the blue points are visible (Color figure online)

The fillet surface is controlled using BGM. The von Mises stress is used to compute the displacement normal to the surface with a threshold value equal to 150 MPa.

In Fig. 9.53 a magnified view of the hot spot is used to show the effect of the BGM sculpting. Results are summarized in Fig. 9.54 where the evolution after 3 iterations is summarized. The first stress map refers to the baseline geometry. It is worth to notice that the best result is gained after two iterations since at the third a small stress increment is noticed.

This could be due to the fact that in this example only the nodes at the fillet are controlled with BGM but the stress redistribution with updated shape moves the hot spot on the aerofoil region. A stress reduction of about 20 MPa (i.e. about 10% with respect to the initial value) is obtained.

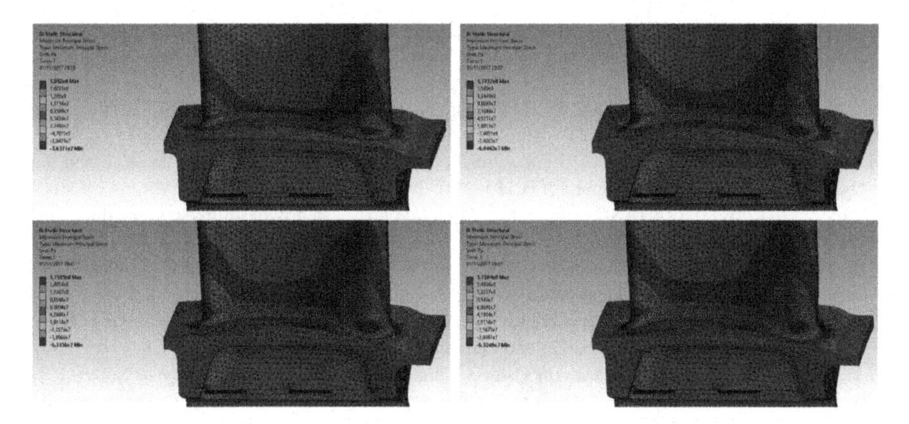

Fig. 9.54 Maximum principal stress at the fillet. From top to bottom: original stress solution and then stress evolution after 3 BGM cycles

References

ANSYS Inc. (2017) ANSYS Fluent User's Guide

AFCEN RM (2007) Design and construction rules for mechanical components of nuclear installations. Section 1—Subsection Z, Appendix A16: Guide for Leak Before Break Analysis and Defect Assessment

Allain E, Paradot N, Ribourg M, Delpech P, Bouchet JP, De La Casa X, Pauline J (2014) Experimental and numerical study of snow accumulation on a high-speed train. In: 49th international symposium of applied aerodynamics, Lille, 24-25-26 March. FP18-2014-ALLAIN

Anderson TL (2017) Fracture mechanics: fundamentals and applications. CRC press

Appiah-Kubi P (2011) U.S Inflight Icing accidents and Incidents, 2006–2010. Master's Thesis, University of Tennessee, 2011

Biancolini ME (2014) How to boost fluent adjoint using RBF Morph. In: Automotive simulation world congress 2014 on 9–10 October 2014 in Tokyo, Japan

Biancolini ME, Chiappa A, Giorgetti F, Porziani S, Rochette M (2017) Radial basis functions mesh morphing for the analysis of cracks. Procedia Struct Integrity 8(2018):433–443, ISSN 2452-3216. https://doi.org/10.1016/j.prostr.2017.12.043

Carpinteri A, Brighenti R, Spagnoli A, Vantadori S (2003) Fatigue growth of surface cracks in notched round bars. In: Proceedings of Fatigue Crack Path. Parma, Italy

Chwalowski P, Florance JP, Heeg J, Wieseman CD, Perry B (2011) Preliminary computational analysis of the HIRENASD configuration in preparation for the aeroelastic prediction workshop. In: International forum of aeroelasticity and structural dynamics (IFASD), IFASD 2011-108, Paris, France

Cole J, Sand W (1991) Statistical study of aircraft icing accidents. In: 29th aerospace sciences meeting and exhibit, Reno, NV, AIAA, AIAA Paper 91-0558, 10 p

Costa E, Biancolini ME, Groth C, Travostino G, D'Agostini G (2014) Reliable mesh morphing approach to handle icing simulations on complex models. In: Proceedings of the 4th EASN association international workshop on flight physics and aircraft design, EASN, Aachen, Germany

European Commission (2013) RBF4AERO Project, http://cordis.europa.eu/project/rcn/109141_en.html

Galland F, Gravouil A, Malvesin E, Rochette M (2011) A global model reduction approach for 3D fatigue crack growth with confined plasticity. Comput Methods Appl Mech Eng 200(5):699–716

Heywood RB (1969) Photoelasticity for designers (chapter 11: improvement of designs). Pergamon Press

Mattheck C, Burkhardt S (1990) A new method of structural shape optimisation based on biological growth. Int J Fatigue 12(3):185–190

NASA (2012a) HiReNASD model. https://c3.ndc.nasa.gov/dashlink/static/media/other/HIRENASD_base_legacy.htm

NASA (2012b) 1st AIAA Aeroelastic Prediction Workshop, https://c3.ndc.nasa.gov/dashlink/static/media/other/AEPW_legacy.htm

Petty KR, Floyd CDJ (2004) A statistical review of aviation airframe icing accidents in the U.S. In: 11th conference on aviation, range, and aerospace, October 2004, Hyannis, MA

Pilkey WD, Pilkey DF (2008) Peterson's stress concentration factors. John Wiley & Sons

Riccio C, Simms NJ, Oakey JE (2017) Predicting deposition onto superheater tubes in biomass and coal-fired combustion system, Conference Paper

Trenker M, Payer W (2006) Investigation of snow particle transportation and accretion on vehicles. In: 24th applied aerodynamics conference, 5–8 June, San Francisco, California, AIAA 2006-3648

Waldman W, Heller M (2015) Shape optimisation of holes in loaded plates by minimisation of multiple stress peaks. Defence Science and Technology Organisation Fishermans Bend (Australia) Aerospace Div, http://www.dtic.mil/docs/citations/ADA618562

Wright WB (2002) User manual for the NASA Glenn Ice Accretion Code LEWICE. Version 2 (2):2

Chapter 10
FSI Workflow Using Advanced RBF Mesh Morphing

Abstract Advanced workflow for fluid structure interaction (FSI) modelling using computer-aided engineering (CAE) tools suitable for the simulation of fluids and structures, that typically are related to computational fluid dynamics (CFD) and computational structural mechanics (CSM) techniques respectively, are demonstrated in the present chapter. In this context, RBF are adopted to interface structure and fluid. In the two-way approach the loads computed using CFD (pressures and shear forces) are transferred to the structure using RBF interpolation for the mapping at surfaces (the mapping topic is further deepened in Chap. 13), whilst deformation computed using CSM are then transferred to the CFD mesh using mesh morphing. The latter approach can be also used to transfer the modal shapes, computed using eigenvalues extraction performed through a finite element analysis (FEA) for instance, on the CFD mesh. The effectiveness of the two possible FSI approaches is demonstrated with practical applications pertaining to aeronautical and motorsport fields. The reported FSI implementation can be used to tackle both steady and transient problems. The chapter is concluded showing how the method can handle vortex induced vibrations of a wing in water and the transient effect due to the separation of a store from the wing of an aircraft.

10.1 Importance of FSI in Technical Applications

The demand for developing multi-disciplinary approaches using high fidelity CAE methods is today strongly rising in a widespread range of technical fields including aerospace, automotive, marine, product manufacturing and healthcare to name a few. This is even more valid with the vision of modern design methods which is strongly oriented to work embedded in reliable numerical optimization procedures. The core of a multi-physics numerical investigation is the coupled-field analysis, which lets analysts to determine the combined effects of multiple physical phenomena as in the case of fluid–structure, thermal–mechanical and electric–thermal interaction.

M. E. Biancolini, *Fast Radial Basis Functions for Engineering Applications*,
https://doi.org/10.1007/978-3-319-75011-8_10

In particular, FSI is the interaction of movable or deformable structures with an internal or a surrounding fluid flow (Bungartz and Schäfer 2006) occurring at different length scales. Such an interaction may be the working principle of the component itself (reed valves action, parachute canopy unfolding, movement of a sheet of paper within a printing device) or may be exploited to finely tune components manufacturing in view of lightening a structure as in case of aircraft design.

FSI is a typical multi-physics phenomenon which computational reproduction implies, at least, the resolution of both the structural and fluid-dynamic task and, when temperature effects are relevant, the thermal one as well. In general, the approaches to accomplish its numerical solution can be roughly grouped depending upon governing equations solution approach (monolithic and partitioned methods) and upon the treatment of meshes (matching and non-matching mesh methods) (Gene et al. 2012). Besides, FSI perspectives (Costa 2012) may vary depending on types of flow fields covered (such as compressible, incompressible, laminar, turbulent), types of applications, structural fields (such as thin-walled, rigid bodies, non-linear material), discretization schemes (such as finite volume, spectral methods, multi-body dynamics), flow modelling assumptions (such as continuum, statistical Lattice Boltzmann distribution) and calculation grid treatment (such as moving grid, fixed grid, immersed boundary).

Whatever the particular scenario of the study, the FSI analysis introduces an high level of complexity in the solution achievement and, as such, either the fluid forces or structural deformations are often neglected. However, since this mechanism turns out to be crucial in many cases such a simplification is not acceptable at all from design point of view and, thus, both physical aspects need to be suitably accounted. Given that, over last decades sensible efforts have been done to safysfy such a requirement.

10.2 FSI Simulation Through RBF Mesh Morphing

Both steady state and unsteady FSI analyses can be efficiently handled through RBF mesh morphing adopting the following approaches:

1. Two-way;
2. Mode-superposition (also termed modal superposition).

The two-way FSI approach foresees the exchange of data between CSM and CFD models. In particular, the loads computed by means of the CFD model, that may comprise pressure, shear stresses and temperature, are transferred to the CSM model, typically through a mapping technique (see Sect. 13.2), in order to determine the displacement field of the deformable structures. Successively, these displacements are used to create the RBF field to morph the CFD model in the respect of the CSM one to compute again the CFD loads in the updated configuration. This loop is commonly repeated for several times to accomplish a steady FSI study, whilst it can be run lots of times for performing unsteady FSI analyses.

On the other hand, the mode-superposition approach is based on a preliminary structural modal analysis from which vibrations data, which consist in a certain number of natural modes and frequencies, are acquired such that they are used in the creation of a morphing criterion for the CFD domain. The mesh is thus made parametric on modal coordinates and updated using a mesh morphing tool during the progress of the CFD computation. An intrinsically elastic numerical model is then enabled and no further iteration with the structural solver is required. During the CFD computing, the modal forces are extracted by integrating the loads on the wall boundaries (wetted surfaces) of the deformable parts of the studied system and employed to define the weights of the morphing action of each of the retained modal shapes.

Both methods have specific advantages and drawbacks that will be deepened in the following sections.

10.3 Two-Way FSI Through RBF Mesh Morphing

10.3.1 Two-Way FSI Theoretical Background and Characterization

The already introduced two-way approach is the one commonly adopted to numerically solve an FSI problem using high fidelity numerical models because it does not impose theoretical limits to such a computational investigation.

On the contrary, the procedure needed to be set to enable such a type of study is rather complex and solver-dependent. As a matter of fact, several technical aspects, related to the data exchange between the numerical methods, introduce complexities and uncertainties in its set-up.

The first element of complication is the need for the bidirectional mapping of data among the common not-matching surface meshes (wetted surfaces) of the CSM and CFD model, topic that will be detailed in Chap. 13. This operation, typically introduces an error in data transfer.

The second problem to handle is that several actions such as solutions extraction, files format conversion, set-up update, run management as well as solutions quality check, are required to couple the CSM and CFD environments. Moreover, if the analysis has to be included in an optimization loop, routines controlling the procedure in an automatic process have to be implemented.

Another aspect to tackle is that the adaptation of the CFD domain to the shape of the deformed model estimated by the structural code, that typically has to be performed for several times for a steady study and even at every time step for unsteady analyses, is a critical process that may degrade the quality of the computational grid. Given that, the necessity to propagate the known deformed surfaces into the volume mesh requires the use of efficient and robust algorithms.

10.3.2 *Two-Way FSI Through RBF Mesh Morphing*

To run the two-way FSI numerical process the capability of mesh morphing to adapt the shape of deformable parts of the CFD model in the respect of the FEM displacement caused by the fluid-dynamics loads, can be fruitfully exploited. In such a manner, it is possible to automate the cyclic exchange of data between the numerical models making them congruent with regards to the deformed shape of the analysed system. This operation is the answer given by RBF mesh morphing to the latter problem identified in the previous section referring to CFD mesh deformation.

Figure 10.1 sketches the typical workflow of a steady and unsteady FSI analysis, based on the two-way approach and performed by means of mesh morphing.

In the two-way FSI the mesh morphing task is accomplished by assigning the displacements of the deformable parts calculated through the CSM model by means of source points that are typically generated using the coordinates of its surface mesh nodes. The calculated displacements need to be arranged in the respect of a format allowing the processing of these data by the morpher tool. To complete the RBF field, other two settings are created in the same RBF morphing solution: one to maintain unchanged the position of mesh nodes of the parts that are requested to remain fixed, and another to delimit the action of morphing in the volume where the deformation will take place. The basic principles for the definition of a well-posed RBF points' arrangement suitable for high quality mesh morphing are covered in Chap. 6, where an overview of its use in multi-physics is given in Sect. 6.5.

Whatever the type of study, namely steady or unsteady, the RBF field has the same framework and only the settings generated from the structural displacements change. Relating to the solution achievement, for the steady state case few FSI cycles are commonly enough to get a converged deformed solution, whilst for the transient case the number of cycles is dictated by the number of time steps of the simulation itself and the frequency selected to update the deformation of the CFD model.

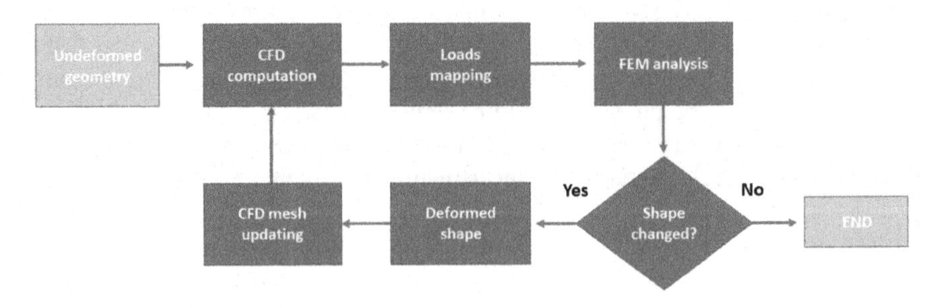

Fig. 10.1 Mesh morphing based workflow of two-way FSI

10.4 Mode-Superposition FSI Through RBF Mesh Morphing

10.4.1 Mode-Superposition FSI Theoretical Background

Modal analysis is a well-established theory of structural mechanics, applicable to both continuum and discrete systems, consisting of the determination of the vibration characteristics of a structure, namely its undamped free vibration modes each identified by a natural pattern (mode shape) and frequency. This framework makes use of the modal coordinates vector $\{q\}$, whose evaluation enables to calculate the deformation expressed in natural coordinates of the studied structure $\{x\}$ exploiting the following relationship (Zienkiewicz et al. 2005):

$$\{q\} = [X]^{-1}\{x\} \tag{10.1}$$

where $[X]$ is the modal matrix whose columns are identified by the eigenvectors of undamped free vibration solution of the dynamic equation normalized with respect to mass.

In fact, assuming that nonlinearities are not present, it can be demonstrated that the dynamic behaviour of a discrete mechanical system can be expressed in terms of modal coordinates as follows:

$$\ddot{q}_i + 2\varsigma_i\omega_{n,i}\dot{q}_i + \omega_{n,i}^2 q_i = \frac{F_i(t)}{M_{ii}} \tag{10.2}$$

where $\ddot{q}_i$ is the modal acceleration, $\dot{q}_i$ is the modal velocity, whilst $\omega_{n,i}$ and ς_i respectively are the modal circular frequency and the damping factor provided by the following formula:

$$\omega_{n,i} = \sqrt{\frac{K_{ii}}{M_{ii}}} \quad \varsigma_i = \frac{C_{ii}}{2M_{ii}\omega_{n,i}}$$

where K_{ii} is the modal stiffness, M_{ii} the modal mass (equal to 1 if modes are normalized with respect to mass), $F_i(t)$ is the modal force that depends on the CFD loads basically related to pressure and shear stresses acting on elastic structures. It is worth to underline that modal mass and stiffness are diagonal matrixes.

In the case the physical phenomenon of FSI can be considered as static (or steady state) the Eq. (10.2) can be written as:

$$\omega_{n,i}^2 q_i = \frac{F_i}{M_{ii}}$$

that, in turn, can be posed as:

$$q_i = \frac{F_i}{K_{ii}} \tag{10.3}$$

Taking into account the Eqs. (10.1) and (10.3), nodal displacement vector can thus be expressed as follows:

$$\{x\} = \sum_{i=1}^{N} \{X_i\} \frac{F_i}{K_{ii}} = \sum_{i=1}^{N} \{X_i\} q_i. \tag{10.4}$$

In other terms, the whole deformation of the structure in steady state conditions can be determined by linearly combining a certain number of modes (retained modes) using their modal coordinates as weights.

In the case that the physical phenomenon of FSI cannot be considered as static, the entire Eq. (10.2) is considered instead. In this design scenario, the solution can be written as follows:

$$\begin{aligned} q(t) =& \mathrm{e}^{-\varsigma\omega_n t}\left[q_0 \cos(\omega_d t) + \frac{\dot{q}_0 + \varsigma\omega_n q_0}{\omega_d}\sin(\omega_d t)\right] \\ &+ e^{-\varsigma\omega_n t}\left\{\frac{1}{m\omega_d}\int_0^t e^{-\frac{b(t-\tau)}{2m}} f(\tau)\sin[\omega_d(t-\tau)]dx\right\} \end{aligned} \tag{10.5}$$

where ω_d are the damped circular frequencies of the system having the following expression:

$$\omega_d = \omega_n\sqrt{1-\varsigma^2} \tag{10.6}$$

and q_0 and $\dot{q}_0$ are the initial conditions (modal coordinates and modal velocities) of the system at the initial instant of computing.

The formulation of the integral of Eq. (10.5), known as Duhamel integral, assumes that the reaction of a system subjected to a force $f(t)$ can be expressed as the summation of the reactions of the system to the single impulses constituting the total force itself. Usually this formulation cannot be applied for a generic numerical analysis. However, given the initial conditions $q(t)$ and $\dot{q}(t)$, if the acting force F is constant within every time step Δt, the Eq. (10.5) has the following form:

$$
\begin{aligned}
q(t+\Delta t) = {} & e^{-\varsigma\omega_n \Delta t}\left[q(t)\cos(\omega_d \Delta t) + \frac{\dot{q}(t) + \varsigma\omega_n q_0}{\omega_d}\sin(\omega_d \Delta t)\right] \\
& + e^{-\varsigma\omega_n \Delta t}\left\{\frac{F(t)}{\omega_d}\left[\frac{4\omega_d}{\varsigma^2\omega_n^2 + 4\omega_d^2}\right.\right. \\
& \left.\left. - e^{-\varsigma\omega_n \Delta t}\frac{2\varsigma\omega_n \sin(\omega_d \Delta t) + 4\omega_d \cos(\omega_d \Delta t)}{\varsigma^2\omega_n^2 + 4\omega_d^2}\right]\right\}.
\end{aligned}
\tag{10.7}
$$

The solution (10.7) can be adopted to update the modal coordinates in a time marching algorithm, a common practice in FEA solvers (MSC 1997). Given that, considering the relationships (10.1) and (10.7), nodal displacement vector in time can thus be expressed as follows:

$$
\{x(t)\} = \sum_{i=1}^{N}\{X_i\}q_i(t). \tag{10.8}
$$

Similarly to the steady case, the whole deformation of the structure in unsteady state conditions can be determined by linearly combining a certain number of modes using their time dependent modal coordinates as weights.

For what described above, once the modal forces F_i are computed in the CFD model, the real displacement of the structure, for both steady and unsteady FSI, can be accordingly calculated using the known data of the CSM modal analysis (natural modes normalized with respect to the mass $\{X_i\}$ and natural frequencies $f_i = \frac{\omega_{n,i}}{2\pi}$).

The mode-superposition approach enjoys several advantages. The first is that only modal data, gained by means of experimental tests, simplified analytical formulations or FEM analyses, are needed from the structural side (loads mapping is not required at all). The second is the improvement in terms of performances its utilisation allows to obtain. This fact is even more relevant for transient analyses for which very high speed up of the whole process can be gained. The main drawback is that the approach cannot manage nonlinearities due to material, contact or large displacements.

Considering the purpose of the mode-superposition technique, just a subset of the first modes (retained modes) is commonly determined and used because the lowest frequency modes have the highest energy levels and are then physically prominent due to the fact that mechanical systems are characteristically low-pass. Such an operation, called modes truncation, allows a favourable reduction of degrees of freedom (DoF) that need to be treated in order to suitably reproduce the vibrational behaviour of the structure. Best practices suggest to involve a number of modes so that the summation of their effective participation factors reaches a threshold value of around 70% of the total mass to properly mimic the actual vibrational behaviour of the entire deformable structure. Such a target is, in general, not necessary for FSI analysis.

10.4.2 Mode-Superposition FSI Through RBF Mesh Morphing

As outlined, mesh morphing can be used to carry out FSI numerical processes based on the mode-superposition approach to mould the surfaces of deformable parts of the CFD model according to the imported vibrational data and the modal forces connected to the fluid-dynamics loads determined during the CFD computing. In the respect of this approach, in a certain sense the CFD numerical model, perfectly rigid by default, is made elastic once the vibrational data are loaded.

In the case a CSM structural solver, ANSYS Fluent and the add-on release of the RBF Morph software are used for the calculation of the vibrational data and the CFD solution, and to generate and apply the RBF fields respectively, the workflow to run mode-superposition FSI through mesh morphing is composed by three stages that are visualised in Fig. 10.2. It is worth to add that such a workflow has a general significance, so particular software can be substituted with others but its validity and consistency remains valid.

In Stage 1 the structural modes and frequencies of the retained modes are evaluated through a FEM model that is requested to have the same position, orientation and units of the CFD one. In the successive Stage 2 the morphing problem is defined. Such a phase foresees that the RBF field for each mode including the imposition of a zero motion to the surfaces considered behaving as rigid is set, and

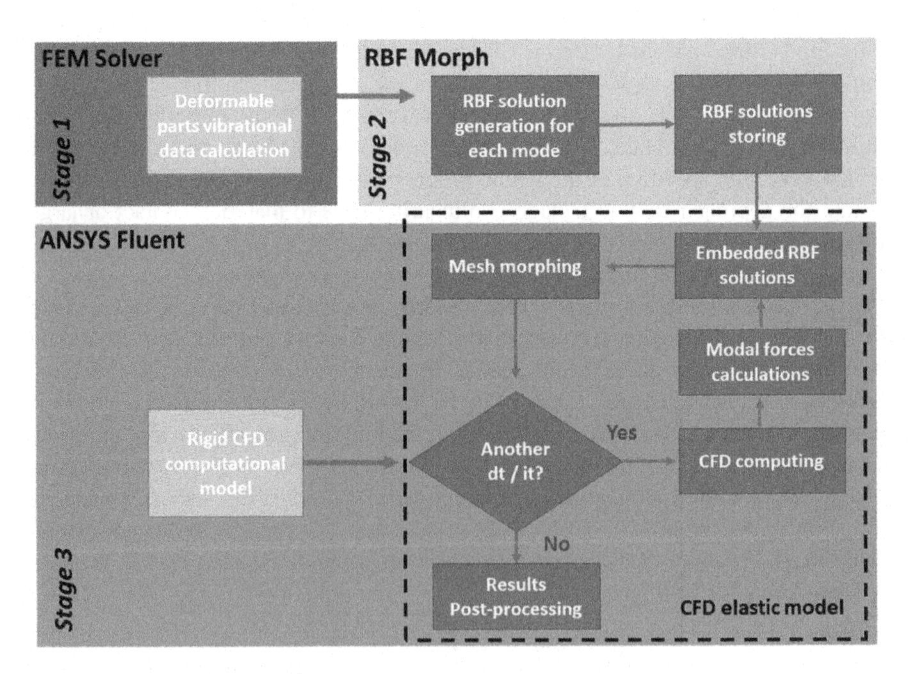

Fig. 10.2 Mode-superposition workflow when a FEM solver, ANSYS Fluent and RBF Morph are used

the generated RBF solutions are saved for the successive use during CFD computing.

In Stage 3 the FSI coupling is enabled and the stored RBF solutions are used for a predefined number of times, depending of the CFD solution (steady or unsteady), to apply morphing.

This crucial operation is straightforwardly executed because the structural response is evaluated directly in the modal space according to Eq. (10.4) for the steady case and (10.8) for the transient one here rewritten as (10.9):

$$\boldsymbol{x_{CFD}} = \boldsymbol{x_{CFD0}} + \sum_{m=1}^{n_{modes}} q_m \boldsymbol{u_{CFD_m}} \tag{10.9}$$

where $\boldsymbol{x_{CFD0}}$ are the positions of the CFD nodes of the undeformed mesh (baseline configuration), q_m are the (unknown) values of modal coordinates and $\boldsymbol{u_{CFD_m}}$ are the modal displacements for the generic retained m-th mode available on the CFD mesh. Since the proposed approach requires the availability of the CFD forces, they are evaluated using nodal conversion; this operation is accomplished by a specific function of RBF Morph which loops over all the faces, calculates the resulting force and accumulates its value at each connected node assuming an uniform distributions. Once nodal forces $\boldsymbol{F_{CFD}}$ are available on the CFD mesh, the integration of the modal forces F results in a simple summation over the nodes (10.10):

$$F_m = \sum_{i=1}^{n_{nodes}} \boldsymbol{u_{CFD\,m_i}} \cdot \boldsymbol{F_{CFD_i}} \tag{10.10}$$

where the m-th modal load is a scalar obtained summing the dot product between the nodal mode displacement and the nodal load of each i node of n_{nodes} nodes of the deformable surface. Considering that a mass normalization criterion was defined for modes extraction, the modal coordinates are computed according to Eq. (10.8):

$$q_m = \frac{F_m}{\omega_m^2} \tag{10.11}$$

and the parametric CFD mesh can adapt its shape on the basis of the actual loads according to relationship (10.12):

$$\boldsymbol{x_{CFD}} = \boldsymbol{x_{CFD0}} + \sum_{m=1}^{n_{modes}} \frac{F_m}{\omega_m^2} \boldsymbol{u_{CFD_m}}. \tag{10.12}$$

The whole approach can be extended to unsteady analyses integrating the complete form of Eq. (10.2) and performing the update at each time step according to (10.7).

10.5 Two-Way FSI Test Case

In the next sections, the application of RBF mesh morphing to perform the steady-state two-way FSI of a test case concerning the aviation sector is described. In particular, the analysed model is the P1xx, a business jet designed and manufactured by Piaggio Aerospace.

*10.5.1 P1*xx Wing

The static aero-elastic analysis described in this section was presented first by Cella and Biancolini (2012) and it was originally performed coupling the commercial Navier–Stokes code CFD++ with the NASTRAN FEM code. A Patran interpolating procedure was adopted to map the aerodynamic wing load onto the non-conformal structural mesh whilst the CFD grid was morphed accounting the FEM grid nodes' displacement by means of the stand-alone version of the RBF Morph software. The same experimental reference was then considered for a second study in Biancolini et al. (2016) where the CFD solver ANSYS Fluent and its built-in mapping procedure were adopted. The latter reference considers both the two-way coupling and mode superposition and it is here considered as a reference.

The wind-tunnel model of the P1xx this test case refers to, was tested in its complete configuration in a transonic facility. During this experimental campaign (see in Fig. 10.3), several corrections were applied in order to obtain the aircraft data in real free-flight conditions.

Fig. 10.3 P1xx case: wind-tunnel model installation

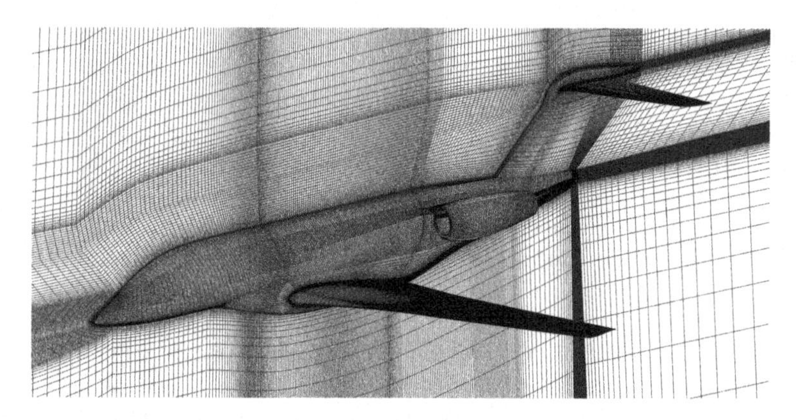

Fig. 10.4 P1xx case: CFD computational grid detail

The multi-block structured hexahedral computational grid has been generated using ANSYS ICEM CFD. All the wall surfaces were set as viscous and adiabatic, and the cells close to the walls were clustered such that y^+ was kept between 0.5 and 1 to completely solve the boundary layer. Figure 10.4 depicts the surface mesh of the model and symmetry plane.

The wind-tunnel static pressure, temperature and velocity were imposed at far field, whilst backpressure was imposed at the domain outlet. The flight conditions were Mach 0.8, Reynolds 4×10^6 and 1.4° for the AoA. The two-equation realizable $k - \varepsilon$ turbulence model was used together with a near-wall correction in the inner boundary layer region where the turbulent viscosity (μ_t) and dissipation rate (ε) were suitably altered and smoothly blended with their high Reynolds number definitions from the outer region.

The FEM structural model, fully steel-made, is limited to the external wing region and it is constrained at the wing-fuselage junction. Its mesh, shown in Fig. 10.5, was created using ANSYS ICEM CFD tool and comprises 27,666 solid elements and 7363 auxiliary shell elements (with negligible stiffness and mass) added on the skin to define the wetted interface for enabling mapping.

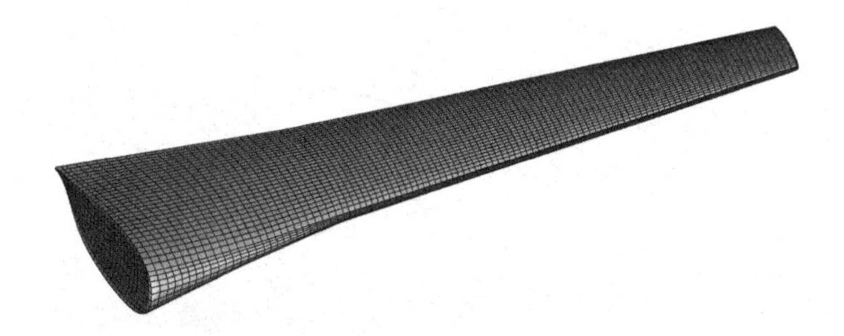

Fig. 10.5 P1xx case: wing FEM model

The interpolation procedure for mapping, performed using zero-th order interpolation, was implemented in the CFD code. A detail of the overlapped visualisation of the FEM (thick solid line) and CFD (points) surface wing meshes is depicted in Fig. 10.6.

The RBF field is defined by selecting a number of source points on the wing and on a cylinder-shaped delimiting domain boundaries shown in Fig. 10.7. The first set comprises the wing FEM nodes positions with their displacements, whilst the latter defines the limits of the fixed domain with a uniformly distributed number of source points.

The mesh morphing is performed in the respect of the two-step approach adopting the linear RBF $\varphi(r) = r$ with global support: at first the CFD wing nodes are morphed in the respect of the FEM nodal displacement and, at second, the morphing action is extended in the volume by smoothing the grid up to the prescribed zero displacement at delimiting cylinder boundaries.

It is worth to specify that usually the linear RBF is the best choice for volume morphing because it guarantees the minimum mesh distortion and it can be accelerated using fast algorithms (see 3.6.1). An high-order RBF, such as the cubic one ($\varphi(r) = r^3$), can be used in the first step to smoothly control surfaces using just a limited number of source points. In this specific example, the FEM data are fine enough and such to assure very good results even if the linear RBF in the first step is employed.

Considering the profiles of the aerodynamic coefficients reported in Fig. 10.8, for this aeroelastic analysis just three iterations (cycles) between the CSM and CFD solver were sufficient to obtain a converged static solution (Fig. 10.9).

Due to mapping procedure, a systematic overestimation of the Z component around 2.5% of the load applied on the FEM model is evaluated. Such a value was determined by calculating the Z component of the integral value of the mapped loads on the wing surface on both the source CFD code and the FEM code after having imported the file of loads.

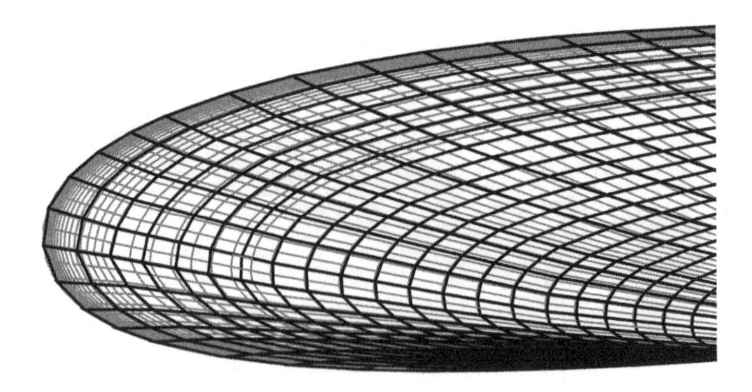

Fig. 10.6 P1xx case: detail of FEM and CFD wing surface mesh

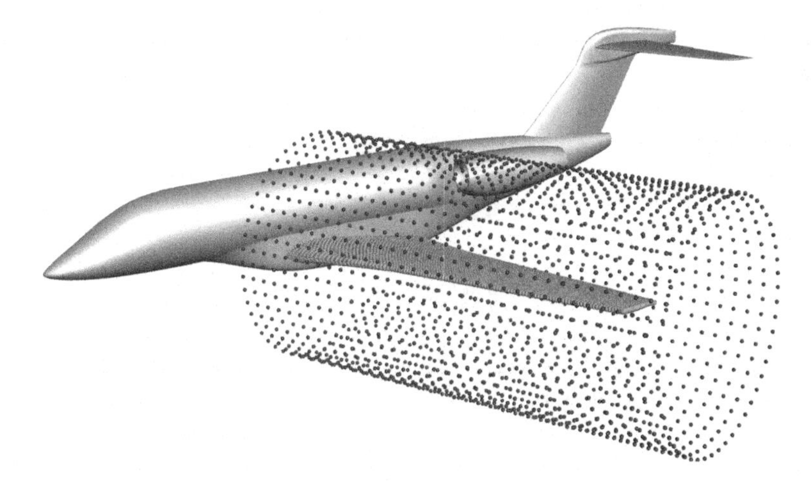

Fig. 10.7 P1xx case: delimiting domain of the morphing field

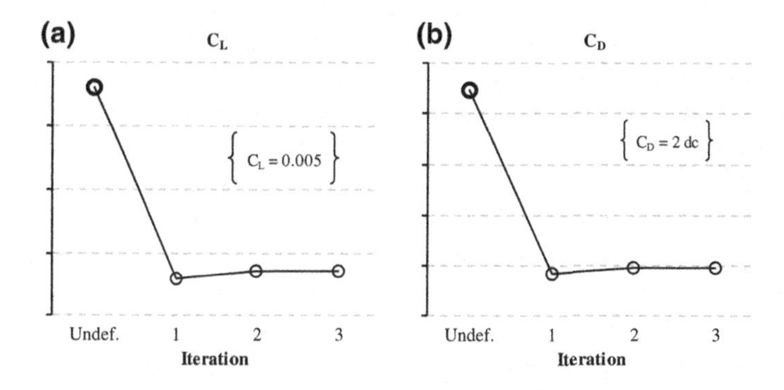

Fig. 10.8 P1xx case: aerodynamic coefficients profiles during the FSI steps

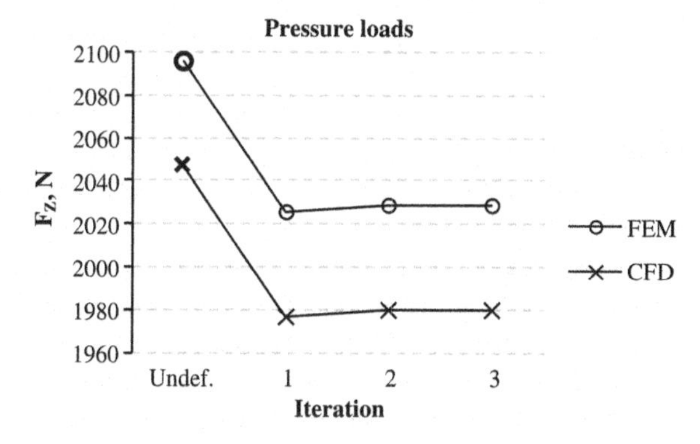

Fig. 10.9 P1xx case: vertical force profiles during the FSI steps

Fig. 10.10 P1xx case: wings deformation amplified 10 times

The wings shape in the final configuration is depicted in Fig. 10.10 amplifying the deformed configuration of 10 times.

10.6 Mode-Superposition for Steady

In the following sections the application of the mode-superposition approach to handle the static aeroelastic analysis of three test cases is deepened. In particular, such test cases respectively concern the P1xx already described, the HIRENASD model that has been previously used for the test case concerning icing simulation detailed in the Chap. 9 and the front wing of an INDY car.

*10.6.1 P1*xx Wing

The modal analysis was performed evaluating up to 20 natural modes of the structure. Figure 10.11 plots the cumulative mass contribution of each mode by the progressive summation of the effective mass participation factor of each mode for the vertical translation and the rotation about the Y axis.

For the FSI analysis only the first six modes are used to populate the modal base. With this number, the percentage of total mass contributing to the vertical translation and rotation along Y direction is respectively 52 and 18%. Although this figure is far from the 70% target suggested in Sect. 10.4.1, the adequateness of this modal base will be confirmed by the results of the modal analysis following reported.

The first six modes' shapes, normalised with respect to mass, are shown in Fig. 10.12 and they respectively refer to the first four bending modes in the lift direction, the torsion mode and the fifth bending mode.

The convergence of the FEM modal analysis calculation was verified by converting all the elements from linear to parabolic till reaching an error less than 1% for all six modes, and successively using the finer model as the reference one. The evaluated natural frequencies are summarised in Table 10.1.

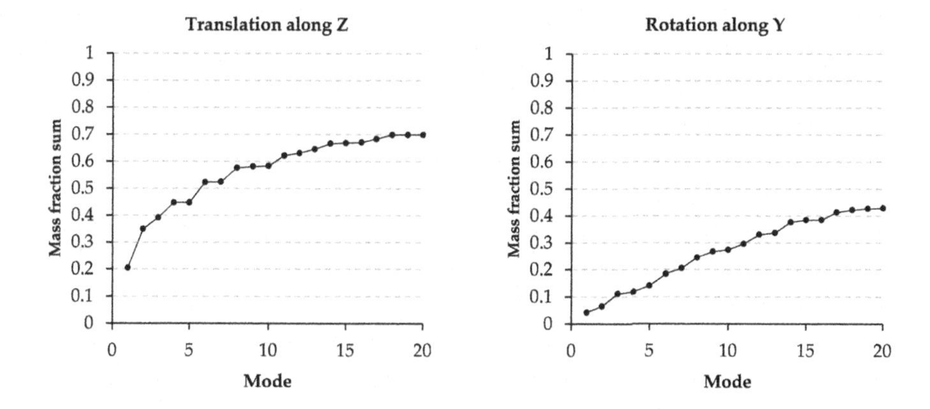

Fig. 10.11 P1xx case: sum of modal mass fraction of first 20 modes

The embedding of the modes onto the CFD mesh has been conducted according to the same set-up detailed in Sect. 10.5.1. Only the FEA displacement field differs: it is the result of subsequent static analyses for the two-way approach, whereas it is the deformed shape of an individual modal shape in the case of mode-superposition.

The pressure coefficient over the wing sections at (a) 20, (b) 40, (c) 60, and (d) 80% of the semi span experimentally evaluated and numerically computed using both the two-way and mode-superposition approach is shown in Fig. 10.13. The numerical solutions accounting for flexibility exactly match and differ from the ones performed on the undeformed shape showing a better match with experimental results especially at cross sections far from the root where the flexibility effect is higher.

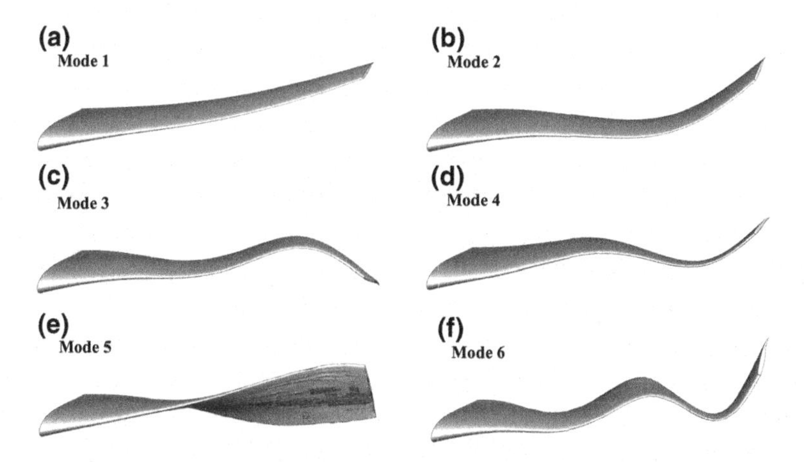

Fig. 10.12 P1xx case: natural modes shapes computed through a CSM solver

Table 10.1 Modes frequencies of the P1xx model

Mode number	Shape	Frequency (linear el.) Hz	Frequency (parabolic el.) Hz	Error %
1	Bending 1	72.89	72.74	0.21
2	Bending 2	251.14	249.93	0.48
3	Bending 3	541.07	537.11	0.74
4	Bending 4	592.96	590.37	0.44
5	Torsion	684.23	683.50	0.11
6	Bending 5	1004.59	995.37	0.93

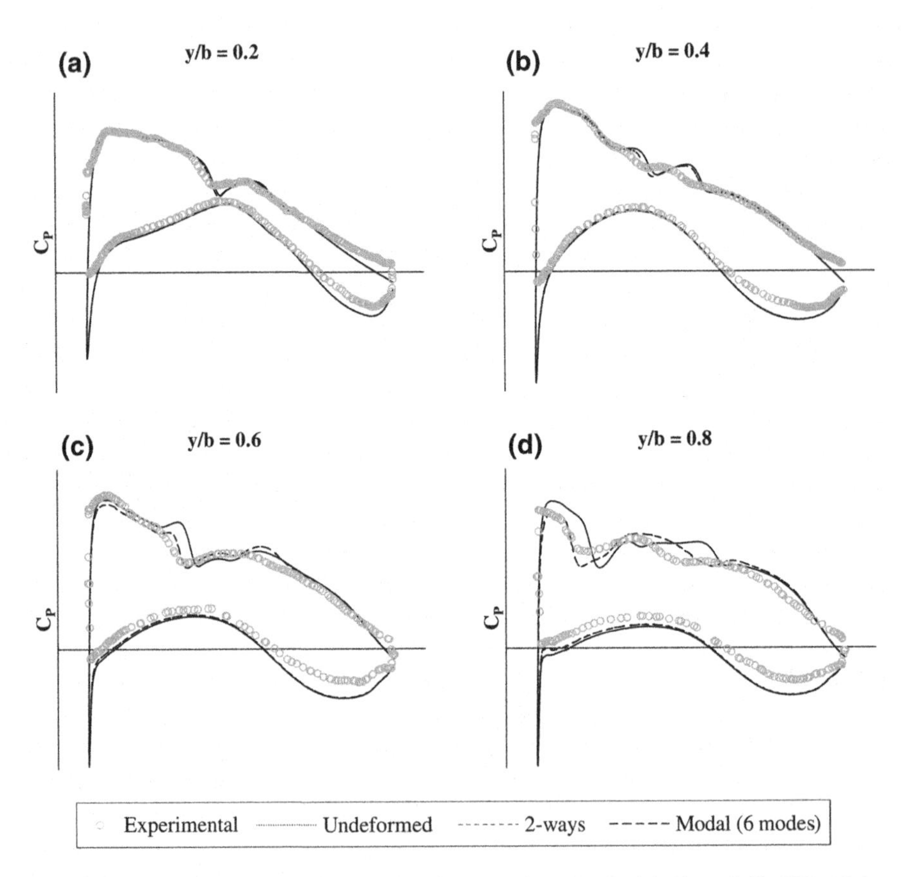

Fig. 10.13 Pressure coefficient at wing sections at (a) 20, (b) 40, (c) 60, and (d) 80% of the semispan

A comparison between the two methods can be performed by inspecting the modes truncation error convergence. The solution computed with a certain number of retained modes, up to six, is compared to the one obtained after the convergence gained with the two-way. Figure 10.14 shows such truncation error as a field: the reference (two-way) and the modal solution displacements are subtracted for each

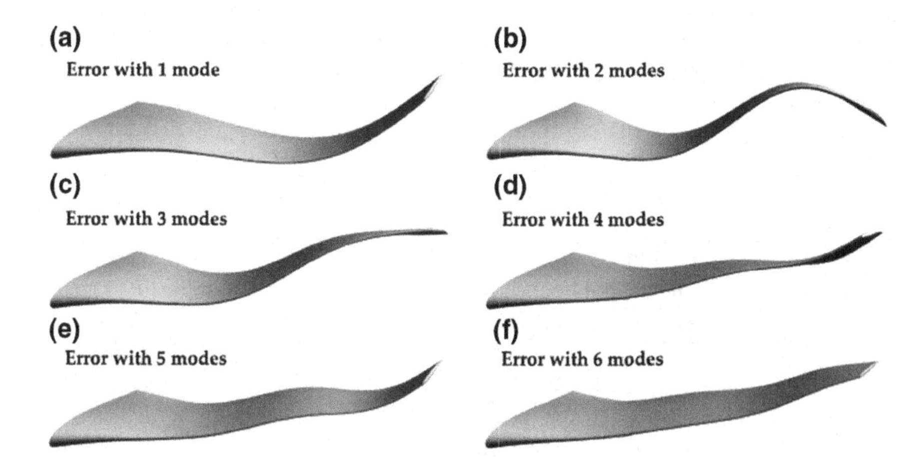

Fig. 10.14 Modal truncation error (computed as the vector between modal computed displacement and reference one) represented as a deformed shape retaining from (**a**–**f**) one to six modes

Table 10.2 Displacements of monitored points

Case	$y/b = 0.4\Delta z$ (mm)	$y/b = 0.7\Delta z$ (mm)	$y/b = 1.0\Delta z$ (mm)
2-way	0.32	1.923	4.852
2-way corrected	0.312	1.875	4.731
1 mode	0.251	1.811	5.188
2 modes	0.307	1.938	4.853
3 modes	0.312	1.933	4.871
4 modes	0.315	1.927	4.890
5 modes	0.318	1.935	4.891
6 modes	0.321	1.933	4.879
6 modes err (%)	2.9	3.1	3.1

node of the wing so that the error can be represented as a field. It is interesting to note that the shape of the error looks similar to the next modes dropped. A quantitative evaluation of the approximation due to mode truncation is provided in Table 10.2 where the vertical displacements at three probe locations are compared; the corrected two-way is modified to compensate the small equilibrium error registered using the FSI mapper of ANSYS Fluent. The difference is very small, that is about 3% retaining six modes.

10.6.2 HIRENASD Wing

Using one of the mesh made available by the Aeroelastic Prediction Workshop (AePW) committee (see Sect. 9.4.2), the EU project RBF4AERO (European Commision 2013) platform was used to automatically perform the CFD simulations of this test case through the SU2 solver. In particular, the hybrid mesh referred to as SOLAR unstructured grid and approximately composed by 1.5 million of mixed cells was employed.

To accomplish the CFD set-up, the flow conditions were computed assuming Mach = 0.8, Re = 7 × 10^6, AoA = 1.5°, a static pressure of 89,289 Pa and a static temperature of 246.9 K, whilst for the numerical solution settings those identified in Sect. 9.4.2 were adopted.

Relating to structural modes data, the ANSYS APDL solver has been used to calculate deformations and natural frequencies starting from the import of the FEM model in NASTRAN format made available by the AePW committee (Chwalowski et al. 2013). Such a model, composed of 357,545 nodes and 225,739 elements of different types, includes the wing, the balance and the wing-balance junction. The mesh of the ANSYS APDL model after the pre-processing accomplishment is depicted in Fig. 10.15. Triangular shells are located on the wing surfaces and are used to monitor the displacements of the FEM case without adding any contribution to the structure stiffness.

In Table 10.3 the value of the natural frequency of the first six modes calculated through the modal analysis are compared to those computed by the workshop committee (NASA 2012). Using 6 modes the effective mass to total mass ratio is about 68% and, in this case, the value is very close to the 70% threshold proposed in Sect. 10.4.1.

In Fig. 10.16 the modes of the HIRENASD model are shown respectively from top left (Mode 1) to bottom right (Mode 6). The AePW classification assigns B to the out-of-plane bending and FA to the in-plane fore-and-aft bending.

Using those data, an RBF solution has been generated for each mode through the stand-alone version of the RBF Morph software. In particular, the RBF solution was a two-step solution. In the first step the displacement of the wing nodes were imposed utilizing the FEM displacements using a file generated ad hoc, whilst in the second step the solution of the first step was imposed for the motion of the wing. The nodes that do not change their position were assigned as fixed and a delimiting domain was utilised to restrict the volume of the morphing action. For the first step of the RBF solution set-up of Mode 1, Fig. 10.17 depicts the preview of the source points' position before (left) and after morphing (right) respectively.

Figure 10.18 (left) shows the source points on the fuselage that maintained unchanged their position during morphing and the source points on the delimiting

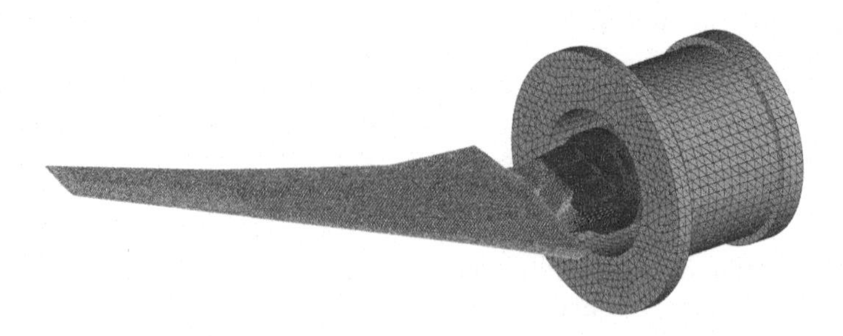

Fig. 10.15 FEM model of the HIRENASD configuration

Table 10.3 Modes frequencies of the HIRENASD model (APDL calculation)

Mode number	Shape	Frequency (Hz)	Frequency ref. (Hz)
1	Bending 1	25.55	25.55
2	Bending 2	80.28	80.24
3	Torsion 1	106.12	106.19
4	Bending 3	160.41	160.34
5	Bending 4	243.14	241.99
6	Torsion 2	252.25	252.22

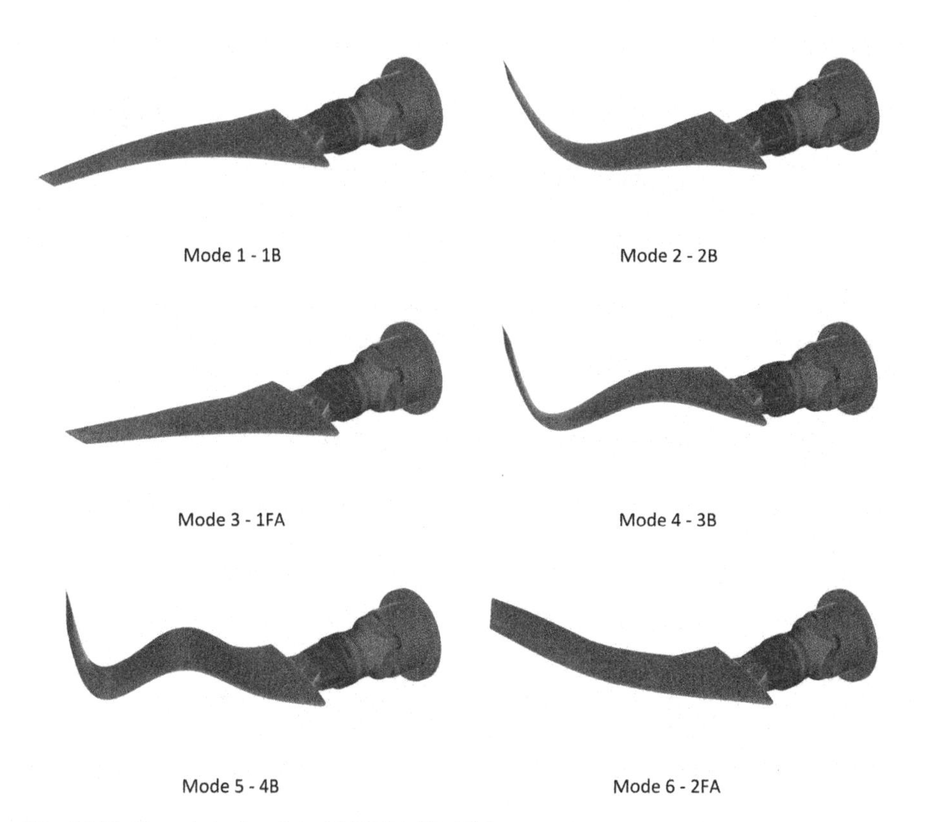

Fig. 10.16 Natural modes of the HIRENASD FEM case

domain (right). Considering the fixed source points distribution, a portion of fuselage near the wing root is left free to deform during morphing because both in the wind tunnel model and the FEM case the fuselage aerodynamic fairing is mechanically uncoupled from the wing root so that a slight motion of the wing root is allowed.

In the FSI study 5000 iterations have been run for the baseline and 400 iterations for each of the four FSI cycles.

In Fig. 10.19 the Cp gained at monitoring wing sections on the CFD model is plotted together with the corresponding value measured during the experimental test

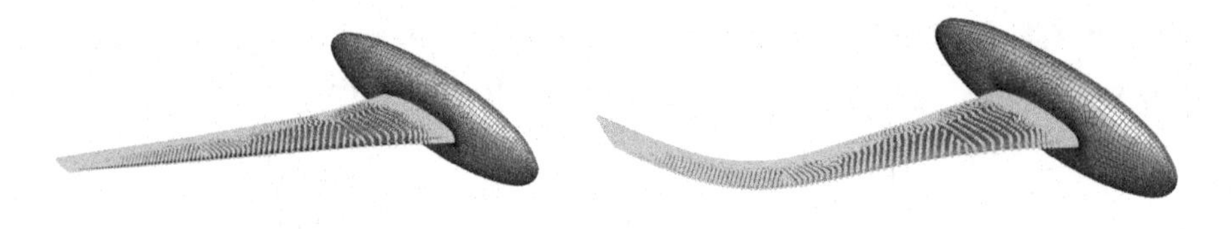

Fig. 10.17 Preview of the source points before and after morphing (Mode 1)

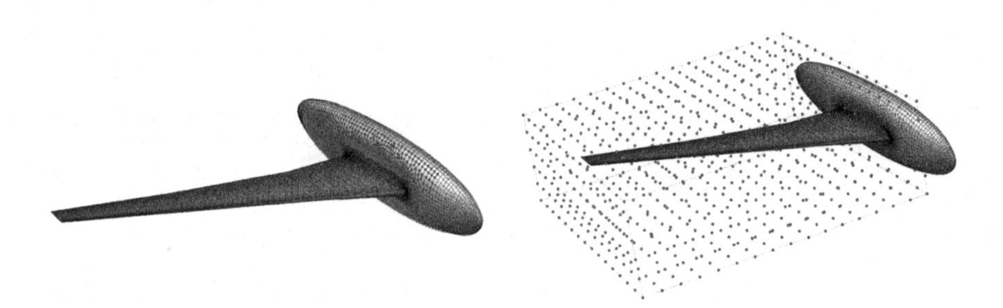

Fig. 10.18 Fixed source points (left) and delimiting domain source points (right)

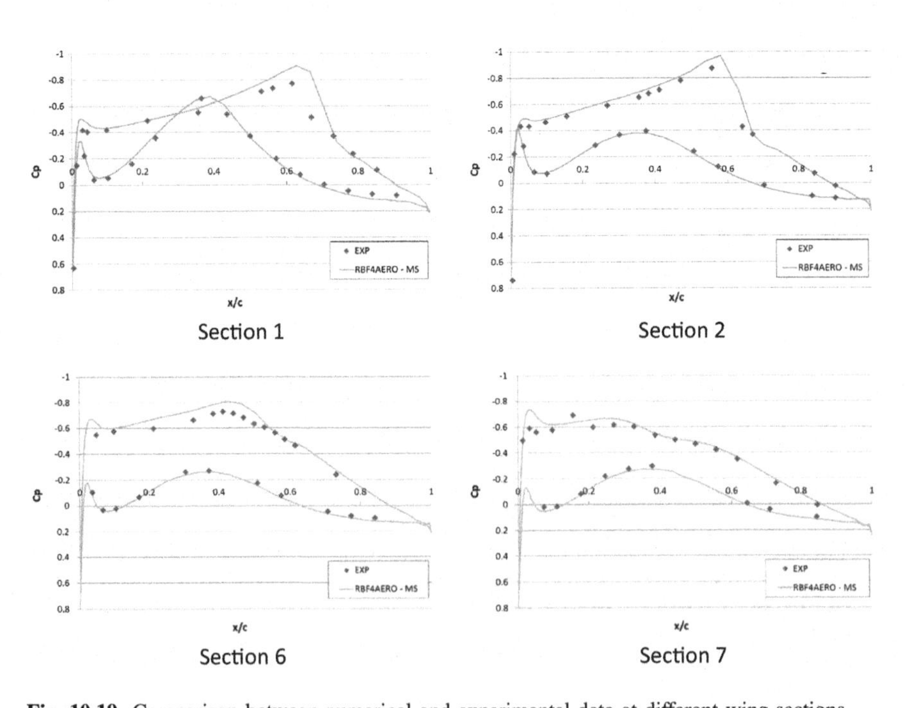

Fig. 10.19 Comparison between numerical and experimental data at different wing sections

(NASA 2012). The reference data available include the pressure coefficient distribution at four monitoring stations of the HIRENASD configuration, located at 14.5% (section 1), 32.3% (section 2), 65.5% (section 6) and 95.3% (section 7) of the total wing span, and the numerical maximum displacement of the trailing edge in proximity of the wing tip. The final value is 14.6 mm and it is close to the 13.75 mm numerically obtained (Chwalowski et al. 2011).

10.6.3 Dallara INDY Race Car

In this section the steady FSI analysis concerning a motorsport application using RBF Morph (add-on version) and ANSYS Fluent and adopting the mode-superposition approach is described (Invernizzi 2013). In particular, an aerodynamic analysis of a Dallara Formula INDY race car, shown in Fig. 10.20, has been performed taking into account the elasticity of the front wing during CFD computing utilising the vibrational data computed by means of the NASTRAN solver. Considering the front wing deformation is an important parameter to accurately evaluate the front loads and the maximum deformation of the front wing itself. In fact, the former value has effects on whole race car performance, whilst the latter influences the structural safety and determine the fulfilment of stringent race regulations.

According to what just defined, in order to obtain the elastic model a modal analysis using a FEM tool has to be carried out for the extraction of the mass normalized modal shapes and their relative frequencies. In doing that, both the CFD and FEM model are generated from the same geometry since the same position, dimension shape and orientation are needed to properly link such numerical models. The starting baseline geometry of the race car model is shown in Fig. 10.21.

The FSI study needs that materials characterization and constraints imposition are suitably defined for each component of the CAD model. In this context,

Fig. 10.20 Dallara INDY race car

Fig. 10.21 CAD model and detail of the front wing

particular attention has been taken in order to correctly represent the carbon properties in terms of lamination and fibres orientation.

The results of the modal analysis of the FEA model showed that the first 5 modes are suitable to describe the deformation of the front wing assembly. Such modes are respectively shown in Fig. 10.22 from top left to bottom, whilst Table 10.4 summarises the natural modes' relative frequencies.

The two-step procedure was adopted to generate the RBF solutions calculating, for each mode, first an RBF solution using the displacements input file and leaving the rigid elements free to deform and, then, applying it only to the surfaces interested by the FEM analysis and constraining the others. To restrict the morphing action and to exclude the points outside the interested portion, a delimiting domain was used. Finally, in order to further improve the quality of the solution and to obtain a better grid smoothing, the tyre was wrapped in a zero motion domain that has an higher source points resolution and finishes in the clearance area between the endplate and tyre itself.

Relating to CFD simulation, the flow field in steady conditions was gained by means of a RANS solution using a turbulence model and imposing a freestream velocity of 50 m/s (180 km/h).

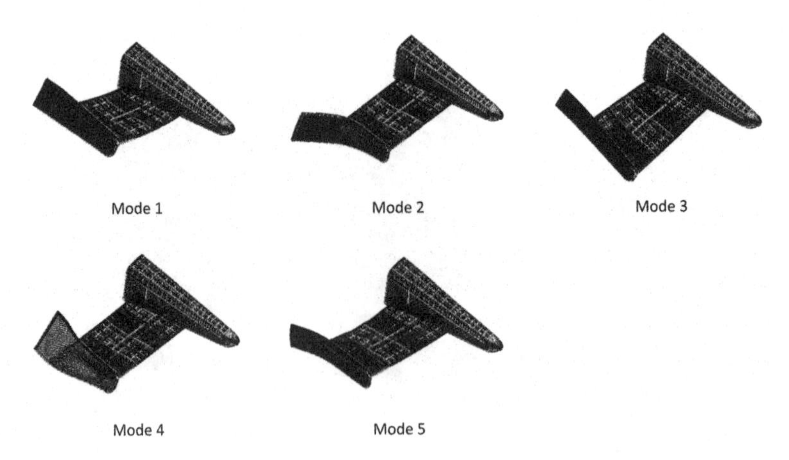

Fig. 10.22 Natural modes of the front wing FEM model

Table 10.4 Modes frequencies of the front wing

Mode number	Frequency (Hz)
1	28.2
2	49.6
3	85.0
4	104.2
5	133.2

An iterative loop to perform the FSI cycles has been implemented to obtain the static aeroelastic equilibrium. During the elastic computing both modal forces and displacements were monitored. Evaluating such profiles, the modal convergence was judged to be achieved just after four FSI cycles. In Fig. 10.23 the obtained deformed shape of the nose-front wing-endplate (red) system is superposed to the original rigid geometry (black) using the first mode only (left) and 5 modes (right) respectively. To ease the comprehension and get a better assessment of the difference, the deformed shape was amplified by a factor 10.

The convergence of the method, using up to 5 modes, is shown in Table 10.5 where the error was calculated with respect to the modal superposition using 5 modes.

As expected, the maximum error of the front wing deformation is reduced by adding modes in the displacement calculation. The influence of the modes introduction can be qualitatively highlighted by plotting the error calculated subtracting the shape obtained with 5 modes to the one using up to 4 modes (see Fig. 10.24); as already observed in Sect. 10.6.1 this visualization allows to understand that the missing information, i.e. the error map at a certain truncation, is shaped as the first dropped mode.

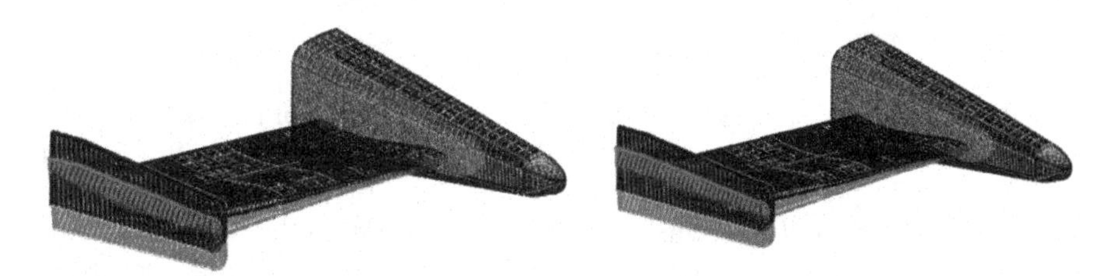

Fig. 10.23 Deformed and undeformed amplified shapes using just the first mode (left) and 5 modes (right)

Table 10.5 Displacement error sensitivity study

Modes used	Maximum displacement (mm)	Maximum difference (mm)	Maximum error (%)
1	5.941	0.4946	8.3
2	5.898	0.3817	6.5
3	5.584	0.1483	2.7
4	5.56	0.07722	1.4
5	5.555	0	0

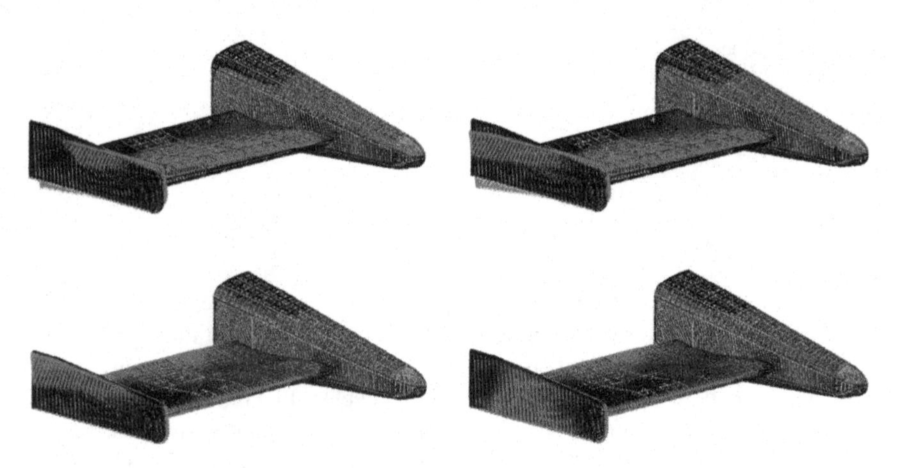

Fig. 10.24 Error of the final front wing deformation using up to 4 modes

10.7 Mode-Superposition for Transient FSI Analyses

The following subsections deal with transient FSI analyses based on the use of the mode-superposition approach. Such method is time consuming because the response of the system is evaluated with a time marching solution in the domain of time. The same calculation framework can be also used for specific transient analyses in which each individual mode is accelerated with a prescribed time history and the signal response is used to build a reduced order model (ROM) suitable for flutter analysis (Castronovo et al. 2017). Direct transient FSI can be used for the study of vortex induced vibrations as well. In the paper of Di Domenico et al. (2017) the experimental study of Ausoni et al. (2012) has been reproduced showing a very good matching in the reproduction of induced vibration. In the same paper an evaluation of natural modes in water, based on fast Fourier transform (FFT) processing of the response of the system in calm water when subjected to an initial displacement, is also demonstrated. In the next subsections two transient FSI studies are presented: the first one refers to the just mentioned paper of Di Domenico et al. (2017) and is used to showcase the just introduced concept, whilst the second is a transient FSI study aiming at determining the wing aeroelastic response caused by a store separation (Reina et al. 2014).

10.7.1 Vortex Shedding Induced Vibrations

In Di Domenico et al. (2017) the numerical accuracy of the proposed FSI method was demonstrated studying the vortex shedding induced vibrations of a literature test case whose experimental data are available (Ausoni 2009). The investigated body is a NACA 0009 hydrofoil operating at zero angle of attack in a uniform high speed flow with $Re_h = 16.1 \times 10^3 - 96.6 \times 10^3$ and with the trailing edge thickness as reference length h. In Fig. 10.25 the hydrofoil geometry and set-up is shown.

The blunt geometry, built in steel with density $\rho = 7850\ kg/m^3$ and submerged in water, is clamped at one side and embedded in a pivot leaving free tip rotation. These particular boundary conditions are necessary to amplify the torsional vibration modes induced by the water flow. The trailing edge is truncated in order to obtain the blunt shape where the flow separation occurs.

The first six modal shapes, obtained in ANSYS Mechanical, were extracted and are shown in Fig. 10.26 together with their shape and frequency. Nodal displacements associated to modal shapes have been exported and employed to set up the RBF problems in RBF Morph, obtaining a shape modification parameter for each mode. Embedding the modal shapes using RBF Morph was then possible to start the transient FSI analysis. As far as boundary conditions are concerned, a pressure

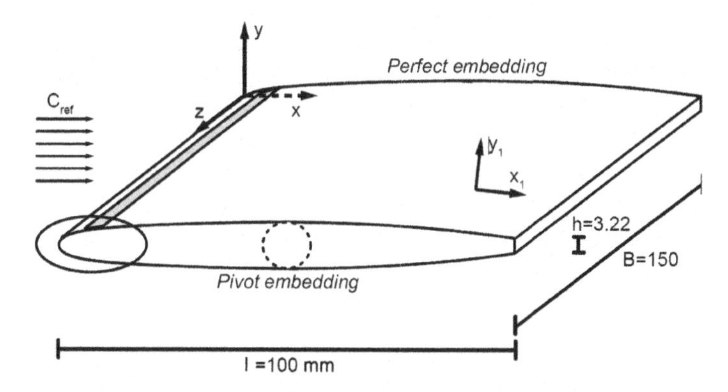

Fig. 10.25 Blunt trailing edge NACA 0009 hydrofoil (Ausoni 2009)

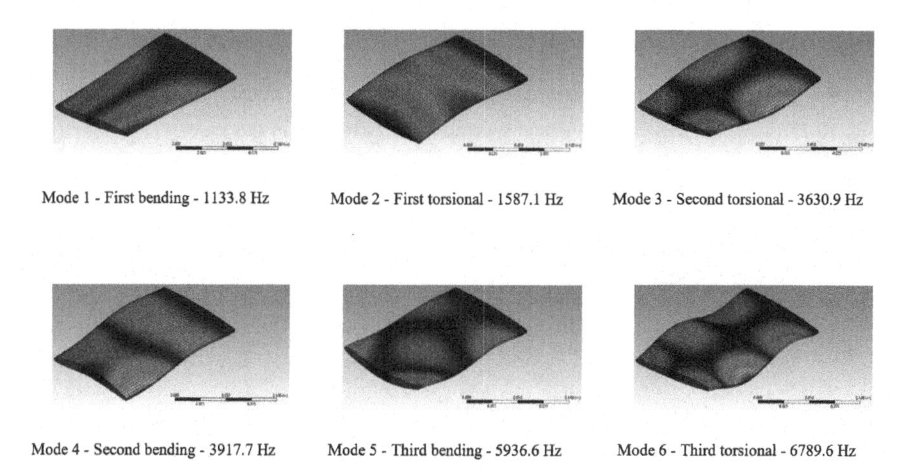

Fig. 10.26 Hydrofoil natural modes

of 101,325 Pa and a temperature of 288.15 K were applied, whilst a kinematic viscosity of $10^6 mm^2/s$ and a density of 998 kg/m^3 were set for the fluid. The simulation was carried out for 10,000 time steps of 2.0×10^{-5} s with five iterations per time step.

Simulations were accomplished with two flow velocity values, 16 and 22 m/s, in order to catch the Lock-In and Lock-Off (Ausoni et al. 2012) conditions. Monitoring a point positioned at the trailing edge, the time-velocity amplitude was registered and the FFT was hence calculated.

In Fig. 10.27 the time-velocity amplitude (left) and the FFT (right) are shown for the 16 m/s flow velocity case. The FFT of the output data shows a good synchronization between the shedding of the vortices and the body oscillation: this particular condition is the Lock-In and it depicts the resonance phenomenon. The frequency of the vortex shedding is really close to the natural one and the amplitude of the velocity deformation signal shows a quite constant value that is typical of the resonance phenomenon. The FFT shows a dominant frequency of 909.91 Hz.

In Fig. 10.28 the time-velocity amplitude and the FFT are shown for the Lock-Off case with a flow velocity of 22 m/s. In absence of synchronization the time development of the signal shows a modulation due to intermittent weak shedding cycles. The spectra for the Lock-Off condition highlights the presence of a major noise in the signal that shows a dominant frequency at 1202.9 Hz. The results are in good agreement with experimental data (Zobeiri et al. 2012) as depicted in Fig. 10.29.

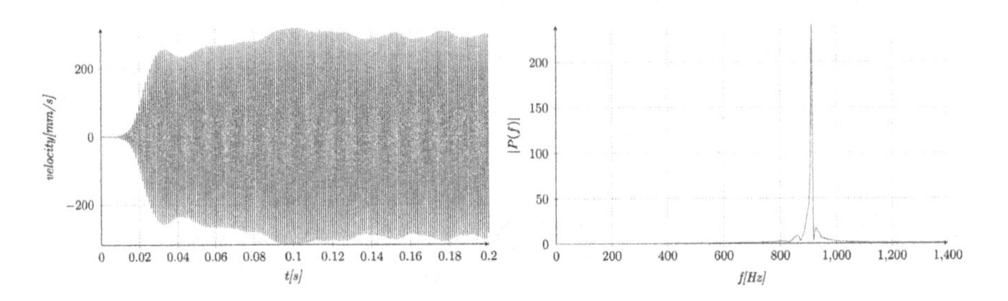

Fig. 10.27 Time-velocity amplitude and spectral analysis for the 16 m/s case

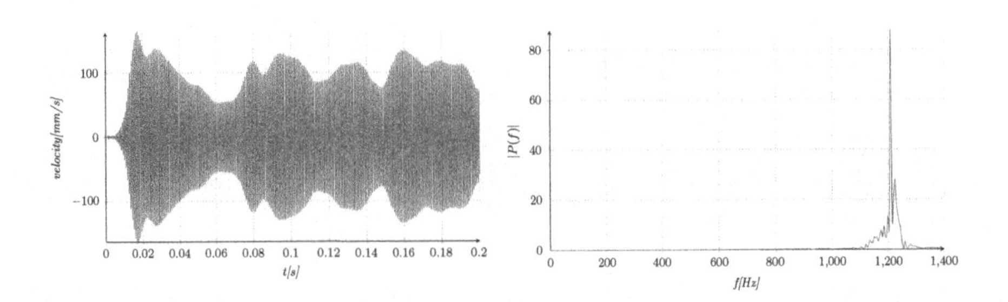

Fig. 10.28 Time-velocity amplitude and spectral analysis for 22 m/s

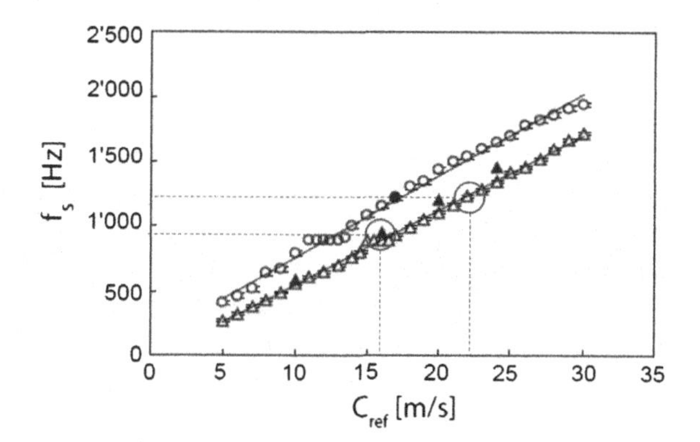

Fig. 10.29 Predicted versus measured frequencies

To further characterize the system behaviour and validate the achieved results, the hydrofoil natural modes in water were computed by employing the same workflow but changing the initial boundary conditions. The transient response of the system under an initial deformation was indeed investigated by extracting the resonant frequencies with an FFT analysis. The initial deformation was obtained by imposing a starting value for each modal coordinate properly tuned to obtain a maximum displacement of 1.0×10^{-3} m. The hydrofoil submerged in calm water was released from this initial configuration and left free to vibrate. The obtained spectral response is shown in Fig. 10.30 and compared to vibration in air.

As water has a density larger than air, the water added mass and damping reduce natural frequencies and velocity amplitudes as predictable. Natural frequencies for the hydrofoil both in air and submerged in water are shown in Table 10.6. The results validate the Lock-In frequency achieved with the 16 m/s flow being the first mode at 891.9 Hz.

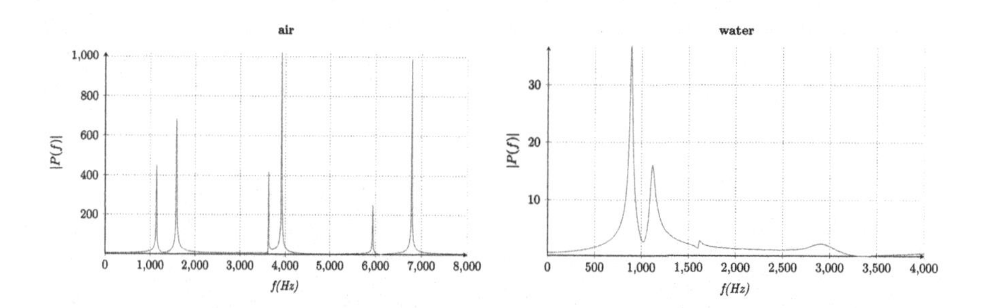

Fig. 10.30 System frequencies in air and in water

Table 10.6 Natural frequencies for the hydrofoil submerged in air and water

Mean	Mode 1 (Hz)	Mode 2 (Hz)	Mode 3 (Hz)	Mode 4 (Hz)	Mode 5 (Hz)	Mode 6 (Hz)
Air	1133.8	1587.1	3660.9	3971.7	5936.6	6789.6
Water	891.9	1118.8	1619.6	2902.7	–	–

10.7.2 Store Separation

In this section the transient FSI study described in Reina et al. (2014) and aiming at calculating the wing aeroelastic response following an event of store separation is described. The store separation problem is particularly felt for the high-performance military attack aircrafts since they typically carry on vast arrays of air-to-air or air-to-ground weapons on wing-mounted pylons.

Assuming the aircraft in flight at a certain subsonic Mach number, the wings are in a condition of elastic equilibrium under the joint action of several loads such as weight, external stores weight and aerodynamic forces as well. If at a certain moment there is the release of one of the external stores, the wing will lose its equilibrium and begins oscillating. The substantial weight and aerodynamic characteristics of these external stores are such to significantly influence the aero-elastic behaviour and structural response of the wings (Chambers 2005). The determination of such characteristics is then of capital importance because the pilot can be made aware and be provided, in advance, with all information necessary to get the right reaction.

The complete numerical methodology adopted for the accomplishment of this study foresees the development of three subsequent steps. In the first step the FEA model is used to extract the natural modes of the wing structure and to perform a static structural analysis to determine the effect of the loads acting on the wing. In the second step a steady aero-elastic analysis is conducted to determine the condition of elastic equilibrium of the wing in the presence of the store. During the third step, that takes as initial conditions the modal coordinates and the flow field resulting at the end of the second one, an unsteady aero-elastic analysis is performed to get the transient response of the wing due to the release of the underwing body. This event is simulated by imposing the disappearance of the store itself making transparent to the flow the wall of the store itself.

The wing is modelled by extruding a NACA 0012 aerofoil for an half wingspan of 2.5 m. The store, whose weight is 300 kg, is modelled joining a cylinder and a hemisphere. The geometry finally created is shown in Fig. 10.31.

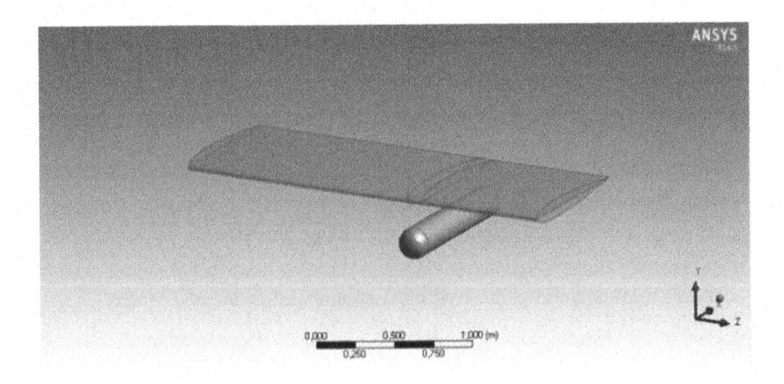

Fig. 10.31 Store separation case: isometric view of the model for the store separation study

The CFD grid is an hybrid unstructured mesh with ten layers of cells inflated from the wing surface and the remaining portion of the simulation volume filled with tetrahedral cells. Moreover, a tetrahedral mesh domain conformal to the outside one is generated inside the store volume to simulate the fall of the store by transforming its wall surface in an internal type surface.

The structural mesh of the wing comprises 1260 hexahedral solid elements for a total number of 6930 nodes. Using such a mesh, two FEM analyses, one modal and one static structural are carried out by means of ANSYS Mechanical. A linear elastic material is selected (Al6061-T6) and the model is constrained fixing all the nodes on the surface at the root of the aerofoil. The first six modes of vibration are shown in Fig. 10.32.

The vibration modes of the wing are exported as a cloud of points by extracting modal displacements from the FEM surface mesh nodes and passed as input to the morphing tool (add-on release of the RBF Morph software). In such a way an RBF solution for every modal shape is finally computed. In Fig. 10.33 the preview of the source points position before and after applying the morphing field of the first natural mode, are highlighted in the ANSYS Fluent graphic window. Such points include both those imported by the FEM solution and those belonging to the delimiting morphing domain.

The release of the store is expected to have an important influence on the transient profile of deformation of the wing structure, especially in the vertical direction. Figure 10.34 shows the profiles of the lift coefficient obtained for the rigid and deformable wing that are respectively visualized in green and blue. It can be seen as the diagram of the deformable wing is characterized by large fluctuations related to the oscillation of the wing after the release of the store, happening at time equal 0, which tend to fade after approximately 0.6 s. Furthermore, it is evident that in the first instants a significant increase of aerodynamic load occurs, whose prediction is very

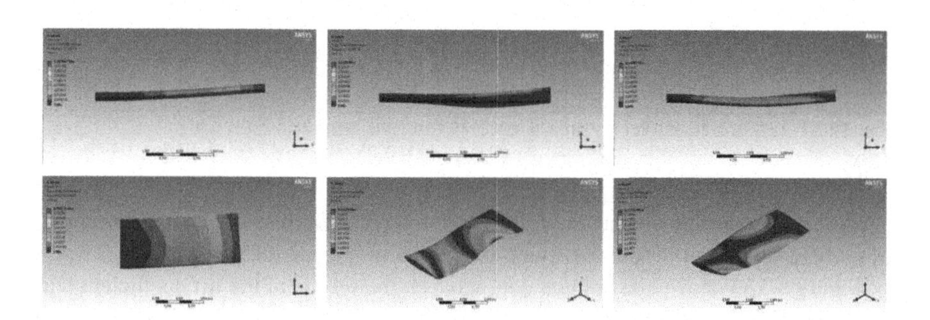

Fig. 10.32 Store separation case: natural modes of the wing

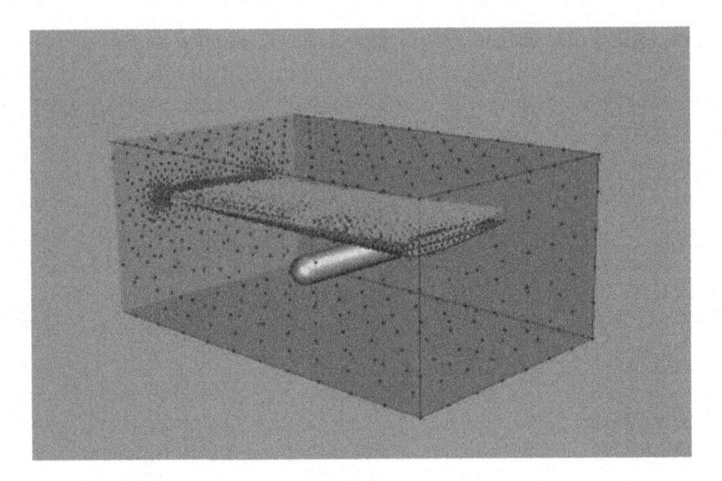

Fig. 10.33 Store separation case: source points

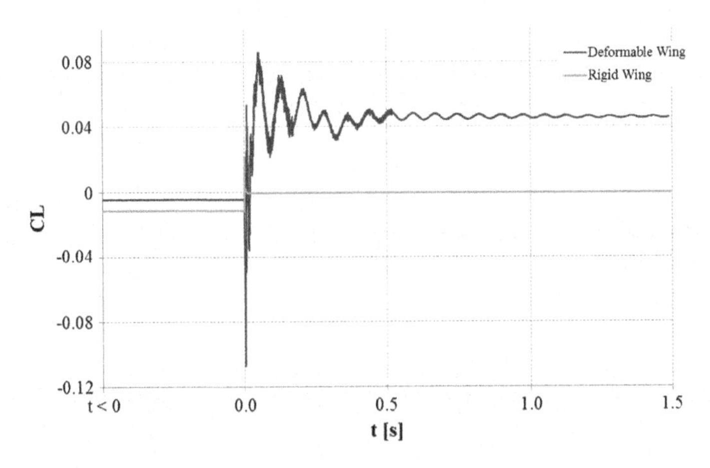

Fig. 10.34 Lift coefficient versus time in cases deformable wing and rigid wing

important to evaluate the structural response of the wing. Such an increase in loading is even more significant if compared with the profile of the case of the rigid wing.

Figure 10.35 shows the comparison between the profile of the lift coefficient and the one of the dimensionless wing tip deflection Y/Yo in which Yo is the tip deflection in presence of the store as a function of dimensionless time $\frac{\omega t}{2\pi}$. It is clearly understandable that the damped oscillatory is similar for both curves.

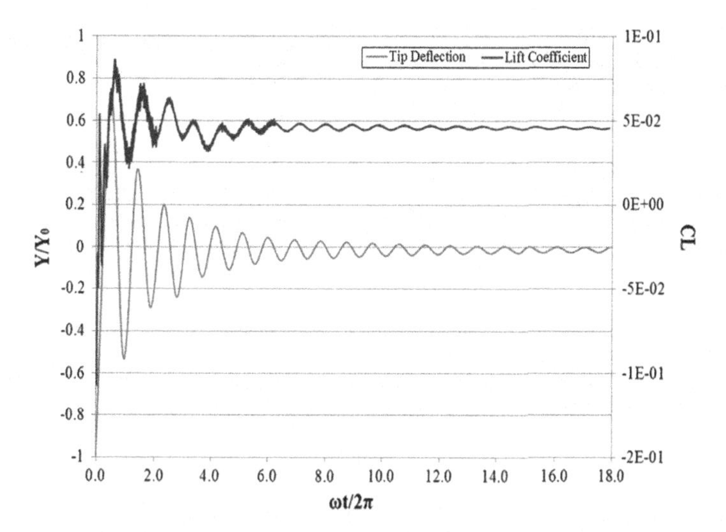

Fig. 10.35 Comparison between lift coefficient and tip deflection versus dimensionless time

References

Ausoni P (2009) Turbulent vortex shedding from a blunt trailing edge hydrofoil. Ph.D. thesis. STI. Lausanne. https://doi.org/10.5075/epfl-thesis-4475

Ausoni P, Zobeiri A, Avellan F, Farhat M (2012) The Effects of a tripped turbulent boundary layer on vortex shedding from a blunt trailing edge hydrofoil. J Fluids Eng 134:051207. https://doi.org/10.1115/1.4006700

Biancolini ME, Cella U, Groth C, Genta M (2016). Static aeroelastic analysis of an aircraft wind-tunnel model by means of modal RBF mesh updating. J Aerosp Eng 29. ISSN: 0893-1321. https://doi.org/10.1061/(asce)as.1943-5525.0000627

Bungartz HJ, Schäfer M (2006) Fluid-structure interaction: modelling, simulation, optimization, lecture notes in computational science and engineering, vol 53. Springer, Berlin

Castronovo P, Mastroddi F, Stella F, Biancolini ME (2017) Assessment and development of a ROM for linearized aeroelastic analyses of aerospace vehicles. CEAS Aeronaut J 8:353–369. https://doi.org/10.1007/s13272-017-0243-6

Cella U, Biancolini ME (2012) Aeroelastic analysis of aircraft wind-tunnel model coupling structural and fluid dynamic codes. J Aircr 49(2):407–414

Chambers JR (2005) Innovation in Flight. NASA

Chwalowski P, Florance JP, Heeg J, Wieseman CD, Perry B (2011) Preliminary computational analysis of the HIRENASD configuration in preparation for the aeroelastic prediction workshop. In: International Forum of Aeroelasticity and Structural Dynamics (IFASD), IFASD 2011-108, Paris, France

Chwalowski P, Heeg J, Dalenbring M, Jirasek A, Ritter M, Hansen T (2013) Collaborative HIRENASD analyses to eliminate variations in computational results. IFASD-2013-1D

Costa E (2012) Advanced FSI analysis within ANSYS fluent by means of a UDF implemented explicit large displacements FEM solver. Ph.D. Dissertation, Program in Environment and Energy—Cycle XXIV, Industrial Engineering Department, University of Rome Tor Vergata

Di Domenico N, Groth C, Wade A, Berg T, Biancolini ME (2017) Fluid structure interaction analysis: vortex shedding induced vibrations, Procedia Struct Integrity, 8(2018):422–432, ISSN 2452-3216, https://doi.org/10.1016/j.prostr.2017.12.042

European Commision (2013) RBF4AERO Project. http://cordis.europa.eu/project/rcn/109141_en.html

Gene H, Jin W, Anita L (2012) Numerical methods for fluid-structure interaction—a review. Commun Comput Phys 12(2):337–377. https://doi.org/10.4208/cicp.291210.290411s

Invernizzi S (2013) Advanced mesh morphing applications in motorsport. In: Automotive simulation world congress 2013, Frankfurt am Main, Germany, 29–30 Oct 2013

MSC (1997) NASTRAN basic dynamic analysis user's guide spiral-bound—June 15. ISBN-10: 1585240036

NASA (2012) HiReNASD model. https://c3.ndc.nasa.gov/dashlink/static/media/other/HIRENASD_base_legacy.htm

Reina G, Della Sala A, Biancolini ME, Groth C, Caridi D (2014) Store separation: theoretical investigation of wing aeroelastic response. Paper presented at Aircraft structural design conference, Belfast

Zienkiewicz OC, Taylor RL, Zhu JZ (2005) The finite element method: its basis and fundamentals, 6th edn, Butterworth-Heinemann, UK

Zobeiri A, Ausoni P, Avellan F, Farhat M (2012) Vortex shedding from blunt and oblique trailing edge hydrofoils. In: Proceedings of the 3rd IAHR international meeting of the workgroup on cavitation and dynamic problems in hydraulic machinery and systems, Brno

Chapter 11
Optimization Workflows Assisted by RBF Surrogate Models

Abstract This chapter presents how RBF can be exploited in the completion of an optimisation workflow. The contents of this chapter are specific of aerodynamics applications for which the parametric shape is obtained according to RBF mesh morphing and can be considered as a companion of Chap. 6. Nevertheless, this is an established use of RBF in the definition of optimization workflows of high cost objective functions and the concepts may be employed regardless of the specific strategy adopted to have parametric evaluations.

A very well established strategy to significantly reduce the number of objective function evaluations is based on the design of experiment (DOE) approach. According to this technique, once the design parameters and variation ranges are defined, first the objective functions are evaluated on a set of design points (DPs). To accomplish this task, several options and explorative methods are available such as full factorial and optimal space filling for instance. Afterwards, an approximate model of the system in the input design parameters space, generally referred to as surrogate model or metamodel, is requested to be evaluated by means of the response surface method (RSM). Also in this case several techniques differing in terms of the order of fitting, accuracy and performances can be adopted. The strategy proposed in the present Chapter foresees the use of the RBF to generate the surrogate model (RSM). The optimization is successively conducted and finalised using the RSM which can be examined using post-processing tools and, optionally, refined to increase the resolution within the most promising areas of the design space. The background of such optimisation approach is out of the scope of this book. The interested reader could refer to the book by Cavazzuti (2013). The RBF based surrogate model is the main focus of this chapter: its implementation is first described with a step by step example and, then, the use and performances of such a model are showcased by describing the optimization of representative industrial cases. Three optimisations performed by means of the support of an RBF-based surrogate model are presented. These cases respectively deal with a modified version of the NACA 0012 aerofoil, an industrial glider (Biancolini et al. 2016) and the trim of a yacht sails (Biancolini et al. 2014a, b).

M. E. Biancolini, *Fast Radial Basis Functions for Engineering Applications*,
https://doi.org/10.1007/978-3-319-75011-8_11

11.1 Optimisation Workflow

In the CAE sector an optimization workflow consists of several components that work exchanging information in order to run computational processes in a wide-spread variety of manners that are characterized by a certain level of automation. There are many optimization tools on the market available as commercial and open-source software. The main components that are typically present in an optimization workflow are the one driving the different implemented numerical procedures, the one that parameterizes the shapes of the models, a metamodel, a component in charge of jobs scheduling and a post-processing tool.

Multi-physics and multi-objective optimisation problems can be tackled adopting advanced CAE platform with an high level of automation. Such platforms allows to exchange information across CAE solvers specialised in different disciplines (or physics as the case of FEA and CFD for coupled FSI analyses), to study variations of the whole design for a new set of the design parameters (as parts' shape or boundary conditions), to extract performance index form individual and coupled physics. The high demand of repeated and intense calculations makes the usage of high performance computing (HPC) a key enabler in many cases. There are many examples of industrial platforms. Among them the ANSYS Workbench software can be considered as a representative example: all the CAE solvers can be driven adopting a common interface and common platform tools (parametric CAD modeller, automatic meshing, and optimisation tool). The user can drive the full workflow using a common dashboard where input and output parameters are summarized. Workbench provides tools for interconnect solutions in multi-physics workflow. It's obviously a clear advantage having all the CAE solver by the same vendor to simplify the integration; similar solutions are provided by other CAE players. However this is not a typical scenario because the standard is to adopt different CAE solvers from different vendors. In this case solutions that are capable to interact with different technologies are required. An example of tool for interconnection between different solvers are the MPCCI libraries that allows to set-up complex multi-physics workflow. An example of platform for optimisation capable to drive multiple CAE tools is modeFRONTIER. Open source solutions are available as well, the Dakota software is an example.

Custom platforms with an high level of automation can be developed for specific projects. The optimisation platform of the RBF4AERO is sketched in Fig. 11.1 (Bernaschi et al. 2016) and has RBF mesh morphing and interpolation technologies adopted for (see Chap. 6) shape optimisation, and multi-physics mapping. The optimisation platform of the RIBES project sketched in Fig. 11.2 (Biancolini et al. 2017) allows to drive shape optimisation of a structure using FEA by a FEA deformation based mesh morphing and RBF metamodeling in the parametric space.

It is worth to specify the role of RBF in the just cited workflows. In the RBF4AERO platform RBF are used to parameterize the computational model through mesh morphing in all the numerical processes it is able to run and to map

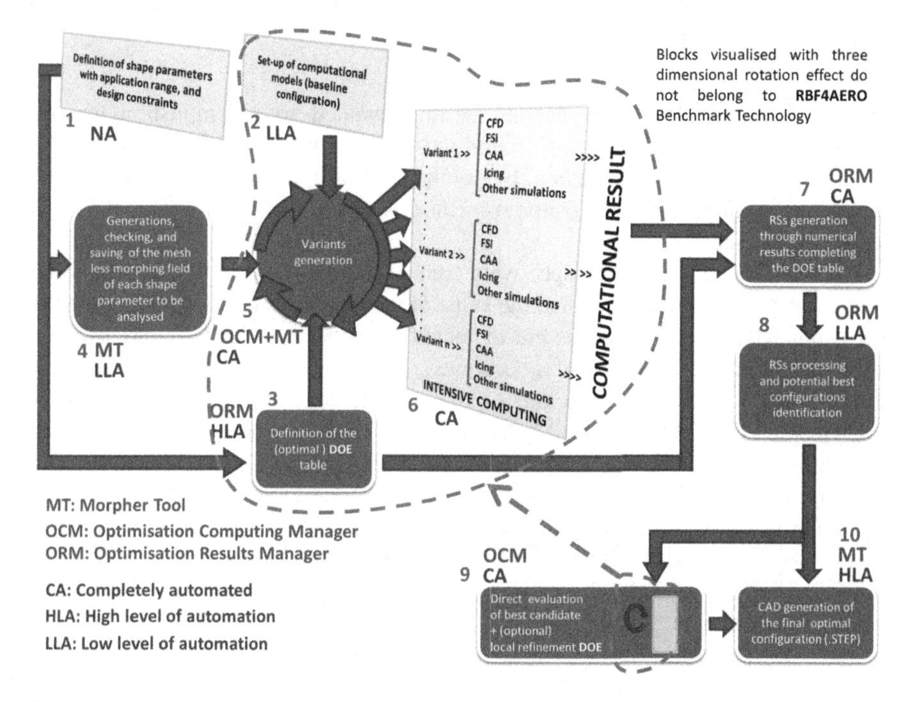

Fig. 11.1 The RBF4AERO platform allows to perform advanced shape optimisation of multi physics and multi objective coupled analyses

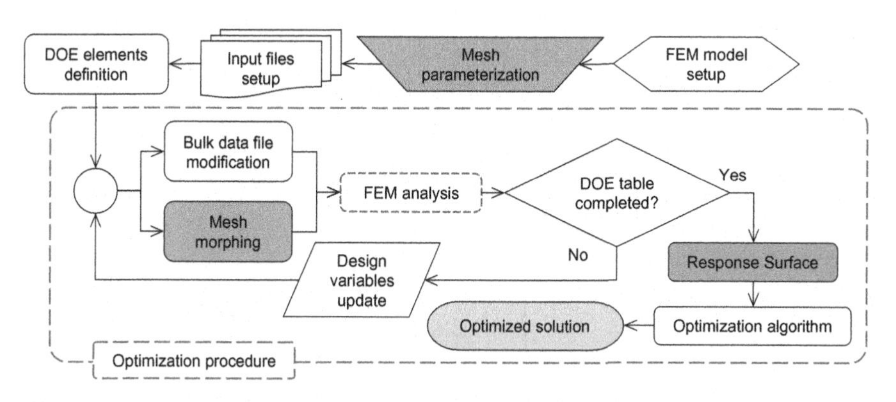

Fig. 11.2 The RIBES optimisation platform automates FEA calculation

the computed CFD loads onto the wetted deformable surfaces of the CSM model when the two-way FSI analysis is enabled. In the RIBES platform RBF are employed to run the RSM model and to enable mapping for FSI.

11.2 RBF Metamodeling

Details about the RBF meta-modelling implemented to accomplish a generic optimization are provided in this section through a step-by-step description of the practical application of the method. In specific, a dataset composed by 45 DOE DPs defined according to the full factorial method is employed. Such a dataset is collected in the Table 11.1.

The columns of the table respectively contain the ID of the DOE case, the value of input parameters p1, p2, p3 and the value of the responses r1, r2. To illustrate the RS using RBF, three problems are extracted from the same dataset. At first the effect of the parameter p1 only is accounted. Then the combined effect of parameters p1 and p2 and finally the effect of all parameters are evaluated. The described implementation is done in the Mathcad tool working environment but it can be straightforwardly extended to different programming languages or design tools. The calculated data are stored in a matrix variable as reported in Table 11.2.

The proper dataset is firstly extracted (11.1). In the case of one parameter only, 5 points are kept and then used to define a set of RBF points (11.2) in an npar-dimensional space (in this case npar = 1). The nres (nres = 2) scalar functions are extracted as well (11.3).

$$\mathrm{DOEres} := \left|\begin{array}{l} \text{for } i \in 1..5 \\ \quad D^{\langle i \rangle} \leftarrow \left(\mathrm{DOEres}^{T}\right)^{\langle i \rangle} \\ \mathrm{out} \leftarrow D^{T} \end{array}\right. \tag{11.1}$$

$$x_k := \left|\begin{array}{l} \text{for } i \in 1..\mathrm{rows}(\mathrm{DOEres}) \\ \quad \left(x_{k_i}\right) \leftarrow \mathrm{submatrix}(\mathrm{DOEres}, i, i, 2, 1 + \mathrm{npar})^{T} \\ \mathrm{out} \leftarrow x_k \end{array}\right. \tag{11.2}$$

$$g_k := \left|\begin{array}{l} \text{for } i \in 1..\mathrm{rows}(\mathrm{DOEres}) \\ \quad \text{for } ic \in 1..2 \\ \quad\quad g_{i,ic} \leftarrow \mathrm{DOEres}_{i,ic+4} \\ \mathrm{out} \leftarrow g \end{array}\right. \tag{11.3}$$

Once the RBF input is defined, the generic RBF tools of Sect. 4.1 can be adopted. In particular, these are the **RBF_fit(·)** (4.1) and **RBF_eval(·)** (4.2) functions. However, in this section all the intermediate calculations are reported according to the Eqs. (2.17) up to (2.21) of Sects. 2.3.2 and 2.3.3.

Table 11.1 Data set used to show a generic optimization with an RBF-based surrogate model

D	p1	p2	p3	r1	r2
1	−10	0	0	8.943	−67.93
2	−5	0	0	9.043	−66.511
3	0	0	0	8.843	−62.817
4	5	0	0	8.511	−59.45
5	10	0	0	8.758	−57.541
6	−10	7.5	0	8.636	−63.683
7	−5	7.5	0	8.669	−61.68
8	0	7.5	0	8.384	−58.691
9	5	7.5	0	8.714	−56.685
10	10	7.5	0	9.283	−55.714
11	−10	15	0	8.88	−58.852
12	−5	15	0	8.765	−56.955
13	0	15	0	8.735	−55.274
14	5	15	0	9.2	−54.106
15	10	15	0	9.88	−53.388
16	−10	0	5	8.387	−72.622
17	−5	0	5	8.341	−69.853
18	0	0	5	8.358	−69.238
19	5	0	5	8.29	−66.436
20	10	0	5	8.007	−62.793
21	−10	7.5	5	8.164	−66.361
22	−5	7.5	5	8.106	−65.282
23	0	7.5	5	8.064	−64.155
24	5	7.5	5	7.872	−60.939
25	10	7.5	5	7.988	−58.648
26	−10	15	5	8.642	−62.619
27	−5	15	5	8.44	−60.62
28	0	15	5	8.11	−58.339
29	5	15	5	8.493	−56.803
30	10	15	5	8.838	−56.183
31	−10	0	−5	9.713	−63.226
32	−5	0	−5	9.547	−59.811
33	0	0	−5	9.427	−56.595
34	5	0	−5	9.866	−54.778
35	10	0	−5	10.422	−54.154
36	−10	7.5	−5	9.399	−59.586
37	−5	7.5	−5	9.459	−57.179
38	0	7.5	−5	9.654	−54.721
39	5	7.5	−5	10.172	−53.23
40	10	7.5	−5	10.715	−52.51

(continued)

Table 11.1 (continued)

D	p1	p2	p3	r1	r2
41	−10	15	−5	9.72	−55.763
42	−5	15	−5	9.783	−54.153
43	0	15	−5	9.941	−52.512
44	5	15	−5	10.427	−51.197
45	10	15	−5	10.948	−50.724

Table 11.2 Matrix variable of CFD calculated data

		1	2	3	4	5	6
DOEres=	**1**	1	−10	0	0	8.943	−67.93
	2	2	−5	0	0	9.043	−66.511
	3	3	0	0	0	8.843	−62.817
	4	4	5	0	0	8.511	−59.45
	5	5	10	0	0	8.758	−57.541
	6	6	−10	7.5	0	8.636	−63.683
	7	7	−5	7.5	0	8.669	−61.68
	8	8	0	7.5	0	8.384	−58.691
	9	9	5	7.5	0	8.714	−56.685
	10	10	10	7.5	0	9.283	...

The RBF function (11.4) is defined (a cubic spline in this case), and it is used to calculate the interpolation matrix **M** (11.5), and the constraint matrix **P** (11.6).

$$\phi(r) = r^3 \tag{11.4}$$

$$\begin{aligned} M := & \text{ for } i \in 1\ldots\text{rows}(x_k) \\ & \quad \text{for } j \in 1\ldots\text{rows}(x_k) \\ & \quad\quad M_{i,j} \leftarrow \phi\left(\left|x_{k_i} - x_{k_j}\right|\right) \end{aligned} \tag{11.5}$$

$$\begin{aligned} P := & \text{ for } i \in 1..\text{rows}(x_k) \\ & \quad P_{i,1} \leftarrow 1 \\ & \quad \text{for } j \in 1..\text{rows}\left(x_{k_1}\right) \\ & \quad\quad P_{i,j+1} \leftarrow \left(x_{k_i}\right)_j \end{aligned} \tag{11.6}$$

After the definition of a zero matrix (11.7) the system is assembled and solved for each scalar response function (11.8).

$$\text{zero\_mat}(m,n) := \left| \begin{array}{l} \text{for } i \in 1..m \\ \quad \text{for } j \in 1..n \\ \quad \quad \text{mat}_{i,j} \leftarrow 0 \\ \text{mat} \end{array} \right. \tag{11.7}$$

$$\begin{aligned} MP &:= \text{augment}\left(\text{stack}\left(M,P^T\right), \text{stack}(P, \text{zero\_mat}(\text{rows}(x_{k_1}) + 1, \text{rows}(x_{k_1}) + 1))\right) \\ g0 &:= \text{stack}(g_k, \text{zero\_mat}(\text{rows}(x_{k_1}) + 1, \text{cols}(g_k))) \\ \gamma\beta &:= \text{lsolve}(MP, g0) \\ \gamma &:= \text{submatrix}(\gamma\beta, 1, \text{rows}(x_k), 1, 1) \\ \beta &:= \text{submatrix}(\gamma\beta, \text{rows}(x_k) + 1, \text{rows}(\gamma\beta), 1, 1) \end{aligned} \tag{11.8}$$

The generic response **ic** can be evaluated with Eq. (11.9). It is interesting to notice that the equations are valid for an arbitrary number of input parameters and output functions. In the particular case of two responses, the compact form (11.10) is used (valid for any number of input parameters).

$$s_{resic}(x,ic) := \sum_{i=1}^{\text{rows}(x_k)} \left[(\gamma_{ic})_i \cdot \phi(|x - x_{k_i}|)\right] + \beta_{ic} \cdot \text{stack}(1,x) \tag{11.9}$$

$$s_{resall}(x) := \sum_{i=1}^{\text{rows}(x_k)} \left[\begin{bmatrix} (\gamma_1)_i \\ (\gamma_2)_i \end{bmatrix} \cdot \phi\left(\left|x - x_{k_i}\right|\right)\right] + \begin{pmatrix} \beta_1 \cdot \text{stack}(1,x) \\ \beta_2 \cdot \text{stack}(1,x) \end{pmatrix} \tag{11.10}$$

The RS can be used as a closed form representation of the system. A general optimisation algorithm can be used to drive the research in the meta-model space. The calculation of the meta-model is so fast that a simple screening method, that foresee to explore the response using a fine distribution of evaluation points. For this single parameter problem it can be evaluated according to (11.11).

$$\text{Bigset} := \left| \begin{array}{l} \text{id} \leftarrow 1 \\ \text{for } \alpha_{\text{deflector}} \in -10, -9.5 .. 10 \\ \quad \left| \begin{array}{l} \text{res} \leftarrow s_{\text{resall}}((\alpha_{\text{deflector}})) \\ \text{Bigsetrow} \leftarrow \begin{pmatrix} \text{id} \\ \alpha_{\text{deflector}} \\ \text{res}_1 \\ \text{res}_2 \end{pmatrix} \\ \text{for icol} \in 1..4 \\ \quad \text{Bigset}_{\text{id},\text{icol}} \leftarrow \text{Bigsetrow}_{\text{icol}} \\ \text{id} \leftarrow \text{id} + 1 \end{array} \right. \\ \text{out} \leftarrow \text{Bigset} \end{array} \right. \tag{11.11}$$

Figures 11.3 and 11.4 show, respectively, the charts of two coupled responses (and this is valid for any number of input parameters) and a single parameter response.

The case of two input parameters is handled using the same dataset but retaining 15 points (11.12).

$$\text{DOEres} := \left| \begin{array}{l} \text{for i} \in 1..15 \\ \quad D^{\langle i \rangle} \leftarrow \left(\text{DOEres}^{T}\right)^{\langle i \rangle} \\ \text{out} \leftarrow D^{T} \end{array} \right. \tag{11.12}$$

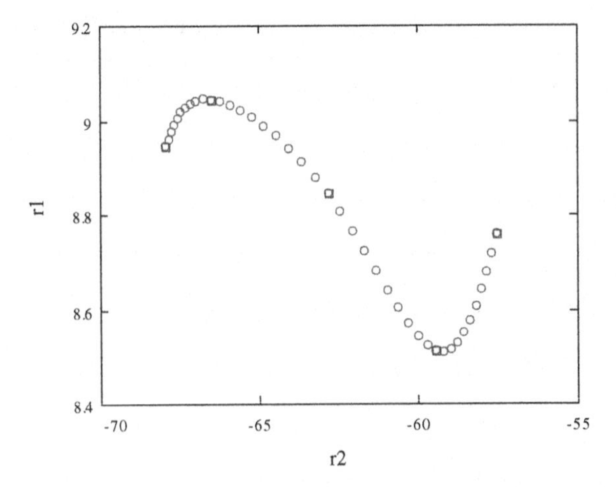

Fig. 11.3 Coupled parameters response plot

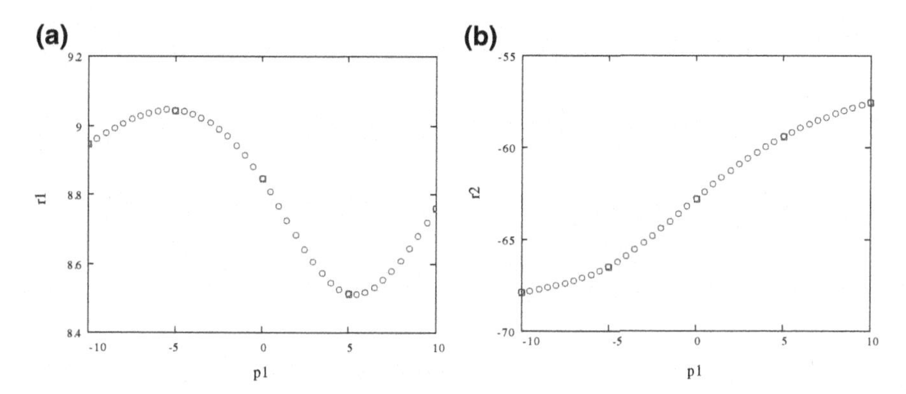

Fig. 11.4 Single parameter response plot: the responses r1 (left) and r2 (right) are evaluated as a function of p1

The same equations of the previous case are still valid with the exception that the big set is now calculated using two parameters (11.13).

$$
\begin{aligned}
\text{Bigset} := \Bigg| \;& \text{id} \leftarrow 1 \\
& \text{for } \alpha_{deflector} \in -10, -9.5 .. 10 \\
& \quad \text{for } \alpha_{driver} \in 0, 0.5 .. 15 \\
& \qquad \text{res} \leftarrow s_{resall}\left(\begin{pmatrix} \alpha_{deflector} \\ \alpha_{driver} \end{pmatrix}\right) \\
& \qquad \text{Bigsetrow} \leftarrow \begin{pmatrix} \text{id} \\ \alpha_{deflector} \\ \alpha_{driver} \\ \text{res}_1 \\ \text{res}_2 \end{pmatrix} \\
& \qquad \text{for icol} \in 1..5 \\
& \qquad \quad \text{Bigset}_{\text{id, icol}} \leftarrow \text{Bigsetrow}_{\text{icol}} \\
& \qquad \text{id} \leftarrow \text{id} + 1 \\
& \text{out} \leftarrow \text{Bigset}
\end{aligned}
\tag{11.13}
$$

The coupling chart is still a simple plot (reported in Fig. 11.5 left) and the responses against a single parameter can be represented as the one in Fig. 11.5 right.

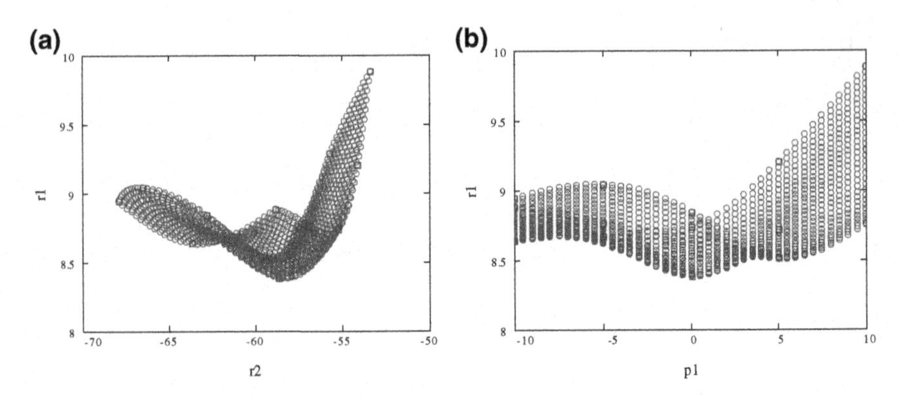

Fig. 11.5 Big dataset response plots: for all the DPs the two responses r1 and r2 are plotted on the left, response r1 is plotted as a a function of the parameter p1 on the right

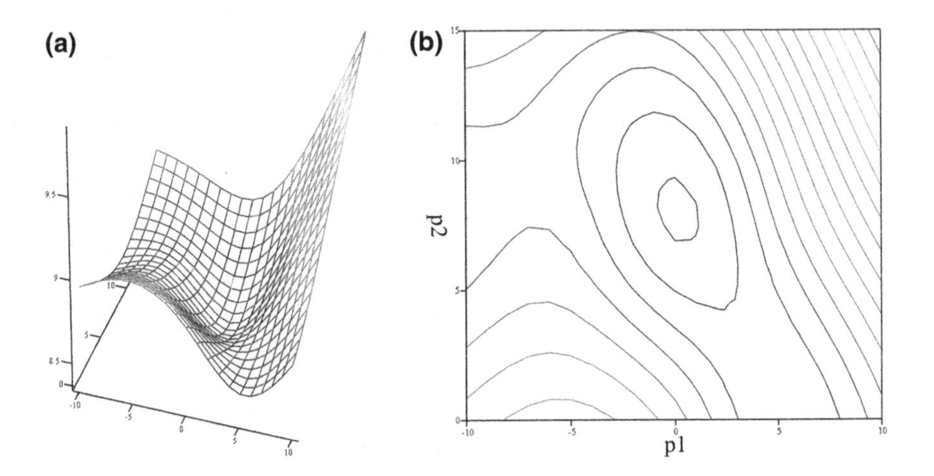

Fig. 11.6 Visualization of response r1 as a function of the two input parameters p1 and p2

The relation between the two input parameters and a single response can also be represented using a surface plot (a) or a contour plot (b) as shown in Fig. 11.6.

The method can be easily extended to the case of three input parameters. In this case the big set is evaluated according to (11.14). The coupling chart depicted in Fig. 11.7 is finally obtained.

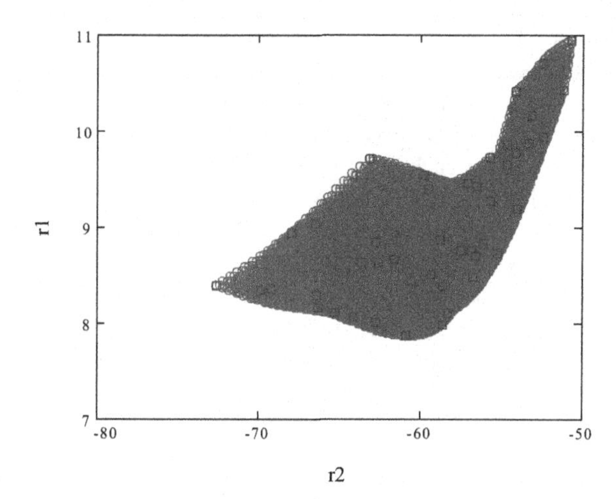

Fig. 11.7 Coupling chart obtained using three input parameters

$$
\begin{aligned}
\text{Bigset} := \; & \text{id} \leftarrow 1 \\
& \text{for } \alpha_{deflector} \in -10, -9.5 .. 10 \\
& \quad \text{for } \alpha_{driver} \in 0, 0.5 .. 15 \\
& \quad\quad \text{for } h_{driver} \in -5, -4.5 .. 5 \\
& \quad\quad\quad \text{res} \leftarrow s_{resall}\left(\begin{pmatrix} \alpha_{deflector} \\ \alpha_{driver} \\ h_{driver} \end{pmatrix}\right) \\
& \quad\quad\quad \text{Bigsetrow} \leftarrow \begin{pmatrix} \text{id} \\ \alpha_{deflector} \\ \alpha_{driver} \\ h_{driver} \\ \text{res}_1 \\ \text{res}_2 \end{pmatrix} \\
& \quad\quad\quad \text{for } \text{icol} \in 1 .. 6 \\
& \quad\quad\quad\quad \text{Bigset}_{\text{id},\text{icol}} \leftarrow \text{Bigsetrow}_{\text{icol}} \\
& \quad\quad\quad \text{id} \leftarrow \text{id} + 1 \\
& \text{out} \leftarrow \text{Bigset}
\end{aligned}
\tag{11.14}
$$

It is very important to notice that relevant points obtained in the response plots can be picked and extracted to know the exact value of responses and input parameters; this is especially needed in response chart as the one of Fig. 11.7. The optimal design region is at the boundaries (Pareto front) and a set of input parameters is retrieved for each point picked. DPs inside the boundaries are defined "dominated" meaning that a DP that is at least better for one of the reponses exists. Optimal points at the bondaries are for this reason defined as "non-dominated" ones. Such DPs have to be recalculated by a single (or a new local set around each point) CFD run for each point. New calculated results are then used to refine the original dataset putting more calculated points in the interesting area. A new RS can be thus fitted to all available data. Usually one refinement step is enough to achieve an acceptable convergence of the method.

11.3 NACA 0012 Optimisation

In this section a CFD optimization of a NACA 0012 aerofoil performed through the OpenFOAM tool is described. In particular, such an extensively used aerofoil represented one of the preliminary computational studies scheduled for testing the feasibility of the numerical optimization procedures required to be implemented during the Seventh Framework Programme project RBF4AERO (RBF4AERO 2013). To accomplish this target, the RBF4AERO platform capabilities in generating a DOE study considering the range of variation of RBF solutions amplifications have been explored, whilst the post-processing of the optimization has been carried out by means of the Mathcad tool.

11.3.1 Baseline Case Set-Up and Results

The baseline model of the 2D fluid-dynamic optimization task is a modified NACA 0012 profile having a trailing edge thickness of 0.6% of the chord length.

As far as flow conditions are concerned, the values of the parameters characterizing the undisturbed field are Mach = 0.753, Re = 3.88×10^6, AoA = 1.95° and T = 273.15 K.

To define the computing scenario of the 2D aerofoil CFD optimization, both geometrical constraints and aerodynamic targets have been specified. In particular, the geometrical constraints are:

1. specified thickness at 95% of chord length: the thickness of the aerofoil profile at 0.95 m from the leading edge has been chosen to be grater or equal to 0.015 m (1.5%);

2. specific thickness at 99% of the chord length: this specific thickness has been set to be equal to 0.006 m (0.6%), namely the thickness of the profile at the trailing edge location;
3. constant aerofoil chord to ensure the coherency of optimization results: the aerofoil length has been imposed to be equal to 1 m.

The aerodynamic targets and constraints of the 2D optimization task are minimum Cd, target Cl and Cm ≤ -0.1 (moment reference point located on aerofoil leading edge).

The optimization task was driven by the aerodynamic efficiency (AE), namely the ratio between Cl and Cd, maintaining approximately the value of Cl gained for the baseline and treating the Cl constraint as an objective function. As such, this constrained single-objective optimization (SOO) problem has been handled as a multi-objective optimization (MOO) problem by selecting from the front of non-dominated solutions the one which satisfies the lift constraint.

Starting from the aerofoil geometry, firstly a 2D structured mesh (see Fig. 11.8) constituted by 8.5×10^3 quadrilateral elements has been generated and, secondly, the resulting quadrangular mesh has been extruded to reach the depth of 0.25 m in the out-of-plane direction. In this way, and in view of satisfying the CFD solver requirements, a three-dimensional computational domain extending around 7 wing chord lengths downstream, 7 upstream and 7 up and bottom constituted by hexahedral elements has been finally obtained. In Fig. 11.8 the resulting extruded aerofoil is depicted together with a detail of the thickness present at trailing edge.

To reach a fully developed solution a simulation time of 0.6 s has been considered using an unsteady compressible solver, the Spalart-Allmaras turbulence model, and the Velocity Inlet, Pressure Outlet and Symmetry boundary conditions.

Table 11.4 reports the main results gained on the baseline configuration of the aerofoil in quasi-steady conditions, whilst the distribution of the Mach number in fully developed conditions over a plane that cuts orthogonally the extruded aerofoil at its middle is shown in Fig. 11.14 (left) where the formation of the shock on the upper part of the profile is well visible.

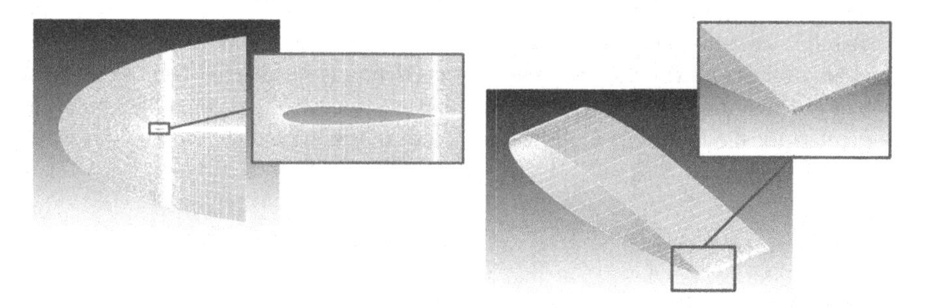

Fig. 11.8 2D quadrilateral mesh for 2D aerofoil optimization (left) and extruded modified NACA 0012 profile (right)

After having registered the appearance of the physical phenomenon characterizing the transonic regime, the objective of the aerofoil optimization is to improve the AE maintaining the value of Cl that is expected to cause a reduction of the extent and intensity of the illustrated shock wave.

11.3.2 Optimization Assumptions and RBF Solutions Set-Up

In order to achieve the targets of the present optimization, before designing the shape modifications, the influence of modifications on the physical behaviour of the aerofoil has been carefully taken into consideration. In particular, those areas are illustrated in Fig. 11.9.

The modification applied in each area has specific consequences. In particular, Area 1 influences the flow acceleration and pressure plateau, Area 2 influences pressure recovery and shock wave intensity, Area 3 influences the drag enhancement and Area 4 supports the maintenance of geometrical constraints. Given that, four shape modifications acting on these areas have been accounted and designed. The RBF set-up of such modifications, respectively termed dispTopFront, dispTopRear, dispBotFront and dispBotRear, has been accomplished through the stand-alone version of the RBF Morph software and they share the same rationale, namely the two-step approach. In the first step of the RBF solution, the portion of the aerofoil profile that needs to maintain its shape is fixed imposing a null displacement to a set of source points extracted from the surface mesh nodes of those areas, whilst in order to assign the aerofoil displacement a set of source points with an imposed displacement is created. In the second-step the RBF solution generated in the first step is imposed to the source points extracted from the surface mesh nodes of the aerofoil which are subjected to displacement, while the remaining surface nodes are kept fixed assigning a null displacement to the source nodes extracted from them. Moreover, one box-shaped domain to delimit the action of morphing is also generated. Among the fixed nodes are those belonging to the tail to satisfy one of the identified geometrical constraints.

All the shape modifications have been created with the aim to lower the aerofoil thickness till the maximum thickness x-location that is maintained fixed to respect one of the identified geometrical constraints concerning the maximum thickness. In particular, for the first three modifiers the mesh morphing at the deformed areas of

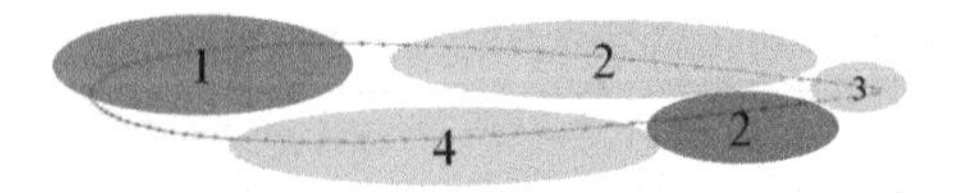

Fig. 11.9 Areas of potential interest for shape modifications of the aerofoil

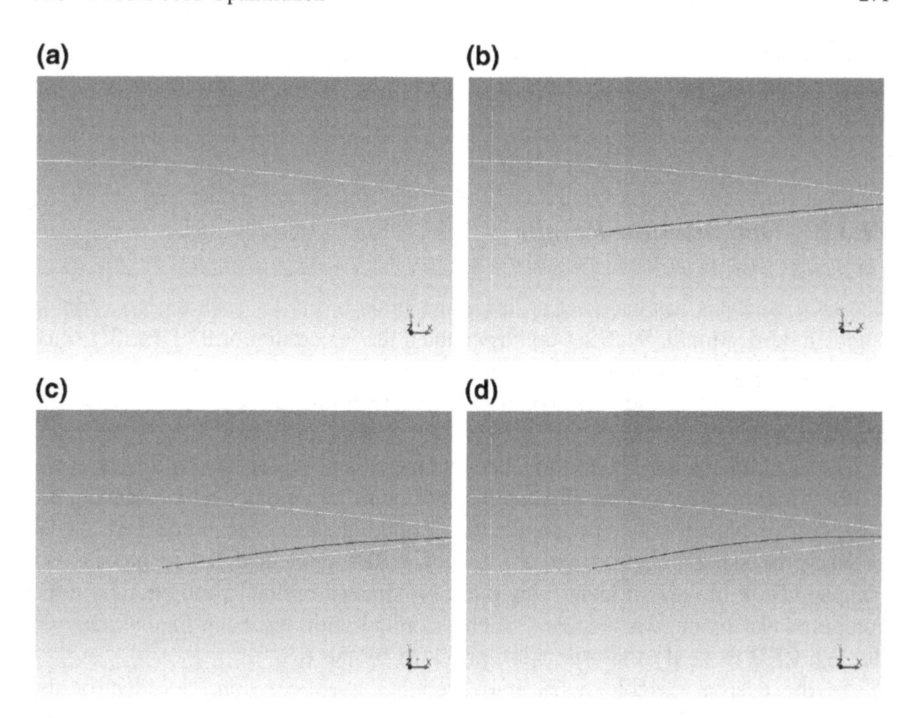

Fig. 11.10 Preview of the effect of dispTopFront RBF solution application at different intensity values

Table 11.3 2D aerofoil optimization case: RBF solution details

RBF solution	Source points of delimiting and moving domains	Source points extracted from surface mesh nodes	Lower bound amplification	Upper bound amplification
dispTopFront	896	290	0.0	1.5
dispTopRear	896	290	−1.5	0.5
dispBotFront	896	228	0.0	1.5
dispBotRear	896	290	0.0	2.0

the profile is applied through the source nodes positioned over the surface of a cylinder-shaped domain to which a predefined displacement is applied, whereas for the fourth is applied by means a source point extracted from surface mesh using an higher order RBF. In Fig. 11.10 the preview of the effect of the application of morphing for this latter RBF solution, termed dispTopFront, for different values of amplification is depicted from a lateral orthographic view by overlapping the baseline and the morphed shapes of the aerofoil.

Table 11.3 summarizes the main data dealing with RBF solutions of the four shape modifications in terms of the number of source points related to the specific used feature of the morpher tool and the amplification value for the lower and upper bound. Referring to these latter, the intervals of values preventing the excessive deterioration of mesh quality have been chosen by evaluating the preview of the morphing action.

11.3.3 Optimisation Results

The DOE analysis has been set up imposing three levels for each design variable (shape modifications). Such set-up has enabled the generation of 81 DPs (3^4) since the space filling has been done according to the full-factorial technique. Relating to objective functions, the aerodynamic coefficients have been monitored and recorded.

The calculations of all DPs have been managed and performed by means of the RBF4AERO platform which filled up the DOE table by automatically retrieving the values of the objective functions once the computing stage is accomplished. As far as post-processing is concerned, the use of a RBF-generated RS to process the complete DOE in view of identifying the approximated optimal configuration of the computational model was adopted. Such a configuration has been finally analysed through CFD to verify the estimation provided by the RS.

As the design variable taken into account have been four, to identify the potential optimal candidate Pareto plots having the drag coefficient in abscissa the aerodynamic efficiency and in ordinate, and the lift coefficient have been generated by the screening method extending (11.14) to the four parameters case. Such results are respectively illustrated in Figs. 11.11 and 11.12.

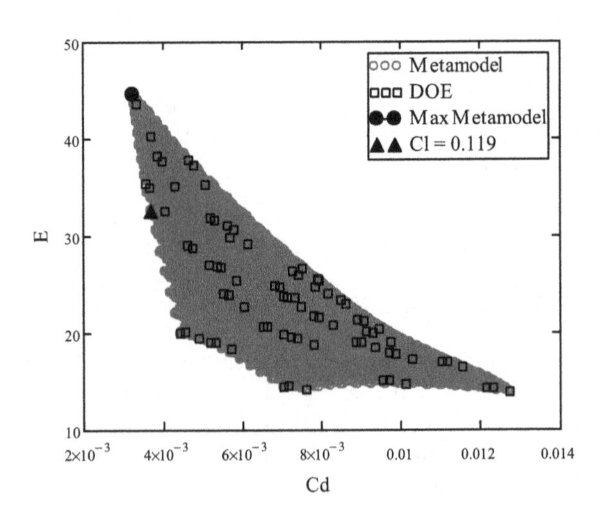

Fig. 11.11 Pareto plot for the 2D aerofoil optimisation concerning Cd and AE

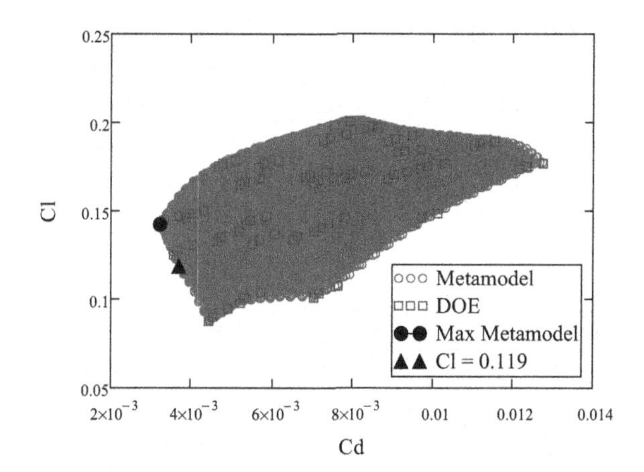

Fig. 11.12 Pareto plot for the 2D aerofoil optimisation concerning Cd and Cl

In those plots squares identify the calculated DPs (indicated with DOE), whilst the cloud of circles are the points of the RBF metamodel (RS). To finalize the optimization, the target is to identify on the Pareto front the configuration that minimizes Cd by guaranteeing, at the same time, a Cl close to the one calculated for the baseline configuration. Such a point is visualized through a triangle in Fig. 11.12.

The identified optimal point sets dispTopFront = 1.425, dispTopRear = −0.4, dispBotFront = 0.0 and dispBotRear = −0.1, and the foreseen Cd and Cl respectively are 3.550×10^{-3} and 1.190×10^{-1}. It is worth to underline that although the potential optimal point is slightly outside the initial defined design space (minimum value for dispBotRear is 0.0), it has been evenly adopted considering the aim of this analysis. The morphed configuration of the mesh corresponding to the above reported amplifications is depicted in Fig. 11.13.

Since that configuration has been identified through the metamodel, and thus it has been provided by an approximated method, the verification of the optimal design point has been conducted through CFD computing. To this end, a CFD solution has been run again with the model morphed according to the amplification values of the optimal configuration. Table 11.4 reports the main results gained on the baseline and optimal configuration of the aerofoil. Taking into account these

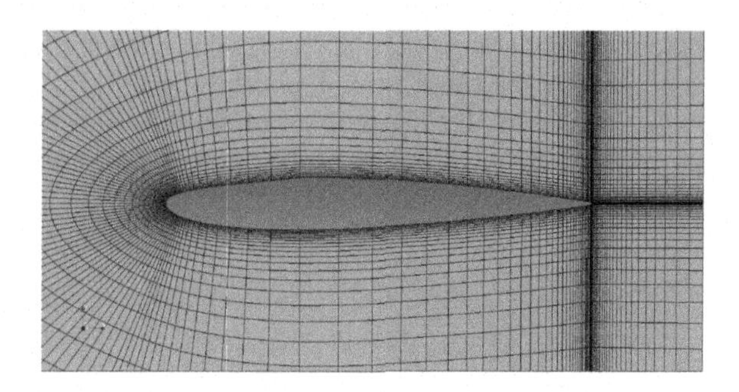

Fig. 11.13 Optimal mesh configuration according to RBF routines based method

Table 11.4 Main results of the aerofoil study in the baseline and optimal configuration

Parameter	Baseline case value	Optimised case value	Relative variation (%)
Cd	6.782×10^{-3}	3.598×10^{-3}	−46.948
Cl	1.197×10^{-1}	1.196×10^{-1}	−0.084
Cm	3.229×10^{-2}	3.394×10^{-2}	5.110
$Mach_{max}$	1.33	1.24	−6.767

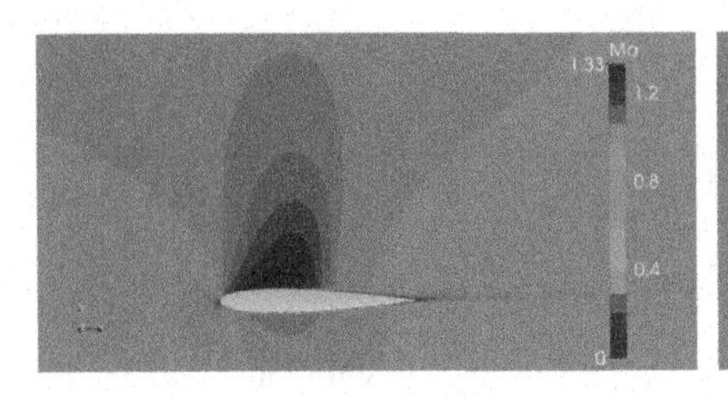

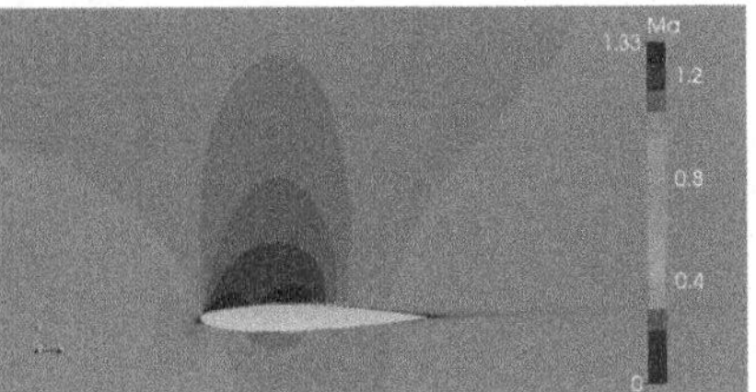

Fig. 11.14 Comparison between the baseline (left) and optimized (right) aerofoil

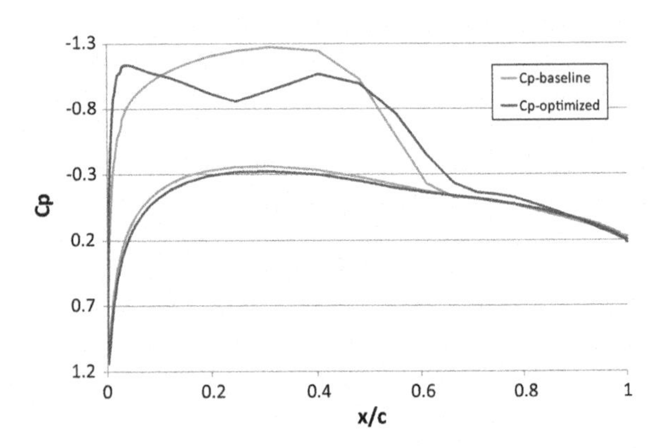

Fig. 11.15 Cp comparison between the baseline and optimized aerofoil

results, the constraint on the Cl value is respected because it has been maintained almost the same of that computed for the baseline.

Figure 11.14 depicts the comparison between the Mach distribution of the baseline configuration (left) and the optimized one (right). As visible in the optimal configuration the Mach peak diminishes by 6%.

In Fig. 11.15 the Cp profiles of the baseline and optimal configuration are plotted and the modification of the pressure plateau is noticeable.

11.4 Glider Optimisation

In this section the glider optimisation previously described in Sect. 8.6.1 is performed adopting a different workflow (Biancolini et al. 2016) that is illustrated in Fig. 11.16 through a block diagram where rectangular blocks identify computational data, circular blocks attain to particular operations or features, whilst blocks concerning the usage of the same software are included in a coloured region.

Such a workflow, managed by means of the ANSYS Workbench environment and the Mathcad tool, starts from the CAD representing the baseline configuration of the model to be analysed, and provides the CAD model of the optimized shapes as final output. In particular, the operations characterised by the highest level of automation are those performed by means of ANSYS DesignXplorer (DX) which acts as driver of the entire optimization procedure. The proposed procedure can be also implemented using different environments and the automation can be steered using either in-house, open source or commercial optimisation tools.

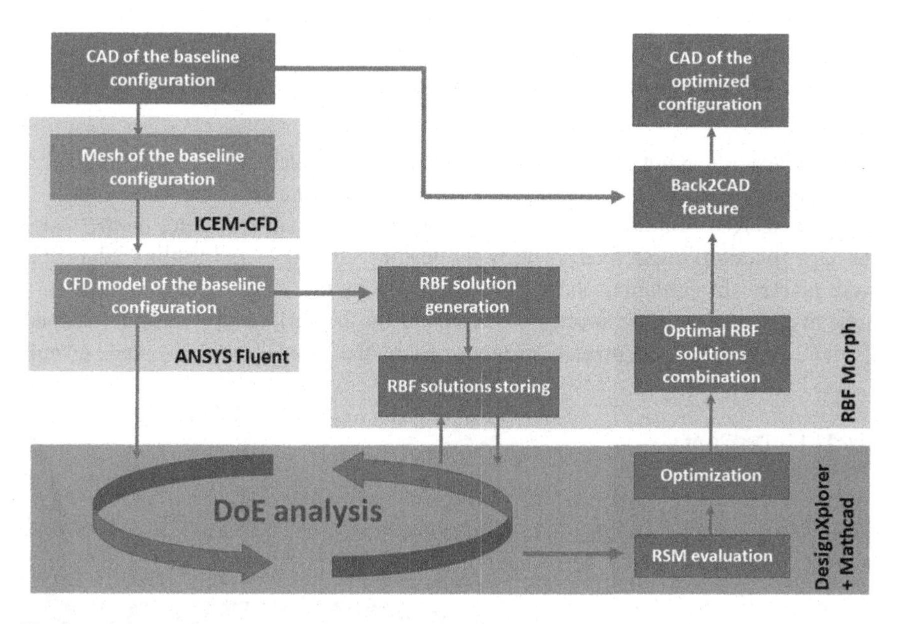

Fig. 11.16 Workflow of the mesh morphing based shape optimization

11.4.1 Baseline Case Set-Up and Results

The mesh consists of a 7 million multi-block structured hexahedral grid covering around 40 wing chords upstream the model, 60 downstream and 50 on the side, and it models just half domain. Figure 11.17 shows the cells distribution on wall surfaces of the case.

The analysis was performed at an AoA of 8°. The flow conditions are Mach 0.08 and Reynolds 1.24 million (based on reference chord of length 0.8 m) that correspond to the flying condition at 2000 m of altitude. A substantial flow detachment is observed at this flight condition (see Fig. 8.19).

A pressure based incompressible RANS solver was used to carry out a steady simulation adopting the two equations $k - \omega$ shear stress transport (SST) turbulence model (Menter 1994), the SIMPLE scheme for the pressure-velocity coupling and the Second Order Upwind scheme for spatial discretization.

Figure 11.22 (left) depicts the visualization of the shear stress streamlines superposed on skin friction coefficient distribution. This output showcases that the flow detachment registered during the experimental flight is numerically reproduced and confirmed also in this case.

11.4.2 Optimisation Assumptions and RBF Solution Set-Up

The optimisation assumptions are the same already detailed in 8.6.1 and, then, also in this case the aerodynamic efficiency ($E = C_l/C_d$) is the objective function.

As far as RBF solutions are concerned, two shape modifiers acting on the wing/fuselage junction, respectively nearby the leading edge (P1) and trailing edge (P2) were created. In particular, such a set-up envisions the assignment a wanted (not null) displacement to the source nodes laying on the surface of a cylinder-shaped domain, located at a proper distance from the model surface, and a null

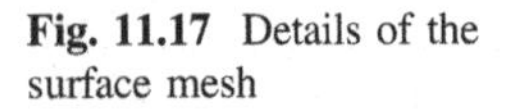

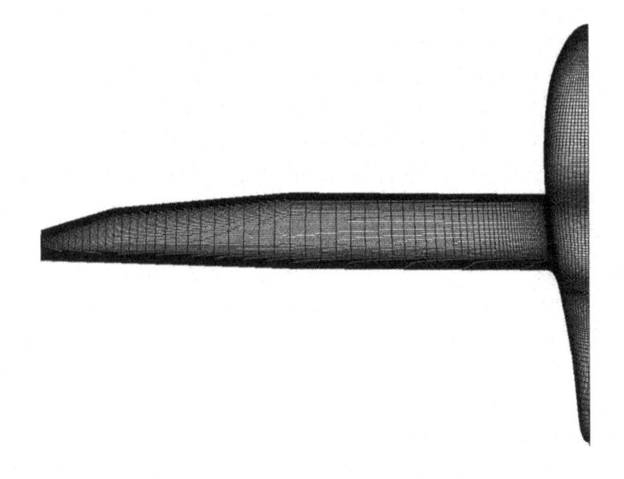

Fig. 11.17 Details of the surface mesh

displacement to the source nodes extracted from surface mesh nodes of the area required to maintain the baseline shape during morphing. Moreover, a box-shape shaped domain to limit the morphing action was also generated. The preview of the location of the source nodes generated by the described set-up is shown in Fig. 11.18, whilst the effect produced on the geometry by the application of the generated solutions is shown in Fig. 11.19.

Similarly to the previous study the limits of the amplifications of both modifiers have been set such to guarantee, even when combined, an acceptable quality of the mesh after the application of morphing which detrimental effects on mesh quality.

Once the RBF solutions have been stored and their bounds have been determined, the computational mesh is made parametric through their amplifications and the CFD computing stage can start.

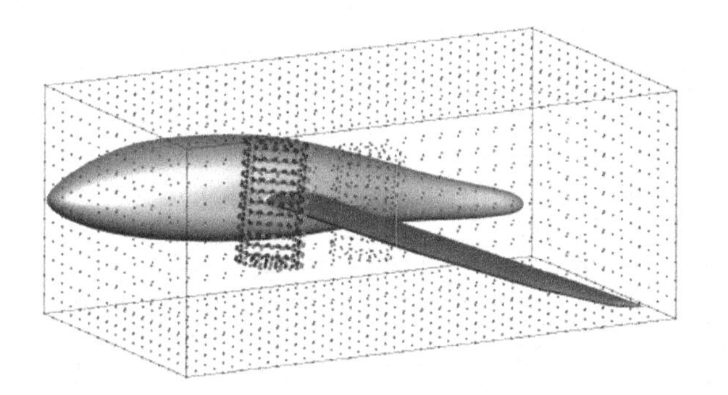

Fig. 11.18 Source nodes position of the RBF morphing problem

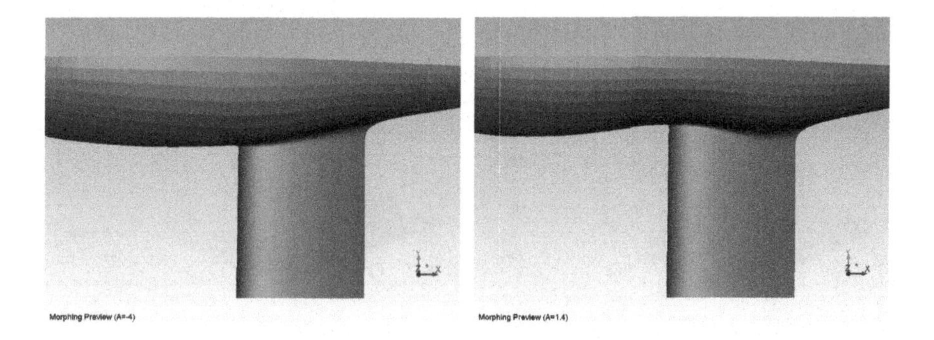

Fig. 11.19 Shape deformation induced by shape modifiers

11.4.3 Optimisation Results

The DOE table was populated with 25 DPs in which the first 20 were defined using a Latin hypercube sampling design method (Costa et al. 2014), whereas other 5 extra points were successively added to enrich the design space in the most interesting area (high values of the aerodynamic efficiency). In order to speed up the DOE table completion, all DPs analyses were ran restarting the computation from the already achieved baseline solution and then morphed according to the imposed input parameters.

Sampling strategy was previously tuned and refined using a coarser mesh for the evaluation of each DP (Costa et al. 2014). Despite the quantitative differences observed when the same point is evaluated using the finer (high-fidelity) and the coarser meshes, as described through the RS shown in Fig. 11.20 generated according to Sect. 11.2 with Mathcad, the relevant design area stays approximatively the same. The squared points on the surface are the values of the objective function of the DOE table. A screening approach on the RBF metamodel has been adopted to generate a finely spaced cloud of DPs (3600 individuals) according to a full-factorial approach. As visible, the trend evaluated in terms of aerodynamic efficiency is similar and the coarser model gives a slight lower value of the maximum aerodynamic efficiency.

The upper design space border exhibits the highest efficiency portion of the RS highlighting a sensitivity of the objective function with respect to the leading edge modifier (P1) higher than the trailing edge one (P2).The high quality and smoothness of the RS was possible applying data normalisation procedure explained in Sect. 4.5.

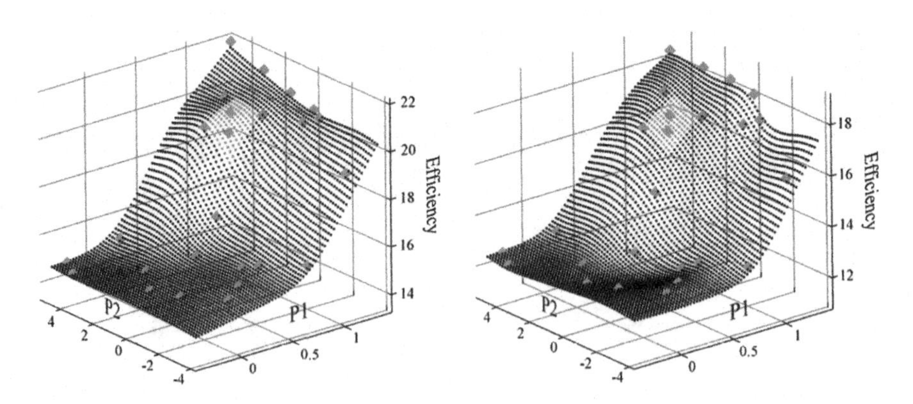

Fig. 11.20 RS obtained using the high-fidelity (left) and coarse (right) model represented as a scatter plot

11.4.4 Optimal Candidate Selection and Optimal CAD Generation

Due to manufacturing constraint, the research of the optimum is focused on the light area of the RS reported in Fig. 11.20. An high accuracy of the RS is expected in this subset where five computed DPs should ensure a good agreement between the metamodel and the objective function.

In the afore-indicated area the first parameter is upper bounded, whereas the second parameter exhibits a quite flat behaviour. A better understanding can be gained cutting the RS at aforementioned bound, as shown in Fig. 11.21. Thanks to this section plot, it can be understood that an area in which the efficiency exhibits an higher rise begins at P2 = 1.5 and reaches the highest efficiency at P2 = 3, after that

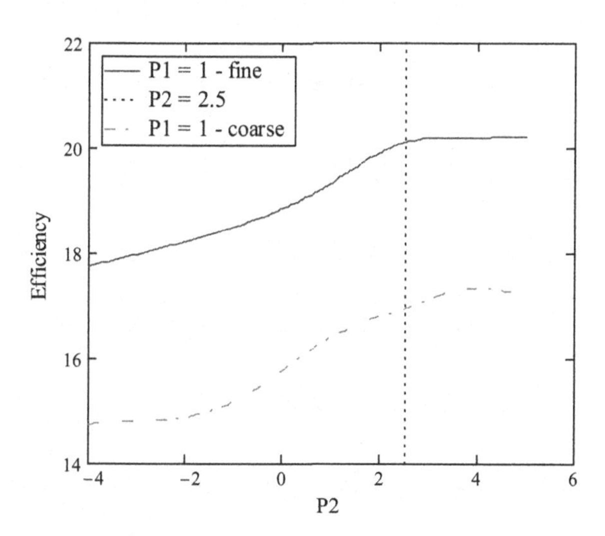

Fig. 11.21 Section of RS at P1 = 1 for the fine (continuous profile) and coarse (dotted) mesh

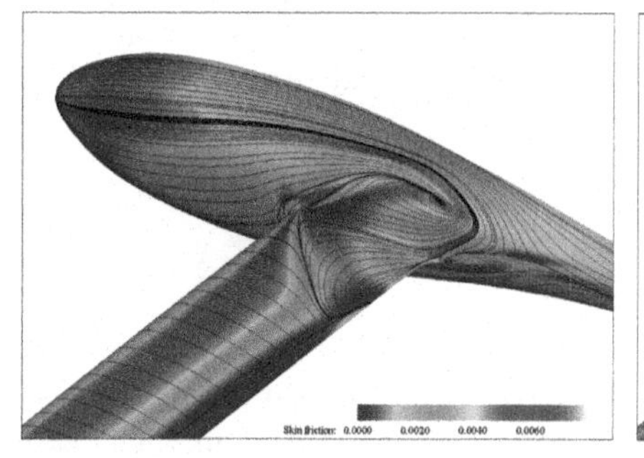

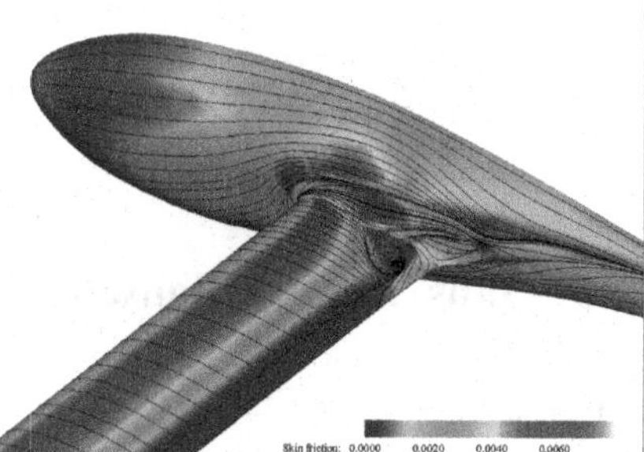

Fig. 11.22 Streamlines superposed on skin friction coefficient distribution for the baseline (left) and optimized (right) configuration

Table 11.5 Main results related to aerodynamic coefficients

Parameter	Baseline	Optimised	Variation (%)
C_D	0.076	0.0605	−20.4
C_L	1.131	1.216	+7.5
E	14.88	20.1	+35.0

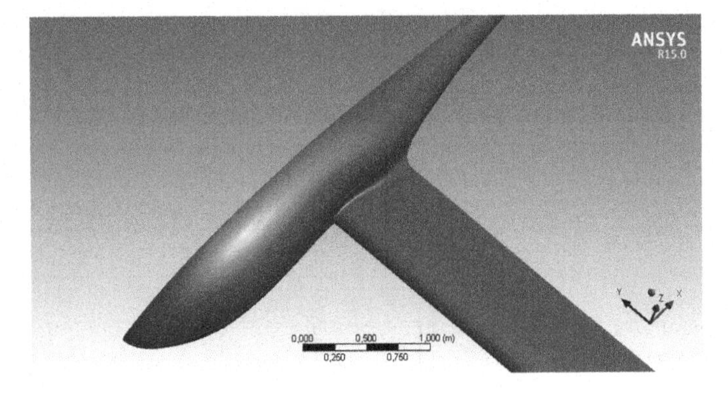

Fig. 11.23 CAD model of the optimized configuration

an almost flat behaviour is observed. For this reason P2 = 2.5 has been selected as optimal candidate (dotted vertical line in Fig. 11.21).

The selected optimum (P1 = 1; P2 = 2.5) has been validated with a new analysis of the geometry generated applying the combination for the amplification values defined by the optimal design. The solution gained through the high-fidelity model is reported in Fig. 11.22 (right), where the significant reduction of separation obtained with respect to the baseline geometry (left) is evident.

In Table 11.5 the aerodynamic coefficients of the baseline and the obtained optimal solution are collected. The aerodynamic efficiency of the optimized configuration is relevantly improved (35%).

As a completion of the design process, the optimal CAD is generated by morphing the baseline CAD model (in STEP format) using the RBF solutions amplified in accordance to the parameters of the optimized solution. The final resulting CAD model is visualized in Fig. 11.23.

11.5 Sails Trim Optimisation

In this section an RBF-based metamodel is applied in the post-processing concerning the optimisation of the trims of yachts sails performed through mesh morphing (Biancolini et al. 2014a) with the objective to find the maximum thrust configuration.

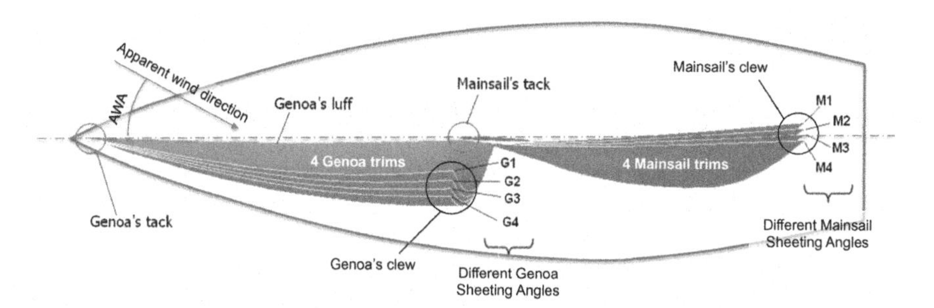

Fig. 11.24 Designation of the wind-tunnel sail positions

In particular, four trims of the fore (genoa) and aft (mainsail) sails of a rigid pressure-tapped 1:15th-model–scale of an AC33-class yacht tested at the University of Auckland wind tunnel (Viola et al. 2011) were numerically simulated and explored. Such a shape modifications are the sheeting angle of both sails, the apparent wind angle (AWA, i.e. the complementary angle between the wind experienced by the yacht and the yacht velocity) and heeling angle. Relating to experimental testing, four genoa sheeting angles (G1–G4) and four mainsail sheeting angles (M1–M4) were tested rotating the sails around the axes defined by the end points of the leading edges by 1.4° and 1° respectively (see Fig. 11.24), for a total number of 16 trim combinations.

The baseline sail trim, G3M2, was chosen as the trim allowing the maximum drive force, namely the global aerodynamic force component along the boat axis.

11.5.1 Baseline CFD Case Set-Up

This experiment was simulated through the CFD solver ANSYS Fluent with a hexahedral computational grid relatively small (1.5 million nodes) and built by means of ANSYS ICEM CFD. Such a grid allowed to get $y^+ < 3$ on the leeward side where regions with separated flow occur, and $5 < y^+ < 25$ on the windward side where the boundary layer is attached.

The steady incompressible RANS equations for Newtonian fluids were solved with a pressure based solver and the $k - \omega$ SST turbulence model with low Reynolds correction enabled. A SIMPLEC scheme was used to couple velocity and pressure, whilst the second order accuracy discretization algorithms were used.

11.5.2 RBF Set-Up

The morpher tool (add-on version of the RBF Morph software) was used to generate the RBF solutions. With regard to the first two sail trim parameters, that are the mainsail sheeting angle and the foresail sheeting angle, a similar RBF set-up was accomplished.

Two clouds of source points extracted from the surface mesh nodes of sails are firstly defined and used to accurately control sail shapes assigning a rigid rotation about one sail axis while keeping fixed the points on the other one (see Fig. 11.25 left). The reference rotation is 1° and, so, the amplification applied to the shape modification of the sail rotation is expressed in degrees. To complete the set-up, RBF centres with a null displacement are added on a cylindrical surface with a spacing defined by the user. Such a spacing has to be tuned to enforce a vanishing RBF field at the border (an optimal value 0.15 m has been used for this study).

The overall RBF problem is composed by 4156 centres and takes about 4 s for fitting (Fig. 11.26 left). The gained shape modifiers are verified imposing a combination of them at the maximum values (i.e. 4.1° for genoa sheeting angle and 3.0° for mainsail sheeting angle) and evaluating the mesh quality degradation. The extent of the combined shape modification, identified by the configuration G4M4, is shown together with the original mesh in Fig. 11.26 (right). In this latter image, in particular, the preview of the volume mesh over an horizontal cutting plane located at 1 m over the base plane, allows to understand how the deformation is propagated in the volume mesh accurately preserving the shape of the volume mesh itself.

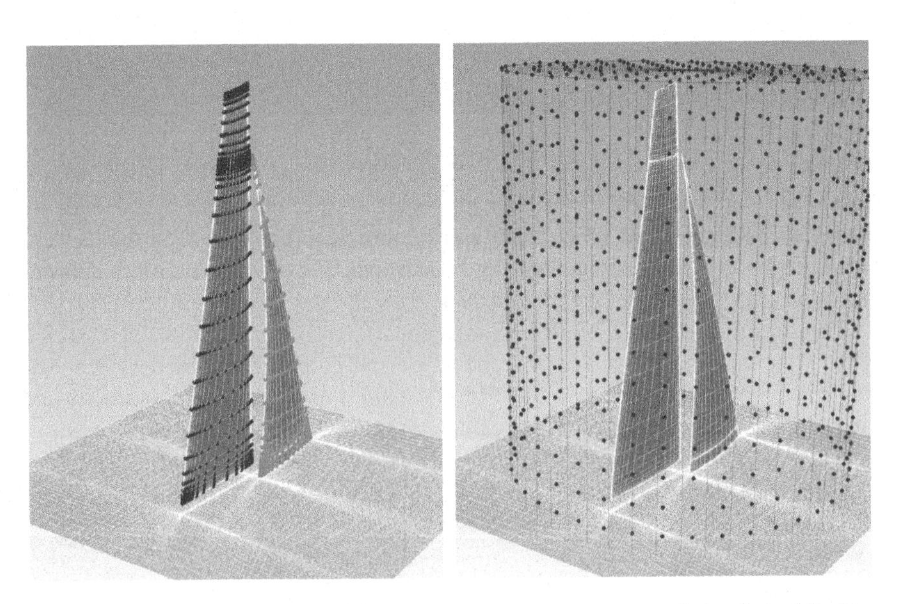

Fig. 11.25 Source points of the genoa and mainsail (left) and on the cylinder-shape delimiting domain (right)

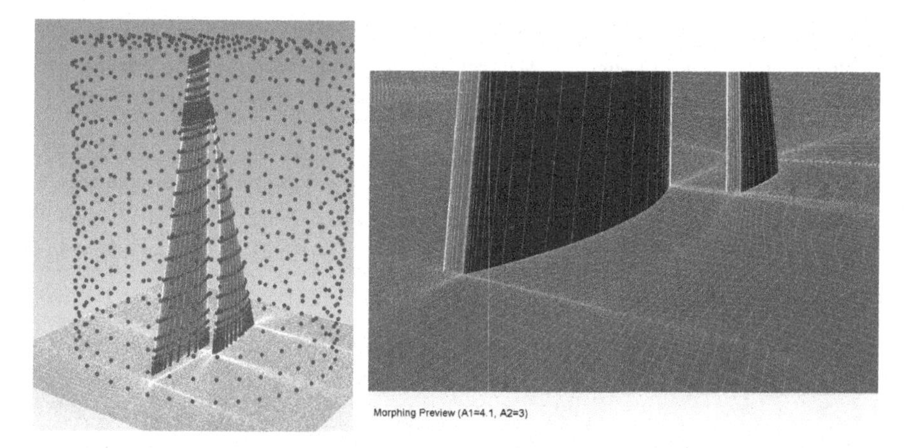

Fig. 11.26 Preview of the source points (left) and morphing preview of G4M4 configuration (right)

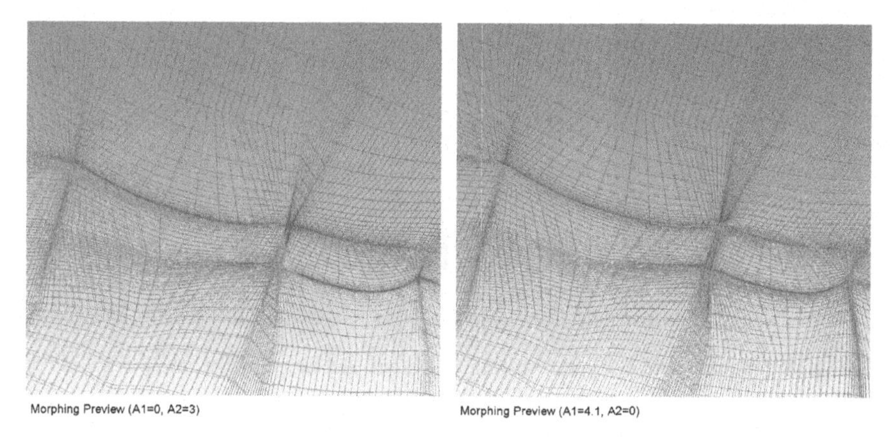

Fig. 11.27 Orthographic view of internal morphed mesh at G1M4 (left) and G4M1 (right) configurations

The morphing effect of the generated shape modifiers acting separately at the maximum height is shown in Fig. 11.27, where the original and the deformed meshes are superposed over the just defined cutting plane, and visualised from the top view and according to a wireframe representation in an orthographic modality.

Using a combination of both shape modifiers, the morphing action takes approximately 90 s in serial manner and approximately 30 s running on 4 cores. Table 11.6 shows that, in the relevant range, the quality of the mesh is very well preserved (the maximum reduction of Minimum Orthogonal Quality is about 13%).

The AWA angle is controlled using the two cylindrical domains, depicted in Fig. 11.28 (left), with a 0.2 m for the spacing so as to finally generate 3300 source

Table 11.6 Quality of morphed versus shape parameters at bounds

Configuration	Genoa sheeting angle (G) (°)	Mainsail sheeting angle (M) (°)	Minimum orthogonal quality	Maximum aspect ratio
G1M1	0	0	1.63E−02	1.555E+03
G4M1	4.1	0	1.38E−02	1.539E+03
G1M4	0	3.0	1.63E−02	1.549E+03
G4M4	4.1	3.0	1.41E−02	1.539E+03

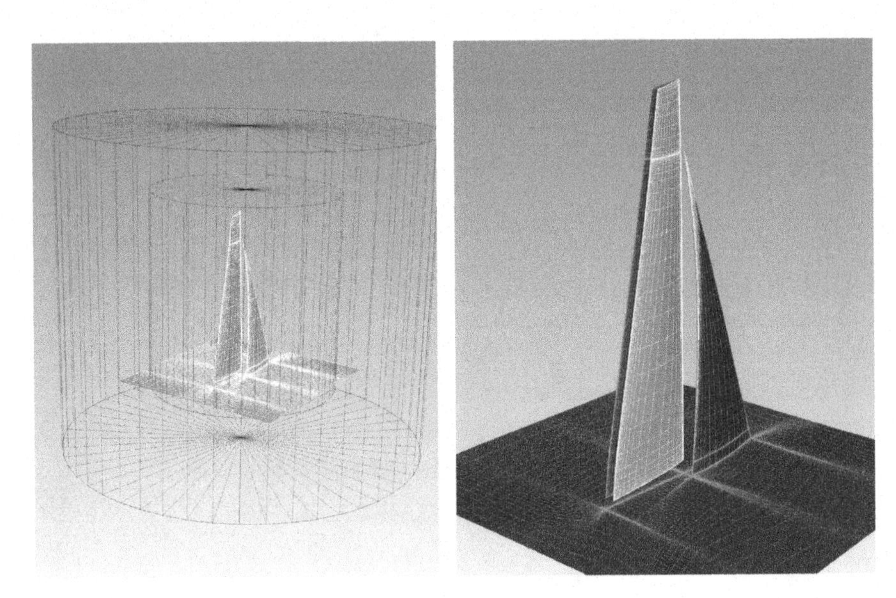

Fig. 11.28 Cylinders used to control AWA modifier (left) and morphing preview at AWA = −5° (right)

points. In specific, the mesh contained in the internal cylinder is subjected to a rotation about the cylinder axis by settings this movement to its source points. The rotation is so propagated outside the domain and it is limited by the external cylinder surface where the RBF points have a null displacement condition so that only the mesh included in the space between the two cylinders is actually deformed as shown in Fig. 11.28 (right). This technique is quite straightforward to build up and it is very effective for any boat configuration.

The heeling angle is more difficult to control and, for this reason, the two-step procedure was employed. In the first step, the morphing of the Genoa surface mesh is managed by imposing a rigid rotation to the source points extracted from the upper part of the sail leaving a deformable buffer to take the mesh deformation, whereas the source points positioned at the intersection between the sail and the

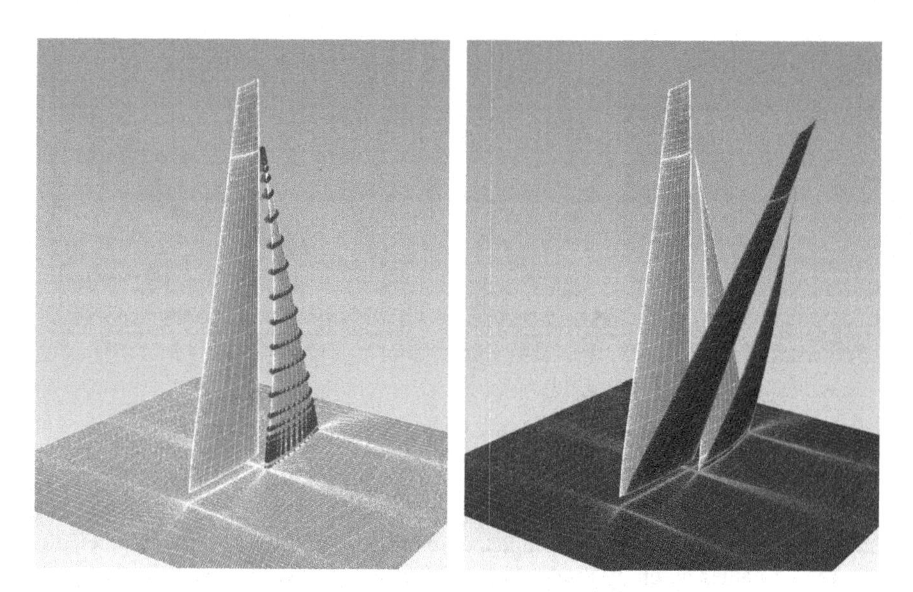

Fig. 11.29 Source points of the first step of the RBF set-up for heeling angle (left) and morphing preview at Heeling Angle = 20° (right)

plate are kept fixed (see Fig. 11.29 left). The second step of the RBF set-up, used for volume morphing, is identical to the one used for sheeting angles, but the movement imposed to the sails is different. In this case, the mainsail is rotated about the heeling axis, the genoa is rotated using the first step result that guarantees the wanted shape at the overall sail, including the deformable part (see Fig. 11.29 right).

The generated RBF fields were verified (Biancolini et al. 2014a) considering, as reference, the remeshing based method previously validated.

11.5.3 Optimisation Results

The RSM method was employed to optimise the combined effect of the four generated shape parameters. In specific, to explore the 4-dimensional design space the optimal space filling approach was adopted using the optimisation software DX (Design eXplorer), whilst the aerodynamic drive coefficient (CX), the aerodynamic side coefficient (CY), and the drag (Cd) and lift (Cl) coefficients were set as objective functions for the optimisation. In order to guarantee a proper density of the design space and a good accuracy in numerical prediction, 100 DPs were calculated through CFD within intervals of the shape modifiers amplifications slightly reduced so as to limit the possible worsening of mesh quality due to the combined effect of the modifications.

Figure 11.30 shows the main results of the optimisation exploiting the RBF-based post-processing. In specific, the CX–CY (left) and Cd–Cl (right) pairs

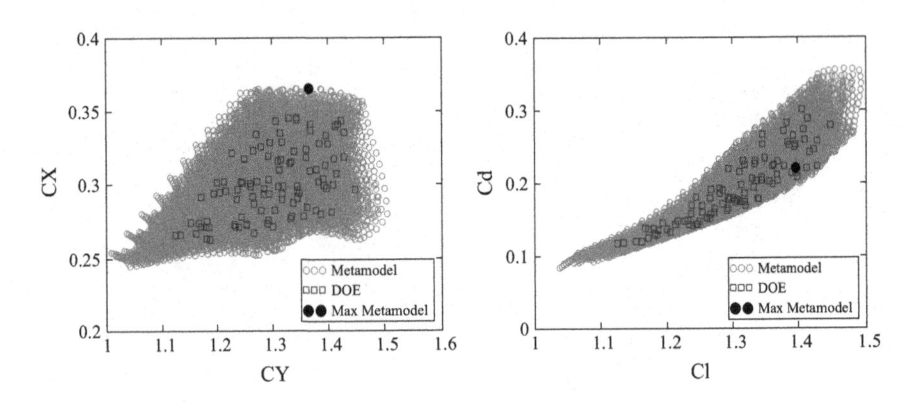

Fig. 11.30 Objective functions values obtained during optimisation

results obtained using CFD, namely those referring to DOE (squares), and employing the Metamodel are depicted (circles). As visible in both plots, the intervals of values given by the Metamodel are larger than the ones of DOE and, in the case of CX–CY results, a quite large range of CY guarantees approximatively the highest value of CX.

A sensitivity analysis of the objective functions with respect to the shape parameters has been conducted assigning 4 values of amplification for each modification. Since in this stage the variation of a single input parameter is applied while other parameters are zero-valued, a larger range for the amplification was affordable because just one modifier stresses the mesh. In specific, 0/+4° for both sheeting angles (mainsail and genoa), 17/29° for the AWA and −1/20° for the heeling were explored. Figure 11.31 shows the results of the sensitivity analysis in terms of CX (left) and CY (right), where RS, in reported legend, corresponds to values that have been obtained through the RSM method.

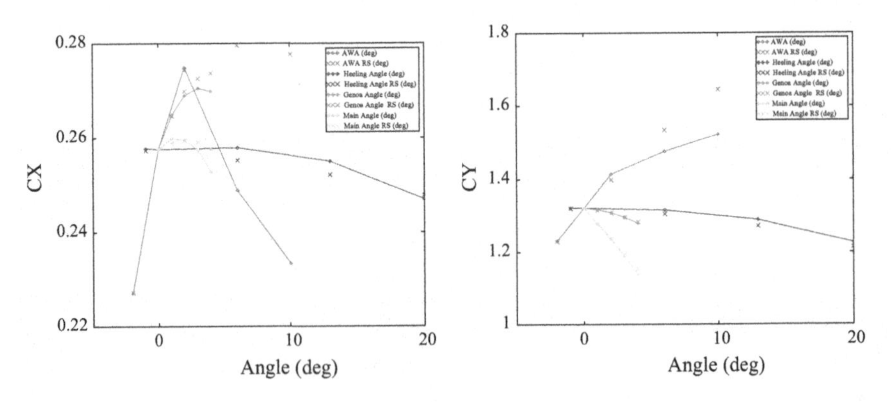

Fig. 11.31 CX (left) and CY (right) for different AWA, heeling, and sheeting angles variations

The Metamodel allows to straightforwardly select a possible optimal solution choosing the point characterised by the highest CX (0.366). Such a point, referred to ad Max Metamodel in Fig. 11.30 (left), is generated by the following combination of the shape modifiers' amplifications: mainsail sheeting angle 3°, genoa sheeting angle 5°, AWA 24° and heeling angle 0°. In such a configuration, the values of CY, Cd and Cl respectively are 1.366, 0.2 and 1.397.

References

Bernaschi MA, Sabellico A, Urso G, Costa E, Porziani S, Lagasco F, Groth C, Cella U, Biancolini ME, Kapsoulis DH, Asouti VG, Giannakoglou KC (2016) The RBF4AERO benchmark technology platform. In: ECCOMAS congress 2016—VII european congress on computational methods in applied sciences and engineering, Crete Island, Greece, 5–10th June 2016

Biancolini ME, Viola IM, Riotte M (2014a) Sails trim optimisation using CFD and RBF mesh morphing. Comput Fluids 93:46–60 (Elsevier)

Biancolini M, Viola I, Ramirez S (2014b) Alla ricerca delle regolazioni ottimali di una vela mediante mesh morphing. A&C. Analisi E Calcolo. ISSN 1128-3874

Biancolini ME, Costa E, Cella U, Groth C, Veble G, Andrejašič M (2016) Glider fuselage-wing junction optimization using CFD and RBF mesh morphing. Aircra Eng Aerosp Technol 88 (6):740–752. https://doi.org/10.1108/AEAT-12-2014-0211

Biancolini ME, Cella U, Groth C, Chiappa A, Giorgetti F, Nicolosi F (2017) Progresses in fluid structure interaction numerical analysis tools within the EU CS RIBES project. In: EUROGEN2017 International conference, Madrid, Spain, 13–15 Sept 2017

Costa E, Biancolini ME, Groth C, Cella U, Veble G, Andrejasic M (2014) RBF-based aerodynamic optimization of an industrial glider. In: 30th International CAE conference, Pacengo del Garda, Italy

Cavazzuti M (2013) Optimization methods: from theory to design scientific and technological aspects in mechanics, Springer, 2012-09-14. ISBN: 9783642311864

European Commission (2013) RBF4AERO Project. http://cordis.europa.eu/project/rcn/109141_en.html

Menter F (1994) Two-equation eddy-viscosity transport turbulence model for engineering applications. AIAA J 32(8):1598–1605

Viola IM, Pilate J, Flay RGJ (2011) Upwind sail aerodynamics: a pressure distribution database for the validation of numerical codes. International Journal of Small Craft Technology, Trans. RINA, 153(B1), 47–58.

Chapter 12
Advanced Field Data Post-processing Using RBF Interpolation

Abstract In this chapter a collection of application faced with an advanced post-processing method based on RBF fields is presented. The idea is that field information known at discrete points in a continuum domain (displacement field of an elastic body subjected to load and deformed, local speed and pressure in a fluid) can be interpolated so that scattered information becomes continuous and, in some cases, differentiable. Examples based on the theory of elasticity are firstly provided: the post-processing of a FEA solution available as a displacement field can be used to compute strains and stresses with the benefit to make them mesh independent (and so ready to be used for the generation of various plot) and better resolved in the space despite the coarseness of the field. Experimental data can be processed according to the same principle so that RBF become an useful tool for strain and stress evaluation by image processing. The chapter then addresses another useful application of RBF mapping suitable for the compensation of metrological data: a complex environment that loads the structure and can adversely affect the quality of the measurement can be subtracted by the inverse mapping of the displacement field achieved by FEA so that measured points can be corrected and positioned in their undeformed configuration for an effective comparison with reference CAD geometry. The chapter is concluded showing how RBF can be used for the interpolation of experimentally acquired flow information related to hemodynamics and showing the potential of an upscaling post processing toward the reduction of the resolution of in vivo acquired flow field (and so the exposition time of the patient).

12.1 Introduction to the Field Data Interpolation Problem

As already explained in the overview of this book, RBF were first adopted as an interpolation tool for scattered data. The first application of multi quadric (MQ) published by Hardy (Editor 1992) for the representation of topographic data dates back to 1971. A review of the evolution of MQ over the following twenty years has been presented in the review by Hardy himself (1990).

M. E. Biancolini, *Fast Radial Basis Functions for Engineering Applications*,
https://doi.org/10.1007/978-3-319-75011-8_12

There are many engineering applications that pose again interpolation challenges similar to the topographic one successfully addressed by Hardy in the past where RBF, not limiting to MQ, can be used. The solution of a physical problem, such as the flow field resulting from a CFD analysis or a deformation field resulting from a FEA analysis, is often computed on a grid and so it is available at discrete positions in the space which can be, for instance, nodal positions or cell centres. A similar scenario occurs for field data obtained experimentally with the extra complexity that, in this case, a noise due to the measurement chain has to be considered as well.

In such situations RBF can be successfully used by scientists and engineers to simplify the analysis of the data acting as a powerful post-processing tool which allows to make meshless the information of interest. In this case, it is worth to notice that the long distance interactions between points that should be captured in mesh morphing applications described in Chap. 6 could be neglected. The information to be reconstructed and interpolated is usually fitted using a local RBF around the observation point. If large datasets need to be processed and the interpolation is required, globally the partition of unity (POU) method, described in Sect. 3.5, allows the user to face the problem with a numerical complexity that scales linearly with the RBF problem size.

The applications hereinafter described include processing of the displacement data for structural analyses as well as flow field data for CFD analyses. Whatever the field of the application, the basic concept is that the output, known at locations on the computational grid, can be interpolated using an RBF so that it can be further processed for various purposes.

In the specific case of FEA analyses, regardless of the complexity and the kind of the analysis (linear static analysis, rigid multibody, dynamic crash analysis, non-linear static plasticity, etc.), if the cell counting is not changed the displacement field can be interpolated and differentiated (see Sect. 2.4) to compose strain components that can be then used to compute stresses that become thus available as a set of continuous meshless functions. These latter, in turn, can be interpolated for advanced plotting features or for the implementation of complex stress assessment procedures. These methods provide the analyst with very useful information suitable for handling important applications such as the optimal positioning of reinforcements (steel in reinforced concrete or fibres in composites material parts) carried out by extracting computed isostatic lines (see Sect. 12.4). The examples reported in this Chapter are based on linear static structural analysis, but further applications can be similarly faced.

Additionally, although the focus is mainly on the processing of numerical results, it can be also applied to experimental data bearing in mind what exposed in Sect. 2.6 about the opportunity of applying a regression instead of an exact interpolation eventually filtering the noise as described in Sect. 2.7. It is worth to comment that the opportunity to use regression instead of the exact fitting was considered from the very beginning in the Hardy (1971) study already cited.

12.2 Evaluation of Equivalent Stresses According to the Implicit Gradient

In this first application, a method for the computation of an equivalent stress quantity based on the implicit gradient theory (IGT) (Maggiolini et al. 2015) is given. According to this method, the local stress field, usually computed using a linear FEA analysis, is processed extracting an averaged stress quantity that is not representative of the stress field but, instead, of its gradient on a scale that is relevant for the material to be assessed. Such an equivalent stress can be used for fatigue analysis at notches, and it works for both singular geometries, that is sharp notches with a theoretically infinite stress, as well as for standard stress raisers. An effective way for computing the equivalent stress consists of the use of a weighted averaging of the local stress field adopting the Gaussian function (12.1) for the computation of the average stress:

$$\sigma_{avg}(x) = \frac{\int_V \psi_G(\boldsymbol{x},\boldsymbol{y}) \cdot \sigma_{loc}(x) dV}{\int_V \psi_G(\boldsymbol{x},\boldsymbol{y}) dV} \tag{12.1}$$

where ψ_G is the Gaussian function (12.2), σ_{loc} the local stress and σ_{avg} the weighted average local stress.

$$\psi_G(\boldsymbol{x},\boldsymbol{y}) = \frac{e^{-\frac{\|x-y\|}{2L^2}}}{2\pi L^2} \tag{12.2}$$

$$L = c\sqrt{2} \tag{12.3}$$

The value of L (12.3) is computed from the constant c that is material dependent and can be set equal to 0.2 mm for a generic steel. The evaluation of the integral becomes straightforward once the data to be integrated are available in a meshless form. It is worth to notice that the constant of (12.2) can be put out of the integral (12.1) and, consequently, it can be neglected.

The RBF integration performed according to the IGT is showcased on the notched geometry sketched in Fig. 12.1 (left), for which the distribution of the first principal stress is extracted as a nodal quantity for a subset of the mesh located at the notch root (Fig. 12.1 right).

12.2.1 Interpolation of the Stresses as Computed by FEA Solver

Using the Mathcad tool as processing means, the stress field, namely the maximum principal stress, is interpolated as a single scalar function employing the sources points extracted from the FEM mesh nodes (Fig. 12.2 left) and adopting the linear

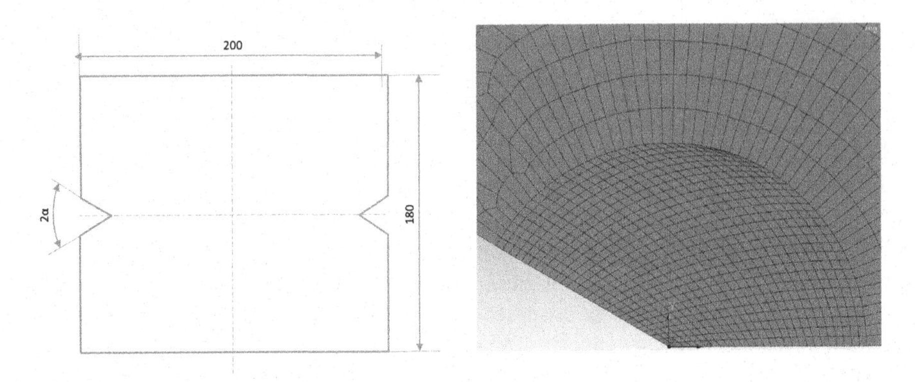

Fig. 12.1 Notched geometry (left) and detail of the mesh at notch (right)

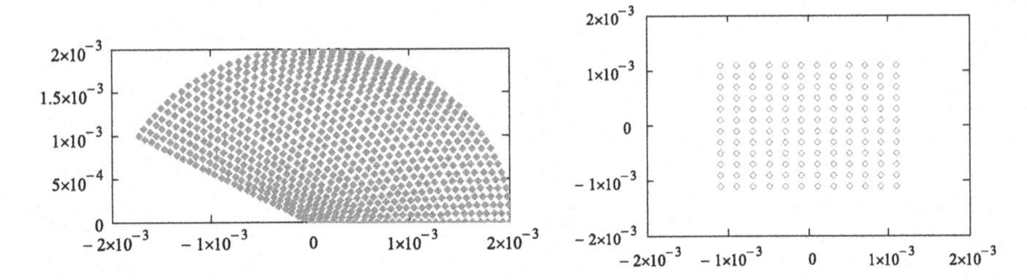

Fig. 12.2 Source points for the RBF fit (left) and uniform spaced square grid for integration (right)

RBF $\varphi(r) = r$. The integration over the field is evaluated on a square grid (Fig. 12.2 right) with adjustable spacing and size. The half-size of the square grid is set equal to $5.5 \cdot c$ that is enough to capture the total decay of the Gaussian function. Grid spacing is defined proportional to c ($1.0 \cdot c$ for the example shown in Fig. 12.2 right).

In this case the stress field is given not just by the usual RBF interpolation, but Boolean conditions have to be added so that the stress is evaluated only for the location inside the structural domain whereas it is set to zero otherwise. Such a condition is assigned through the Eq. (12.4), that express the RBF interpolation of the first principal stress to be averaged. It is worth to underline that, in this specific example, the domain is identified using 3 equations, two lines and a circle, whilst for advanced and more complex applications a similar information is available directly inquiring the geometrical modeller.

$$\sigma_1(x) := \begin{cases} x_{RBF}(x)_1 & \text{if } x_2 \geq 0 \wedge x_2 \geq (-x)_1 \cdot \frac{1}{2} \wedge |x| \leq \frac{2}{1000} \\ 0 & \text{otherwise} \end{cases} \tag{12.4}$$

The weights to be used for integration do not depend on the specific location where the integration is performed and they can be thus computed just once:

$$\begin{aligned} \psi_c := \ & \text{for } i \in 1..\text{rows}\left(P_{1234_1}\right) \\ & \quad \psi_{c_i} \leftarrow \psi_G\left[\left(P_{1234_1}\right)_i, \begin{pmatrix} 0 \\ 0 \end{pmatrix}\right] \\ \lambda_P := \ & \psi_c \cdot \delta_{xy}^{\ 2} \end{aligned} \tag{12.5}$$

The vector P_{1234} contains four vectors of points, one for each of the four quadrants (for symmetry reason just one needs to be computed as the others are identical). The Gaussian function ψ_G takes as input two points, and sampled results are stored in the vector ψ_c. The weight vector is completed by multiplying them by the integration area (the squared grid spacing). It is worth to notice that the grid is shifted so that there are no points lying along the axes (Fig. 12.2 right).

$$\begin{aligned} \sigma_{1e}(x) := \ & \sigma_e \leftarrow 0 \\ & \lambda_{sum} \leftarrow 0 \\ & \text{for } i \in 1..\text{rows}\left(P_{1234_1}\right) \\ & \quad \text{for } iq \in 1..4 \\ & \quad\quad \sigma_{1p} \leftarrow \sigma_1\left[x + \left(P_{1234_{iq}}\right)_i\right] \\ & \quad\quad \sigma_e \leftarrow \sigma_e + \sigma_{1p} \cdot \lambda_{P_i} \\ & \quad\quad \lambda_{sum} \leftarrow \lambda_{sum} + \lambda_{P_i} \ \text{ if } \sigma_{1p} > 0 \\ & \frac{\sigma_e}{\lambda_{sum}} \end{aligned} \tag{12.6}$$

The averaged value at a generic location can be computed using the algorithm (12.6). The integrals at numerator and denominator of Eq. (12.11) are numerically performed by accumulating the data in the variable $\boldsymbol{\sigma_e}$ and $\boldsymbol{\lambda_{sum}}$ that are first initialized to zero. Nested **for** cycles allow to loop on all the points of the grid (visiting a full quadrant with index **i** and the four quadrants with index **iq**), while the value to be weighted $\boldsymbol{\sigma_e}$ is computed by summing the position of the probe point and the one of the integration grid; the normalisation factor at denominator $\boldsymbol{\lambda_{sum}}$ is computed at the same time taking care of the sign of the stress $\boldsymbol{\sigma_e}$ so that only points in the integration volume are used. This action can be thought as a lens that moves over the domain and collects just the data that fall inside the near field. As already remarked, it is important to note that the check of belonging to the integration domain is performed by inspecting the value of the stress: when outside, both the averaged stress and area terms are neglected.

The accuracy of the integration is evaluated by performing a sensitivity study on the integration grid that is the sensitivity of averaged stress at notch root versus the

Table 12.1 Sensitivity analysis of averaged stress at notch root versus the integration parameters

	3.5	4.5	5.5	6.5
1.000	–	7.92958	7.9277	–
0.500	8.25979	8.18858	8.17994	8.17928
0.250	–	8.14207	8.1331	–
0.125	–	–	8.12652	–
0.075	–	–	8.12423	8.12364

integration parameters. The results of this analysis are collected in Table 12.1 where the half-width of the domain is reported by columns, whilst the resolution by rows (units are in c). The width and the resolution are both varied around a baseline guess that foresees to adopt an half-width equal to $5.5 \cdot c$ and a resolution equal to $0.5 \cdot c$. The output shows that an half-width of $4.5 \cdot c$ and a resolution of $0.5 \cdot c$ are enough for the achievement of a good convergence: 9 points along the square are used in this case, and the integration grid of each quadrant comprises 81 points (324 in total). A very good matching with the literature reference value by Maggiolini et al. (2015), obtained using a numerical averaging over the FEA grid (8.14847), is gained.

Figure 12.3 (left) shows the comparison between the local stress and the averaged stress at notch cross section visualised as a Cartesian plot (left). It is worth to notice how plots are easily constructed thanks to the meshless nature of the available information. In particular, a parametric equation of a segment that enters the notch for a width of 1 mm is used; the variable xx is zero at notch surface where the stress peak occurs and expresses the depth under the surface in meters. A comparison between the local stress and the averaged one is given showing how the latter converges to a finite value whilst the other, which should converge theoretically to infinite, grows to a mesh dependent "large value" that is useless for strength assessment.

A similar comparison in the whole field is shown on the right of Fig. 12.3 (right) as a scatter plot that has nodes position as x, y values and the stress as the z one. It can be noticed how the singular stress field is transformed into a smooth one. Such

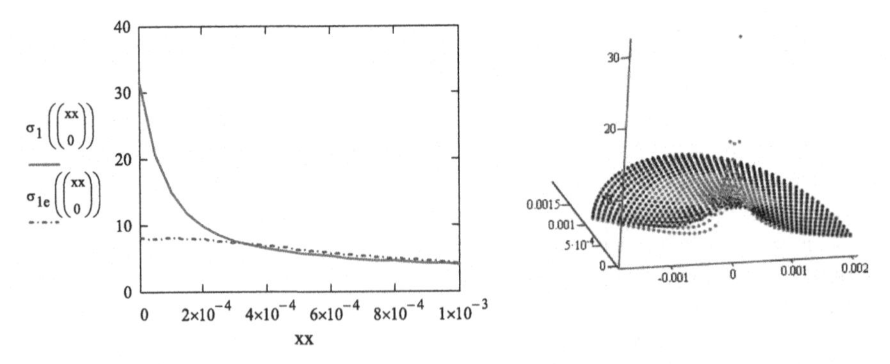

Fig. 12.3 Nominal stress versus equivalent stress comparison at notch cross section (left) and on the nodes (right)

an equivalent field does not depend on the mesh size if the mesh is properly refined so that the characteristic length c can be resolved.

12.2.2 Interpolation of the Stresses Derived by FEA Displacements

In this case just the displacement field coming from FEA analysis is used as the starting point for the stress calculation. Once the field is transformed in a meshless function, strain and stresses are obtained as closed form functions. The full 3D case is demonstrated in Sect. 12.3.1, for the 2D case the two components of the displacement field are first fitted adopting the cubic RBF $\varphi(r) = r^3$, and then the components of the gradient are computed according to the Eq. (2.27) of Sect. 2.4. In particular, the gradients of each component of the stress field are computed according to (2.29) and then paired in a single matrix:

$$\nabla_{\mathrm{RBF}}(\boldsymbol{x}) = \begin{pmatrix} \frac{\partial u(\boldsymbol{x})}{\partial x} & \frac{\partial v(\boldsymbol{x})}{\partial x} \\ \frac{\partial u(\boldsymbol{x})}{\partial y} & \frac{\partial v(\boldsymbol{x})}{\partial y} \end{pmatrix} \tag{12.7}$$

Strain components can be thus computed by combining the components of the gradient as follows:

$$\varepsilon_{xy}(x) := \left| \begin{array}{l} gg \leftarrow grad_{RBF}(x) \\ eps \leftarrow \begin{pmatrix} gg_{1,1} \\ gg_{2,2} \\ gg_{1,2} + gg_{2,1} \end{pmatrix} \end{array} \right. \tag{12.8}$$

Under the plane stress hypothesis assumed for this specific example, the stiffness matrix of the material can be computed as follows:

$$Q_{xy} := \begin{pmatrix} \frac{E_1}{1-\nu_{12}\cdot\nu_{21}} & \frac{\nu_{21}\cdot E_1}{1-\nu_{12}\cdot\nu_{21}} & 0 \\ \frac{\nu_{12}\cdot E_2}{1-\nu_{12}\cdot\nu_{21}} & \frac{E_2}{1-\nu_{12}\cdot\nu_{21}} & 0 \\ 0 & 0 & G_{12} \end{pmatrix} \tag{12.9}$$

whilst the stress field becomes a meshless point function and can be defined by the following relationship:

$$\sigma_{xy}(x) := Q_{xy} \cdot \varepsilon_{xy}(x) \tag{12.10}$$

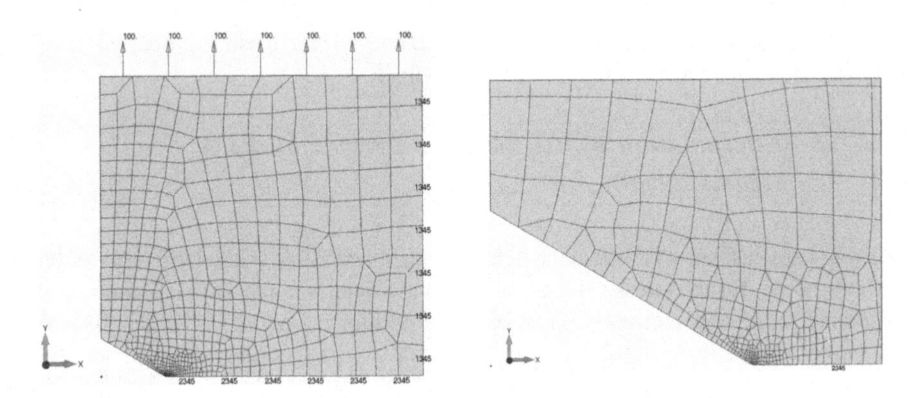

Fig. 12.4 FEM model for the notch example (left) and a detail of the mesh refinement at the notch tip (right)

In this example three new FEM models have been generated (see Fig. 12.4) by means of the Femap pre-processor using a quick non-structured meshing approach that is more representative when industrial scenarios have to be tackled. Only one quarter of the geometry is modelled. Symmetry conditions are suitably applied and a nominal stress of 1 MPa is imposed at the top of the domain. Keeping the same node counts on the edges, the mesh is refined at the notch by simple changing the Bias Factor of the Geometric Bias. The obtained sizes of the smallest elements at the tip for the three models are 0.06, 0.10 and 0.17 mm.

For each mesh a meshless RBF representation of the stress is obtained using as input nodal positions and displacements:

$$\sigma_1(x) := \begin{cases} \text{if } x_1 \geq 0 \wedge x_1 \leq 0.1 \wedge x_2 \geq 0 \wedge x_2 \leq 0.09 \wedge x_2 \geq 0.011547 - (x_1)\cdot\dfrac{0.011547}{0.02} \\ \quad \left| \begin{array}{l} \sigma \leftarrow \sigma_{xy}(x) \\ \dfrac{\sigma_1 + \sigma_2}{2} + \sqrt{(\sigma_3)^2 + \left(\dfrac{\sigma_1 - \sigma_2}{2}\right)^2} \end{array} \right. \\ 0 \ \text{otherwise} \end{cases} \tag{12.11}$$

The first principal stress is computed according to (12.11) where the geometrical conditions are imposed as Boolean operations using the equation of domain boundary segment in the Cartesian plane. The equivalent stress is then computed, as in Sect. 12.2.1, according to (12.6). It is worth to notice that the full domain mesh is composed of about 700 nodes, whilst in the example of Sect. 12.1 about 800 nodes are used just for the local refinement around the notch and the size of the smallest element at the tip is 0.074 mm.

The sensitivity analysis concerning the effect of the local mesh size on the maximum principal stress at the notch root is reported in Table 12.2. In specific, the table columns respectively report the local mesh size, the stress peak at the notch obtained by the FEM solver NX Nastran 8.5.1 and then post-processed by Femap

Table 12.2 Sensitivity analysis of maximum principal stress at notch root versus local mesh size

Size (mm)	FEM (MPa)	RBF (MPa)	AVG (MPa)
0.17	16.59	19.38	8.374
0.10	21.97	24.31	8.114
0.06	29.93	31.24	8.046

(averaged using corner data), the stress peak at the notch computed using the meshless RBF stress field and the averaged RBF stress field adopting an half-width equal to $5.5 \cdot c$ and a resolution equal to $0.25 \cdot c$.

RBF extrapolation improves the local stress value. In Sect. 12.3 this effect is explored on 2D and 3D cases noticing that the RBF derived stresses exhibit a quicker convergence with respect to mesh size if compared with FEM stress recovered on the same mesh using the standard approach. As far as the average value is concerned, the results look to be quite stable with mesh size as the coarser spacing is just 4% apart from the finer one. A good agreement with the reference value (8.148 MPa) described in Sect. 12.2 is found.

12.3 Upscaling of FEA Results at Stress Raisers

Shape functions, also called basis functions, play a fundamental role in FEM structural analysis because they provide the deformation field inside a single finite element processing the displacements computed at its nodes. It is well-known that the application of shape functions entails a continuity issue related to the fact that the displacements derivatives are not required to be continuous between two adjacent elements (Zienkiewicz et al. 2005).

In this section, the use of RBF as FEM structural results post-processing means such to interpolate the nodal displacements in order to obtain continuous analytical functions, is demonstrated adapting the results of Biancolini et al. (2015). Adopting this approach, both the gained strain and stress field prove to be smooth throughout the whole computational domain because they come from analytical derivatives.

In order to verify the effectiveness and the accuracy, the proposed technique is applied to a set of simple 2D and 3D test cases extracting the nodal displacements computed by means of the ANSYS APDL software and processing them with the Mathcad tool. These test cases, characterized by geometrical singularities that induce a concentration of stress whose theoretical value is well known from literature, respectively concern a rectangular plate with a single centred hole, a plate with a set of aligned holes and a solid pipe with a circular hole on the lateral wall.

Two typologies of analyses are specifically carried out (Biancolini et al. 2015): the first aims at evaluating the interpolation accuracy by comparing the value of the stress intensity factor K_t obtained through the RBF-based technique with the one computed by the ANSYS APDL solver. To make more consistent such an evaluation, the theoretical value is taken as reference and, additionally, the dependence of the mesh discretization in proximity of the geometrical singularity is also considered. On the other hand, the second analysis has the purpose to determine the discrepancy between the full field of stress distribution calculated by means of the proposed interpolation technique on four meshes, selected among those generated to develop the first analysis, and the finest one considered to be high-fidelity. In particular, the four meshes are termed coarse, medium, fine and very fine, whilst the finest one is referred to as gold standard (GS). In all cases, the cubic RBF $\varphi(r) = r^3$ is employed.

In the following sections, before describing in sequence the outputs and findings of the aforementioned test cases, a brief introduction of the working environment for the verification testing as well as of the mathematical background of the proposed technique are provided.

12.3.1 Verification Procedure Steps and Mathematical Background of the Proposed Technique

The verification procedure of the proposed technique foresees the accomplishment of the following operations:

- development of set of FEM analyses performed on meshes with increasing accuracy in order to determine nodal displacements in function of the computational grid refinement;
- generation of the RBF interpolant of nodal displacements (scattered data) in view of obtaining a continuous and differentiable analytical function;
- combination of analytical partial derivatives of displacements obtained through RBF to calculate the deformation field in the whole computing domain (mesh);
- application of elasticity laws to determine the stress field in the whole computing domain;
- evaluation of the stress intensity factor to be compared to the one available in literature to finally assess the accuracy of the proposed post-processing numerical means.

The analytical method to calculate the deformation and the stress field starting from the displacement field has been already introduced in Sect. 12.2.1 for a plane stress case. The three dimensional case is developed according to the same concepts and is here presented for the sake of completeness.

$$s = \begin{Bmatrix} u \\ v \\ w \end{Bmatrix} \tag{12.12}$$

Known the displacement vector (12.12) the components of the first order deformation vector are expressed by (12.13).

$$\varepsilon_x = \frac{\partial u}{\partial x} \quad \varepsilon_y = \frac{\partial v}{\partial y} \quad \varepsilon_z = \frac{\partial w}{\partial z}$$

$$\varepsilon_{xy} = \frac{\partial u}{\partial y} + \frac{\partial v}{\partial x} \quad \varepsilon_{yz} = \frac{\partial v}{\partial z} + \frac{\partial w}{\partial y} \quad \varepsilon_{xz} = \frac{\partial u}{\partial z} + \frac{\partial w}{\partial x} \tag{12.13}$$

If the material is elastic, linear and isotropic, and considering the elasticity law, the stresses can be written according to (12.14) where the stiffness matrix $\boldsymbol{Q} = \boldsymbol{H}^{-1}$ is computed as the inverse of $\boldsymbol{H}$ (12.15), that is the compliance matrix of the material, and where E and ν respectively are the Young's modulus and the Poisson's coefficient, whilst $G = \frac{E}{2(1+\nu)}$ is the shear modulus.

$$\begin{Bmatrix} \sigma_x \\ \sigma_y \\ \sigma_z \\ \sigma_{xy} \\ \sigma_{yz} \\ \sigma_{xz} \end{Bmatrix} = \boldsymbol{Q} \begin{Bmatrix} \varepsilon_x \\ \varepsilon_y \\ \varepsilon_z \\ \varepsilon_{xy} \\ \varepsilon_{yz} \\ \varepsilon_{xz} \end{Bmatrix} \tag{12.14}$$

$$\boldsymbol{H} = \begin{bmatrix} \frac{1}{E} & -\frac{\nu}{E} & -\frac{\nu}{E} & 0 & 0 & 0 \\ -\frac{\nu}{E} & \frac{1}{E} & -\frac{\nu}{E} & 0 & 0 & 0 \\ -\frac{\nu}{E} & -\frac{\nu}{E} & \frac{1}{E} & 0 & 0 & 0 \\ 0 & 0 & 0 & \frac{1}{G} & 0 & 0 \\ 0 & 0 & 0 & 0 & \frac{1}{G} & 0 \\ 0 & 0 & 0 & 0 & 0 & \frac{1}{G} \end{bmatrix} \tag{12.15}$$

12.3.2 Rectangular Plate with Single Centred Circular Hole

As previously introduced, the first test case deals with a rectangular plate with a single centred circular hole. The plate length, depth and thickness respectively are equal to 40, 30 and 1, whist the hole's diameter is 10 (dimension can be expressed without units). Since the model is characterized by a geometrical symmetry with respect to two planes, the ANSYS APDL case reproduces only one quarter of the whole plate (see Fig. 12.5) exploiting the assignment of proper symmetry conditions.

The computational mesh is discretised by means of triangular elements (SHELL41) characterized by an in-plane rigidity and by a thickness of 1. The plate is loaded with a longitudinal traction load of 1000 that is applied at the shorter lateral edge of the model. Going from top left to bottom right, the images of Fig. 12.6 respectively depict the mesh of the coarse, medium, fine and very fine (GS) mesh for the plate with one centred hole.

The stress intensity factor, namely the ratio between the maximum and the nominal stress, can be computed using the relationships present in the scientific

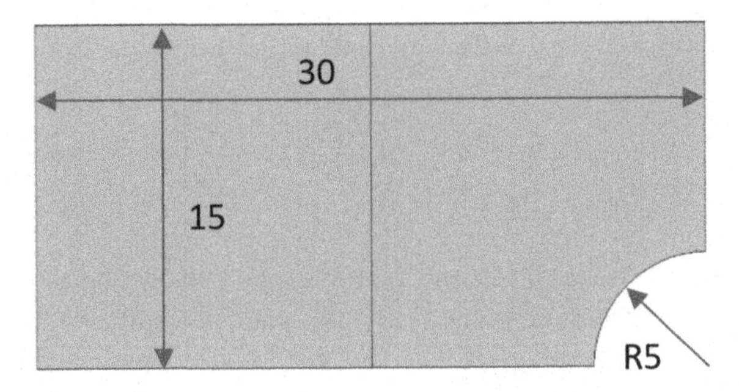

Fig. 12.5 APDL case geometry for the plate with one centred hole. Dimension of the quarter are shown, plate length is 60, width 30 and hole dimeter 10

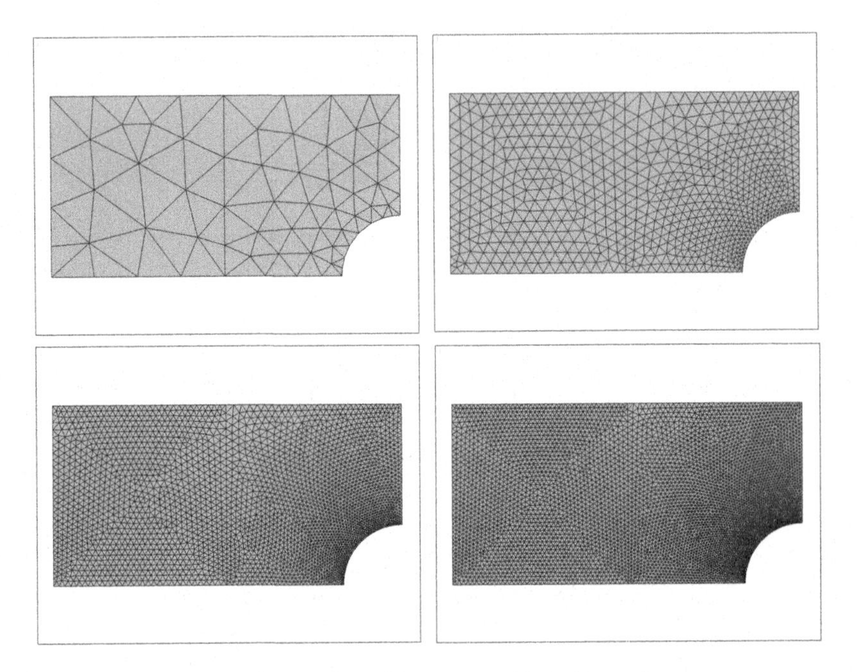

Fig. 12.6 APDL case mesh for the plate with one centred hole (4 cases)

literature (Howland 1930). Its value, which depends on the ratio between hole's diameter and the width of the plate, is equal to 3.47. The results of the mesh sensitivity study keeping the just reported analytical value of the stress intensity factor as reference, are summarised in Table 12.3. In this table the refinement level, defined by the number of subdivisions along the contour of the hole, is reported in the first column, whilst in the second and in the third column the stress intensity factor computed through ANSYS APDL (K_{tg} APDL) and the post-processing (PP) RBF interpolator (K_{tg} PP-RBF) are respectively reported. As it can be straightforwardly

Table 12.3 Convergence of K_t versus mesh spacing, as computed from direct FEM output (APDL) and after RBF post processing (PP-RBF) for a plate with a single hole

Subdivisions of the hole's edge	K_{tg} APDL	K_{tg} PP-RBF
4	2.57	3.31
6	2.80	3.19
8	3.20	3.45
10	3.28	3.50
20	3.44	3.48
30	3.46	3.43
40	3.48	3.44
50	3.47	3.46
60	3.48	3.45
100	3.48	3.46

judged, the advantage of using the RBF-based interpolator is evident just for low refinement meshes, whereas it decreases when finer meshes are taken into account.

Figures 12.7 and 12.8 show the σ_x obtained through FEM analysis and RBF interpolation respectively, in the case that 8 subdivisions are used to discretize the hole's edge.

The accuracy of RBF upscaling on the full domain can be verified evaluating the stress on the GS mesh (100 divisions), for which the maximum fidelity is available, the interpolated stress field coming by processing with RBF the displacements field

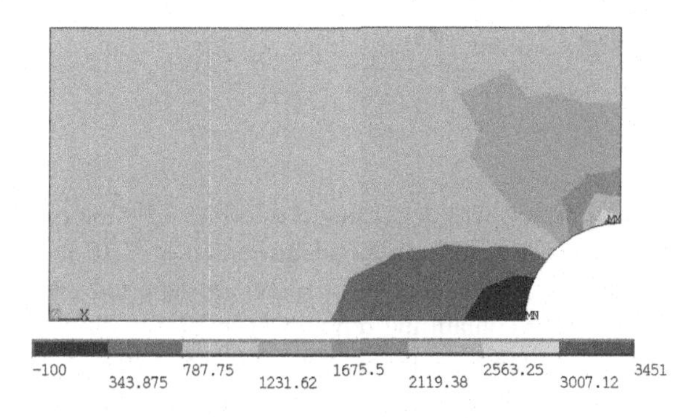

Fig. 12.7 Plate with a single hole case: σ_x obtained by means of FEM analysis

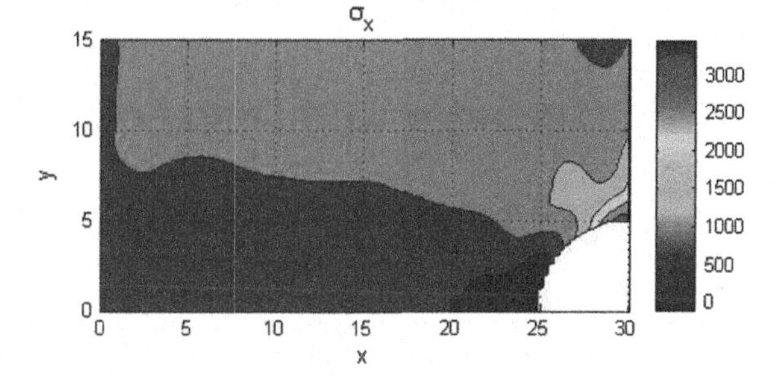

Fig. 12.8 Plate with a single hole case: σ_x obtained by means of RBF interpolation

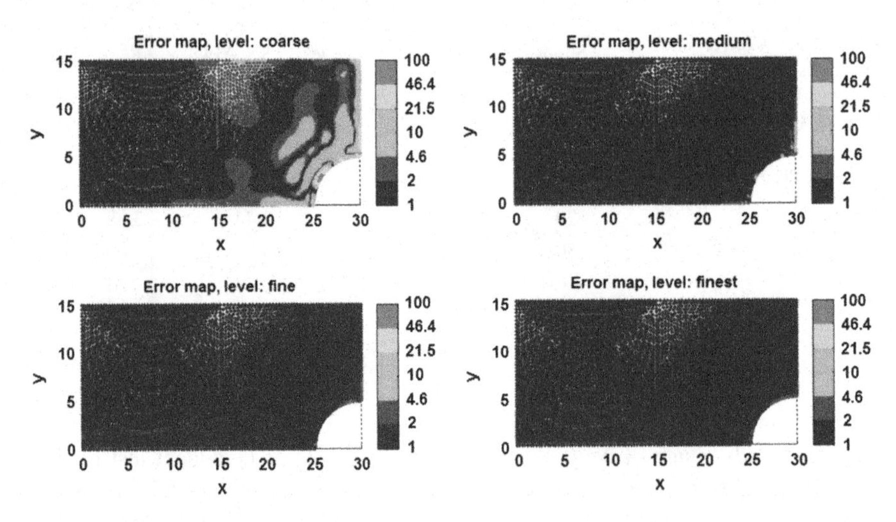

Fig. 12.9 Error distribution after RBF post processing versus mesh refinement (coarse, medium, fine, finest) for the plate with one centred hole

Table 12.4 Error distribution after RBF post processing versus mesh refinement for the plate for the plate with one centred hole

Refinement level	Subdivisions of the hole's edge	Err% max	<1%	1% ÷ 5%	>5%
Coarse	8	74.6	38.3	40.3	21.4
Medium	30	10.9	91.2	8.3	0.5
Fine	60	5.1	97.8	2.2	0.0
Very fine	80	4.7	98.3	1.7	0.0

coming from low fidelity models. In Fig. 12.9 the result of this study is represented as an error map where the adequateness of RBF interpolated stress is highlighted.

Results of the same comparison are reported in Table 12.4 where the columns respectively contain the refinement level, the number of discretisation intervals of the edge of the hole, the maximum value of the relative error in terms of von Mises stress normalised with respect to the nominal value, and the percentage error connected to the indicated intervals.

12.3.3 *Plate with a Set of Centred Aligned Holes*

The second case deals with a plate with a set of five centred aligned holes. The plate length is 100, width 30 and has five equispaced holes with diameter 10. Given the geometrical symmetry with respect to one plane, the ANSYS APDL case is generated accounting only half the model as illustrated in Fig. 12.10.

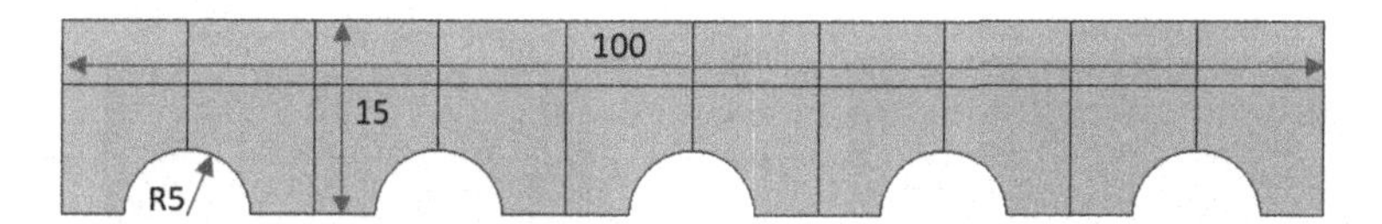

Fig. 12.10 APDL case geometry for the plate with a set of centred holes. Dimension of the half model are shown, plate length is 100, width 30 and hole dimeter 10

Table 12.5 Convergence of K_t versus mesh spacing, as computed from direct FEM output (APDL) and after RBF post processing (PP-RBF) for a plate with multiple holes

Subdivisions of the hole's edge	K_{tg} APDL	K_{tg} PP-RBF
4	2.765	3.086
6	2.869	3.015
8	2.911	3.013
10	2.933	3.005
20	2.957	2.976
30	2.958	2.963
40	2.957	2.956
50	2.956	2.952
60	2.955	2.950
100	2.953	2.945

The computational mesh is generated through quadrangular elements (SHELL41) and the loading condition is the same of the first test case. The same mesh spacing of Sect. 12.3.2 demonstrated in Fig. 12.6 is used for the repeated module of the full mesh.

The stress intensity factor of the current case evaluated at the mid (third) hole is 3.049 (Schultz 1941). Similarly to what already done for the first test case, Table 12.5 summarizes the outputs of the first part of the study. In this case, however, the interpolation based on RBF is applied in the central area of the model in order to lower the computing demand.

Also in this case the accuracy of upscaling in the full domain, collected in Fig. 12.11 and in Table 12.6, are evaluated taking as reference the GS discretisation consisting of 100 subdivisions.

12.3.4 Tube with a Single Transversal Hole

The third test case concerns the 3D model of a tube having a transversal hole and subjected to a twist moment $T = 1000$. The main dimensions of the tube are the external diameter of 1, the internal diameter of 0.8, the hole's diameter of 0.2 and the length of 8. The geometrical model of the case is reported in Fig. 12.12.

The tube is discretised by means of tetrahedral elements (SOLID85). The four meshes of the tube are shown in Fig. 12.13.

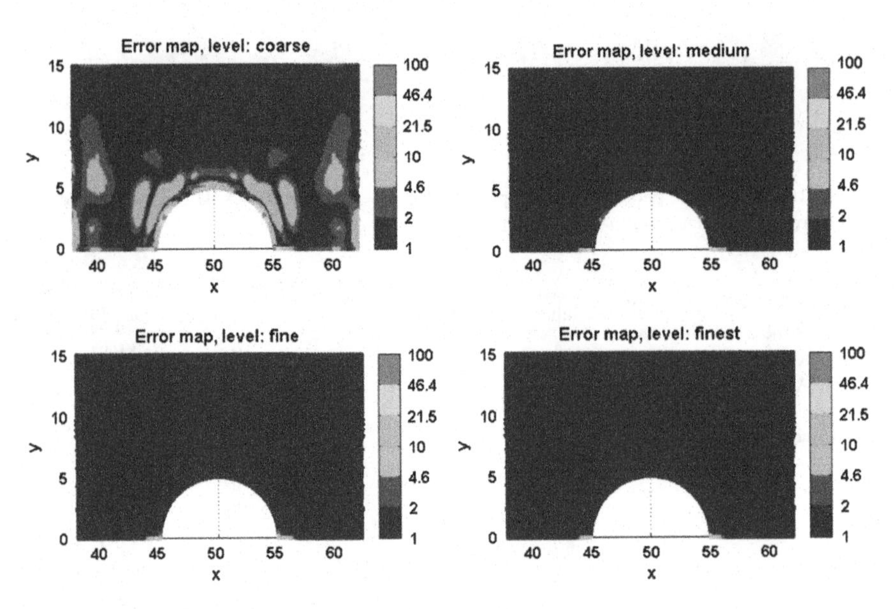

Fig. 12.11 Error distribution after RBF post processing versus mesh refinement (coarse, medium, fine, finest) for the plate with a set of centred holes

Table 12.6 Error distribution after RBF post processing versus mesh refinement for the plate with multiple holes

Refinement level	Subdivisions of the hole's edge	Err% max	<1%	1% ÷ 5%	>5%
Coarse	8	52.6	35.7	49.5	14.8
Medium	30	47.0	96.8	2.8	0.4
Fine	60	23.7	98.3	1.2	0.5
Very fine	80	15.8	99.1	0.6	0.3

Fig. 12.12 APDL case geometry for the tube with a single transversal hole

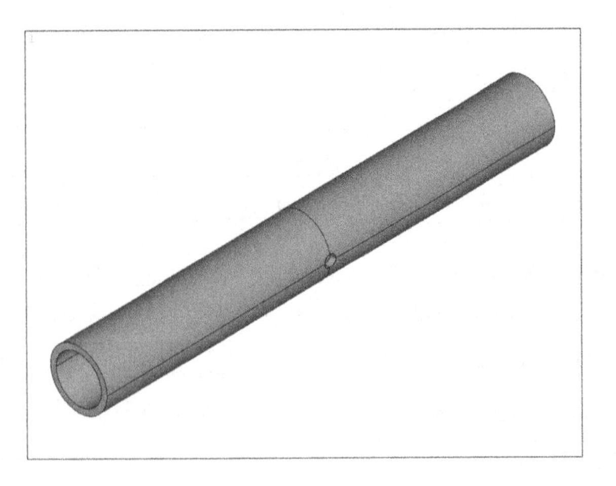

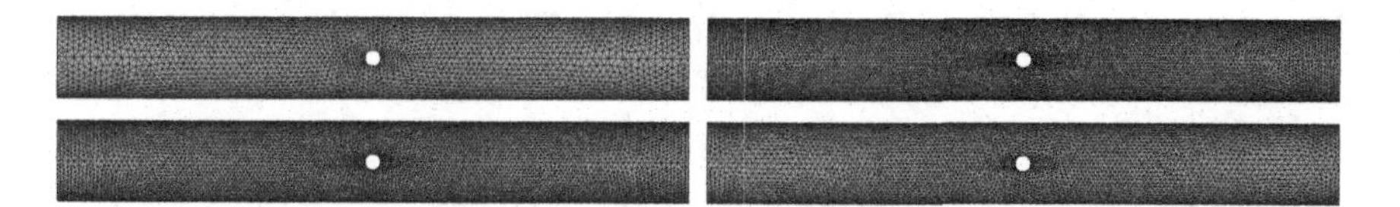

Fig. 12.13 APDL case mesh for the tube with a single transversal hole (4 cases)

The stress intensity factor is given by the ratio between the maximum and nominal tangential stress. This latter, in particular, is given by the following relationship of continuum mechanics:

$$\tau_{nom} = \frac{16 \cdot T \cdot D}{\pi(D^4 - d_i{}^4)} = 8626.28 \tag{12.16}$$

As already done for the second test case, the RBF interpolation is performed in the central area of the model to limit the computing demand.

The reference value of the stress intensity factor for the holed tube is 2.055 (Jessop et al. 1959). According to what already presented for the first two test cases, Tables 12.7 and 12.8 summarise the results gained with RBF post processing.

Table 12.7 Convergence of K_t versus mesh spacing, as computed from direct FEM output (APDL) and after RBF post processing (PP-RBF) for the tube with a single hole

Subdivisions of the hole's edge	K_{tg} APDL	K_{tg} PP-RBF
4	1.450	1.745
5	1.636	1.746
6	1.678	1.749
7	1.779	1.827
8	1.791	1.908
9	1.884	1.899
10	1.922	2.018

Table 12.8 Error distribution after RBF post processing versus mesh refinement for the tube with a single hole

Refinement level	Subdivisions of the hole's edge	Err% max	<1%	1% ÷ 5%	>5%
Corse	5	133.3	17.1	53.1	29.8
Medium	6	96.0	22.4	56.8	20.8
Fine	7	100.1	35.6	48.8	15.6
Very fine	8	79.0	37.4	52.4	10.2

12.4 Isostatic Lines Construction for Planar Problems

The RBF based post-processing of FEA results can be used to extract from the model useful information that could be difficult to be evaluated using the mesh based available solution. In the example of the present section a plane stress analysis of a beam with a uniform load is conducted. Boundary conditions and resulting von Mises stresses are represented in Figs. 12.14 and 12.15, in which FEA

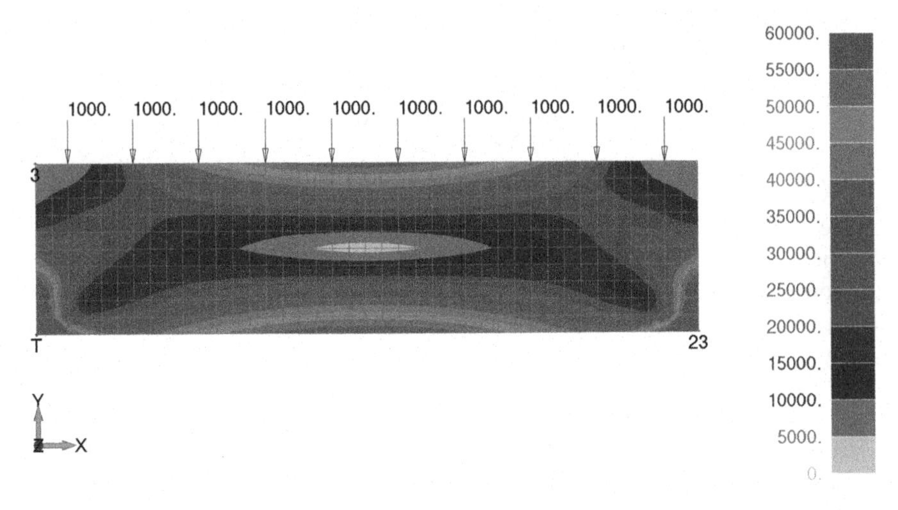

Fig. 12.14 von Mises stress of a beam under uniform load pressure constrained at the corners

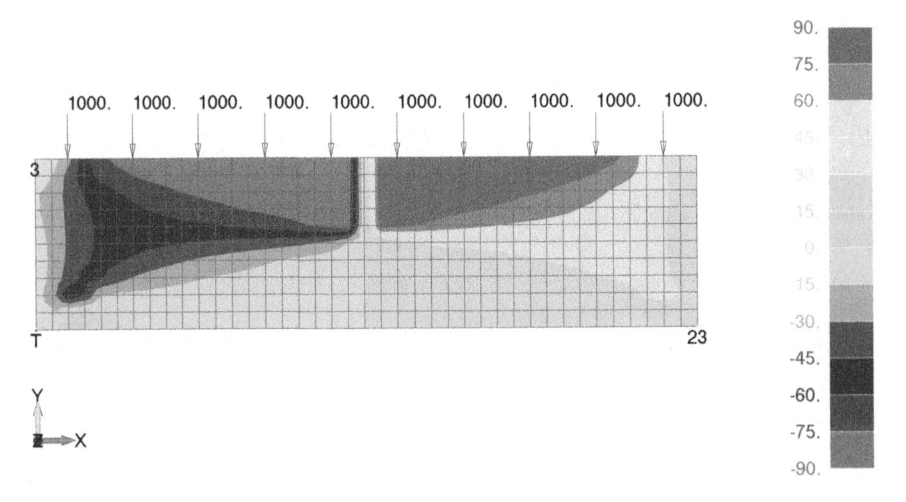

Fig. 12.15 Angle of the maximum principal stress direction of a beam under uniform load pressure constrained at the corners

elements, load and constraint conditions are shown together with the map of stresses that are, respectively, the von Mises stress and the angle of the maximum principal stress.

This latter output, in particular, provides the analyst with useful information about the potential direction for the introduction of reinforcements. A convenient arrangement is therefore provided by plotting the isostatic lines (Halpern et al. 2013) that are always tangent to the principal stress and form two families of curves that are mutually orthogonal. From each point in the structure four isostatic semi-lines can be originated in the positive and negative directions of the principal reference system. It is worth to observe that the stress is not constant along the isostatic lines but it is always of pure traction/compression. A notable application of isostatic lines concept was given by Nervi[1] to define the geometry of ribs reinforced slabs. Details about structural performances of this structure are given in Tarisciotti and Biancolini (2013) where Nervi's approach is compared with topological optimisation.

The meshless stress field is here computed according to the approach of Sects. 12.2.2 and 12.3.1: displacement results computed by FEM are processed and an RBF field with a cubic function $\varphi(r) = r^3$ is defined; differentiation of the RBF allows to compute the strains, so stresses are then computed using the isotropic material model. The construction of the isostatic line is based on the principal reference system angle that is computed according to the Eq. (12.17) that gives as output the four directions to be used for originating isostatic lines as a point function.

$$\varphi(x) := \left| \begin{array}{l} \text{sigma} \leftarrow \sigma_{xy}(x) \\ \sigma \leftarrow -(\text{sigma}_1 - \text{sigma}_2) \\ \tau \leftarrow \text{sigma}_3 \\ a \leftarrow \dfrac{\sigma}{2\cdot\tau} \\ b \leftarrow \sqrt{\dfrac{\sigma^2}{4\cdot\tau^2} + 1} \\ \begin{pmatrix} \text{atan}(a - b) - 90\,\text{deg} \\ \text{atan}(a - b) \\ \text{atan}(a + b) \\ \text{atan}(a + b) + 90\,\text{deg} \end{pmatrix} \end{array} \right. \tag{12.17}$$

[1] Brevetto N° 455678, Ingg. Nervi & Bartoli Anonima per Costruzioni a Roma che ha designato quale autore dell'invenzione Aldo Arcangeli, Perfezionamento nella costruzione di solai, volte, cupole, travi-parete e strutture portanti in genere a due o tre dimensioni, con disposizione delle nervature resistenti lungo le linee isostatiche dei momenti o degli sforzi normali, data del deposito 23 luglio 1949, data della concessione 9 marzo 1950.

The plot of the isostatic lines is generated by a space marching algorithm detailed in (12.18) that takes as input the previous angle (out of the four possible), the previous location and the spatial increment. The four local directions are scanned to select the one closest to the previous one, to compute the unit vector and to finally give the new position and angles supplied as the output:

$$x\varphi_{new}(x_{old}, \varphi_{old}, \delta s) := \left| \begin{array}{l} d\varphi_{min} \leftarrow 1000 \\ \varphi_{vec} \leftarrow \varphi(x_{old}) \\ \text{for } i \in 1..4 \\ \quad \text{if } \left|\varphi_{old} - \varphi_{vec_i}\right| < d\varphi_{min} \\ \quad\quad \left| \begin{array}{l} i_\varphi \leftarrow i \\ d\varphi_{min} \leftarrow \left|\varphi_{old} - \varphi_{vec_i}\right| \end{array}\right. \\ \varphi_{new} \leftarrow \varphi_{vec_{i_\varphi}} \\ n \leftarrow \begin{pmatrix} \cos(\varphi_{new}) \\ \sin(\varphi_{new}) \end{pmatrix} \\ x\varphi_{new} \leftarrow \begin{pmatrix} x_{old} + \delta s \cdot n \\ \varphi_{new} \end{pmatrix} \end{array}\right. \qquad (12.18)$$

For the simple investigated structure, twelve isostatic lines are originated in the two horizontal directions (positive and negative) from six source points located along the symmetry line at the middle of the beam span. Figure 12.16 depicts the isostatic lines obtained by means of the proposed RBF-based method together with the nodes of the FEA mesh represented in the same plot. It is interesting to observe how plotted lines intersect each other with right angles.

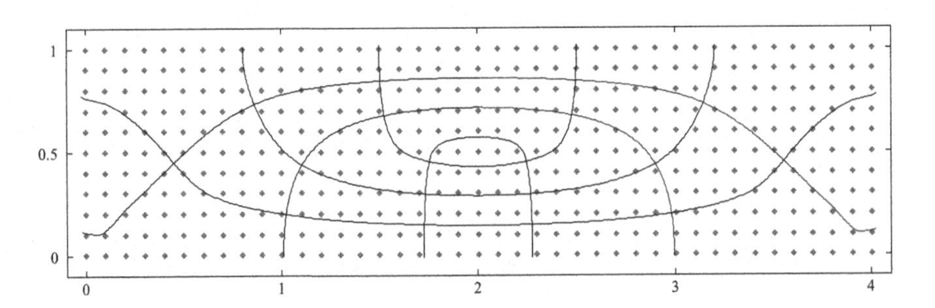

Fig. 12.16 Isostatic lines are plotted using the proposed method

12.5 Post-processing of Images Analysis of Structural Deformations

In recent years electronically controlled sensors, such as digital cameras based on charge-coupled device (CCD) or complementary metal-oxide-semiconductor (CMOS) technologies, have enjoyed lots of interest and have experienced an astonishing headway due to the low price they reached and the high resolution they guarantee.

In the structural analysis context, plenty of approaches (Sutton et al. 2009) have been proposed for the application of digital images to characterize the progress of the deformation of loaded structures through the comparison of registered sequential frames. Most of these methods, based on the recognition of individual points on the pictures through correlation functions (Sutton et al. 1991), are very fast but the resulting displacement field can be affected by some discontinuities. To avoid this unwanted effect, smoothed analyses have been recently conducted with the application of FEM mesh on the acquired regions (Sutton et al. 2009; Amodio et al. 1995; Besnard et al. 2006). Such techniques are able to reach convergence even if large displacements occur by using, before running the CSM analysis, algorithms able to identify some points of the region of interest to be taken as reference points (Flusser et al. 2009). However, one of the limits of these methods is related with the need for managing an high number of controlling points, namely the mesh nodes of the CSM model, that increases the computing time while affecting the solution convergence.

The study presented in this section, here adapted from Biancolini and Salvini (2012), shows the application of RBF in structural deformation analysis using image processing. In particular, the processing concerns synthetic images generated by a FEM model reproducing the starting and the deformed configuration of the structure of interest, and it is conducted processing the images pixels by means of the Mathcad tool. The aim is to showcase the capability of RBF to predict the target deformed image of the CSM mesh using the known displacements of a small subset of source points properly extracted from the mesh nodes of the CSM model. As a matter of fact, in principle RBF should be able to considerably reduce the control points through a choice accomplished in an optimal way, while describing the displacement field accurately. Furthermore, an important additional advantage in using RBF is the capability to increase the control points in the region where important gradients of displacement are expected, so that the local approximation can be accordingly reduced.

In specific, the synthetic images deal with the starting and the deformed triangular mesh of a 2D CSM model of a square (1 m side) with a circular hole (0.2 m radius) in its centre. The deformed configuration is gained imposing a rigid motion (0.1 m) to the hole's nodes along one of the principal axes, while the nodes of the external edges are kept fixed. Figure 12.17 depicts the original (left) and deformed mesh (right) images generated through FEM.

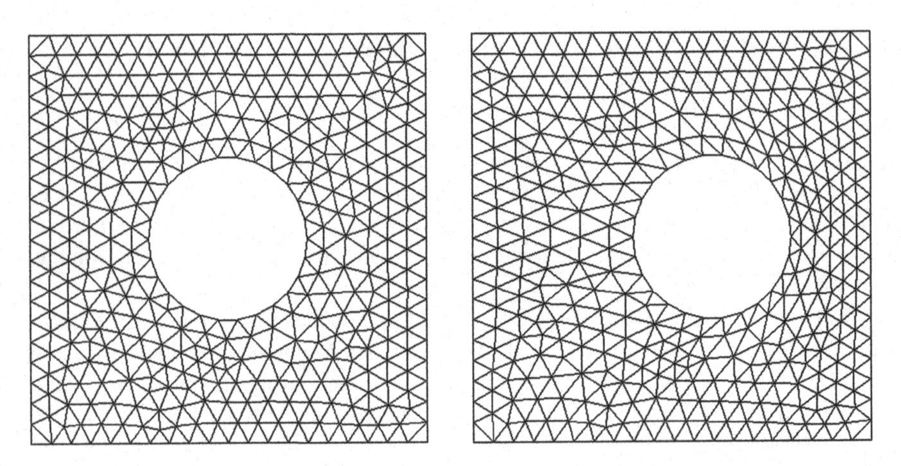

Fig. 12.17 Original mesh (left) and deformed mesh (right) images generated through the FEM model

Figure 12.18 (left) shows the result of the first phase of pixels data processing concerning the import of mesh nodes position and the successive extraction of the RBF source points. As visible, 20 RBF source points (blue empty diamonds) are extracted among the 409 CSM grid nodes (red filled diamonds) such to have 8 source points along the boundaries of the domain, 8 in the bulk and 4 along the hole boundary.

The CSM displacements of the extracted nodes, are employed to impose the unknown values of the image matching problem. In particular, the inverse RBF method (Arad et al. 1994) is used to define a conformal $\mathbb{R}^2 \rightarrow \mathbb{R}^2$ transformation such to map the deformed nodes distribution onto the starting one. This operation is mandatory needed to guarantee image treatment consistency, namely to establish the

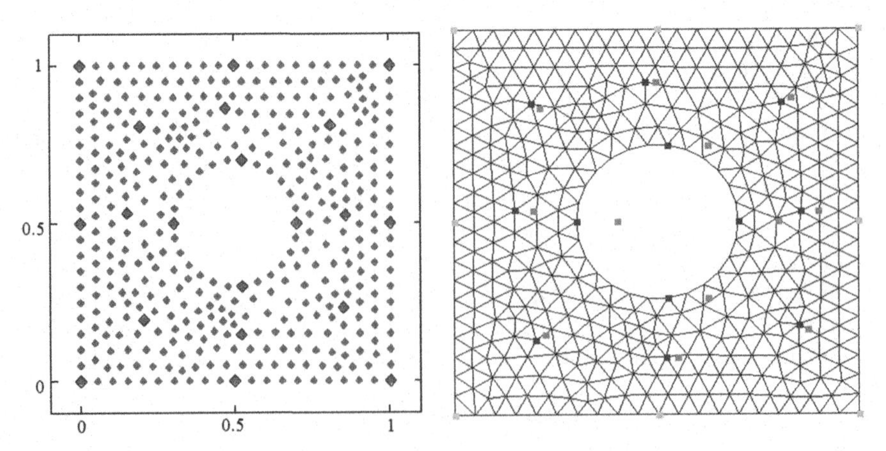

Fig. 12.18 Position of RBF source points and of FEM grid nodes before (left) and after applying morphing (right)

correspondence between the features detected in the deformed image through pixels processing, and those detected in the undeformed (reference) image. The pixels processing, in particular, also foresees that colour values of source nodes pixels at original positions are averaged employing the data of their neighbour pixels. In such a manner the deformed image is generated pixel by pixel picking the data from the original image using the transformation. Figure 12.18 (right) shows the starting image together with the RBF points in both the starting and target position.

Once gamma and beta (RBF sought coefficients) of the RBF system are computed, the RBF fit can be used to compute a new deformed image transforming the original undeformed one. The position of each pixel of the new image to be computed is processed by the inverse RBF to retrieve the corresponding position in the original one. The colour value is suitably computed averaging the values at the four surrounding pixels at each location and assigned to the new one.

It is important to say that the main advantage of using the inverse RBF method is that the new image can be generated visiting all its pixels that are arranged on a regular grid, a local colour interpolation around the RBF computed point on the original one can be easily conducted. The adoption of the RBF direct field would produce scattered locations of the colour information on the destination image and a further global interpolation stage to match the colour on the destination pixels should be considered to complete the task.

The final gained output is shown in Fig. 12.19 (left) where the comparison between the deformed image computed using RBF field and the target image (right) is reported through a superposed visualization.

To quantify the interpolation error of the RBF field, the RBF field is applied to all FEM nodes of the synthetic model and the error *err* is calculated as follows:

$$err = \frac{|\delta_{RBF} - \delta_{FEM}|}{|\delta_{FEM}|} \cdot 100$$

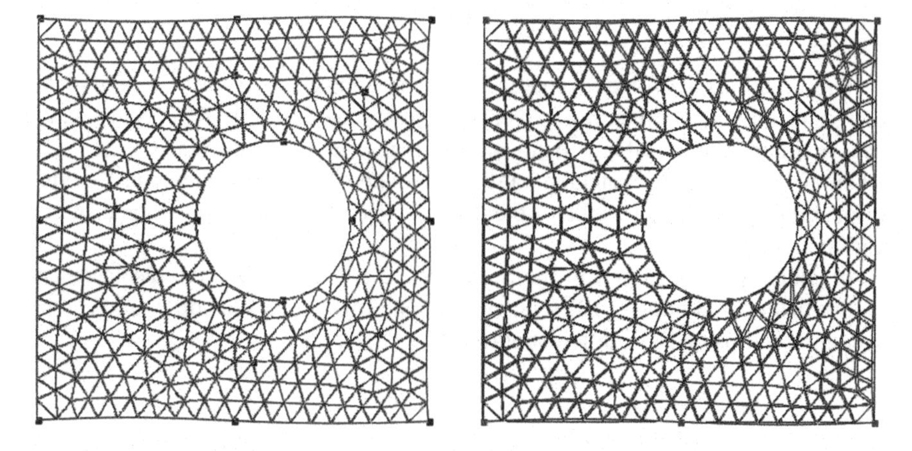

Fig. 12.19 RBF deformed image (left) compared with the target image (right)

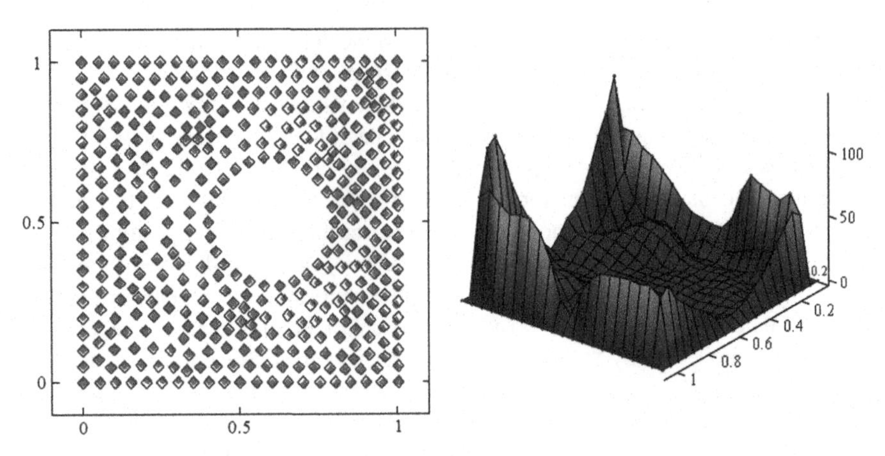

Fig. 12.20 Nodes position computed through the FEM model and the RBF field (left) and the error map (right)

where δ_{FEM} is the FEM nodal displacement, whilst δ_{RBF} is the nodal displacement calculated through the RBF field. Figure 12.20 shows the FEM nodes in the deformed position as computed by FEM (empty blue diamonds) and as interpolated using the RBF field (red filled diamonds) on the left side, whilst the evaluated map error is plotted on the right side.

Considering in depth the error distribution, about 75% of the points have an error lower than 20%, whilst a poor interpolation is achieved for the remaining ones that are basically located on the domain contour. The distribution of the computed error percentage is, in detail, as follows:

- 42.8% below 5%;
- 17.4% in the range 5–10%;
- 14.9% in the range 10–20%;
- 13.9% in the range 20–50%;
- 11.0% over 50%.

The described RBF-based approach can be adopted to process experimental images of deformed structures as well. As already outlined, the RBF field quality can be improved by adding extra source points in the constrained areas and, additionally, the image matching can be automated looking for the control points displacements that minimise the difference between the parametric warped image and the target one.

In addition, as described in the Sects. 12.2.2 and 12.3, once the displacement field is determined through RBF, the evaluation of the strain and stress field can be accomplished using RBF differentiation.

12.6 Compensation of Metrological Data

A problem that is particularly felt in the metrological field is the occurrence of variable, and not avoidable, environmental conditions that change the shape of the component to be accurately measured. The application described in the current section aims at showing how RBF can be employed to reliably and accurately compensate the metrological data pertaining a cantilever beam (Semeraro 2017).

A cloud of (x, y, z) points, acquired at a very high precision onto the real component, is compared with the baseline CAD model in order to check if the actual manufactured part geometry belongs to the design space. This latter, in particular, is defined by the volume comprised between the maximum and minimum offset of the nominal surfaces, being the offset prescribed in terms of design manufacturing tolerance. A constant temperature can be easily accounted if the component to be measured is made of an isotropic homogenous material and is not over constrained.

The conversion required to transform the measured cloud in the reference ambient condition is given by the equation:

$$P_{exp} = P_{ref}(1 + \alpha \Delta T) \tag{12.19}$$

where P_{ref} is the nodal position of a point in the CAD model and P_{exp} is the nodal position of the same point as evaluated during the experiment, α is the thermal expansion coefficient and ΔT is the temperature offset. Further transformations to align the surveyed points to the CAD model are made by best fitting the cloud data on the CAD model. This function optimizes the cloud data position and rotation by searching the least mean square of the distances from the surfaces (Δ) using the following formula:

$$\sqrt{\frac{1}{N-1} * \sum_{1}^{N} \Delta^2} \tag{12.20}$$

where N is number of the points of the cloud. More sophisticated procedures taking into account the relevance of the point, as well as the presence of mechanical constraints, are available in metrology software such as SpatialAnalyzer (New River Kinematics 2017) and Polyworks (Innovmetric 2017) for instance.

There are more complex scenarios such as those characterised by the presence of large part subjected to a temperature field that is not constant, by anisotropic components at a temperature different from the baseline one and by components deflected by the dead load, that make the compensation of experimental data a very challenging task. All the aforementioned cases require a CAE analysis specifically conceived for the prediction of the actual environment occurring during the measurement campaign. The availability of high-fidelity structural model based on FEA is quite common. Nevertheless, the use of such models for the measurement compensation is not obvious for many reasons including that FEA model could be related to a de-featured version of the full CAD model and that FEA results can be easily retrieved at nodal positions that usually are not coincident with measurement points. The standard ISO/IEC GUIDE 98-3:2008 "Uncertainty of measurement—Part 3: Guide to the expression of uncertainty in measurement" provides a complete guideline to address this topic.

An RBF based approach can be adopted to overcome the aforementioned issues. Two options are envisioned. The first one consists of the adaption of the original CAD model so that it is moved to the measured shape. The major drawbacks are that the CAD morphing could be a difficult and time consuming operation and the need for generating many CAD variations of the same part if more than one environment exist. The second approach, presented in this section, is very similar to the strain analysis implemented in Sect. 12.5. It basically foresees the correction of measured data by subtracting the environment effect: all the corrected measured points can be referred to the same baseline CAD geometry.

The RBF interpolation, as that described in Sect. 12.5, is based on the inverse transformation. First the points are collected with a metrology instrument like a Laser Tracker or a Photogrammetry camera. Those points are then elaborated and aligned to the coordinate frame of the FEM or CAD model with the tools available in the standard metrology software.

For the RBF solution framework set-up, the inverse RBF transformation is constructed using all, or just a part, of the FEA nodes as input. FEA data and results can be computed according to a generic FEA solver and are usually available as nodal positions (P_{FEM}) and nodal displacements (δ_{FEM}).

The inverse field, needed for the RBF interpolation, requires first the computation of nodes positions $\mathrm{P_{FEM\_d}}$ in the deformed configuration. Such data can be computed as follows:

$$\mathrm{P_{FEM\_d}} = \mathrm{P_{FEM}} + \delta_{FEM} \tag{12.21}$$

Using this information, the computation of the field δ_{FEM_inv}, that brings the nodes back, can be finalised according to what following reported:

$$\delta_{FEM\_inv} = \mathrm{P_{FEM\_d}} - \mathrm{P_{FEM}} = -\delta_{FEM} \tag{12.22}$$

An RBF interpolation is so defined using the full cloud $\mathrm{P_{FEM\_d}}$ interpolating the inverse displacement δ_{FEM_inv} using an optional distance based decimation (as the sub sampling strategy that is detailed in Sect. 4.3). The positions of an experimental points corrected in the reference CAD conditions is consequently gained using the RBF interpolation:

$$
\mathrm{P}_{\mathrm{ref}}(\mathrm{P}_{\mathrm{exp}}) = \begin{pmatrix} \sum_{i=1}^{N_{FEM}} \gamma_i^x \varphi\left(\left\|\mathrm{P}_{\mathrm{exp}} - \mathrm{P}_{\mathrm{FEM\_d}_i}\right\|\right) + \beta_1^x + \beta_2^x \mathrm{P}_{\mathrm{exp}_x} + \beta_3^x \mathrm{P}_{\mathrm{exp}_y} + \beta_4^x \mathrm{P}_{\mathrm{exp}_y} \\ \sum_{i=1}^{N_{FEM}} \gamma_i^y \varphi\left(\left\|\mathrm{P}_{\mathrm{exp}} - \mathrm{P}_{\mathrm{FEM\_d}_i}\right\|\right) + \beta_1^y + \beta_2^y \mathrm{P}_{\mathrm{exp}_x} + \beta_3^y \mathrm{P}_{\mathrm{exp}_y} + \beta_4^y \mathrm{P}_{\mathrm{exp}_y} \\ \sum_{i=1}^{N_{FEM}} \gamma_i^z \varphi\left(\left\|\mathrm{P}_{\mathrm{exp}} - \mathrm{P}_{\mathrm{FEM\_d}_i}\right\|\right) + \beta_1^z + \beta_2^z \mathrm{P}_{\mathrm{exp}_x} + \beta_3^z \mathrm{P}_{\mathrm{exp}_y} + \beta_4^z \mathrm{P}_{\mathrm{exp}_y} \end{pmatrix} \tag{12.23}
$$

where γ^{dir} and β^{dir} are the sought coefficients obtained by fitting the 3 RBF problems at $\mathrm{P}_{\mathrm{FEM\_d}}$ related to the three components of the displacement field $\delta^{dir}_{FEM\_inv}$, with $dir = x, y, z$. All the measured points can be processed according to that equation. It is worth to remember that the RBF interpolation is a point function available everywhere.

In the framework of the metrology task of the ITER fusion reactor (Poncet et al. 2015) the soundness of the proposed approach has been evaluated in (Semeraro 2017) by a specific experiments illustrated in Fig. 12.21. The surface of a cantilever beam, made of steel, is clamped on one edge and loaded on the opposite one with a weight of adjustable amount and position (left). With regard to instrumentation, about 200 markers suitable for accurate photogrammetric measurements are attached onto the surface of the component.

The photogrammetry system to collect measurements is based on a Dynamo D12 camera coupled with the software V-Stars 4.9 by Geodetic System. Experimental data processing was performed by the software Spatial Analyzer Rel 2017.01.12_21716. To enhance the environmental effects, the cantilever was loaded with a known weight (25 kg). Many load conditions have been considered to represent the effect of an undesired environment which is responsible of elastic deformations that move the part apart away with respect to the reference

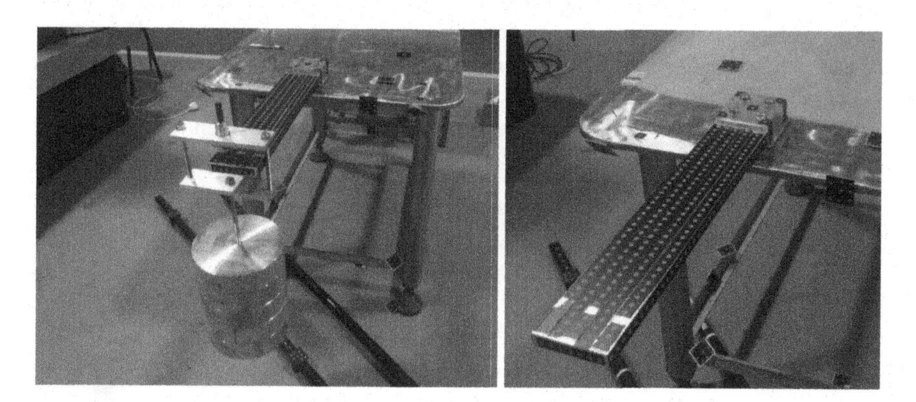

Fig. 12.21 Experimental test arrangement of the cantilever beam: with load applied (left) and undeformed (right)

undeformed geometry. In this section we refer to two conditions: the most aggressive one, namely maximum weight at maximum offset, that produces bending and torsion in the beam and the unloaded one that, due to the high stiffness of the cantilever, allows to neglect the effect of its dead load and can be considered representative of the part free of deformation.

Figure 12.22 shows the FEM model generated to reproduce the experimental layout. Such a model, characterised by a mesh composed by 10800 nodes and 2000 second-order hexa elements (left), was prepared by means of the ANSYS APDL software. Such a set-up allows to capture with an high level of fidelity the displacement field (right) due to the applied loading and constraint conditions.

The positions of FEA nodes for the undeformed model and for the deformed one are then exported in a text file by means of an ANSYS APDL script according to three strategies: all FEA nodes (fine dataset), all FEA nodes on the surface (medium dataset) and a minimal set of FEA nodes which includes only those belonging to seven specific cross sections (coarse dataset).

A Mathcad worksheet was defined to set-up a prototype of the workflow of the proposed method. FEA data are imported as tables and the cloud of points to define the RBF problem is created by converting the table of the deformed FEA positions. Finally, the inverse field is obtained subtracting the two tables, which is the undeformed one minus the deformed one.

An RBF using a cubic kernel ($\varphi(r) = r^3$) is fitted using the direct method (see Table 2.1 of Sect. 2.2). Experimental positions of the markers are available in both the undeformed and deformed configurations as text files. These files contain, in columns, the name of the marker and the x, y, z positions. Since some of the markers are not valid for both configurations, a specific parser able to link consistent locations by name matching has been developed. It is worth to notice that such a matching is required only for validation purposes, whereas in the sought workflow only the deformed positions are available and their processing is quite straightforward.

Figure 12.23 depicts the exported FEM data including both the undeformed (cloud of black points) and the deformed (cloud of red points) mesh. The coarse dataset (left) includes only the nodes on seven cross sections, whereas the medium one (right) includes all the nodes on the component surface.

The RBF corrected cloud is computed by the worksheet and written back adopting the same file format of experimental input to enable further processing.

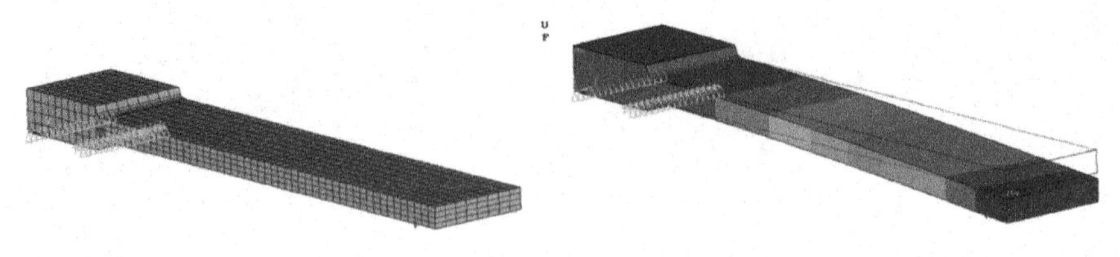

Fig. 12.22 Cantilever case: mesh (left) and displacements (right) obtained through the FEM

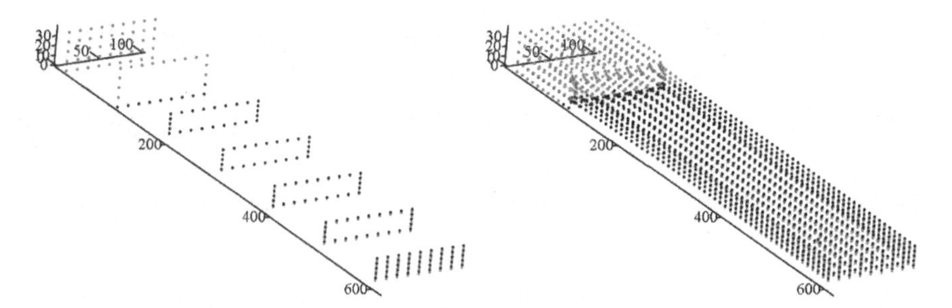

Fig. 12.23 FEM data are exported as clouds of points for the coarse (left) and medium (right) dataset

The quality of the correction has been evaluated computing the distances of the reference CAD versus the experimental cloud for all the available data (i.e. undeformed, deformed and deformed after the RBF compensation).

In the case that no compensation is applied, the effect of the environment (in this case the eccentric load produced by the adjustable weight) induces a substantial mismatch between the real component and its baseline CAD (Fig. 12.24). Given the extent of the discrepancy from the baseline desired shape (2.28 mm), the manufactured component would be judged to have a not conformal shape and, thus, not compliant with its mechanical requirements.

The effective accuracy of the actual geometry can be assessed according to two approaches: by removing the environmental effect, performed by comparing the CAD with the experimental cloud acquired on the undeformed part as depicted in Fig. 12.25 (top), or by compensating the measurement with RBF subtracted with

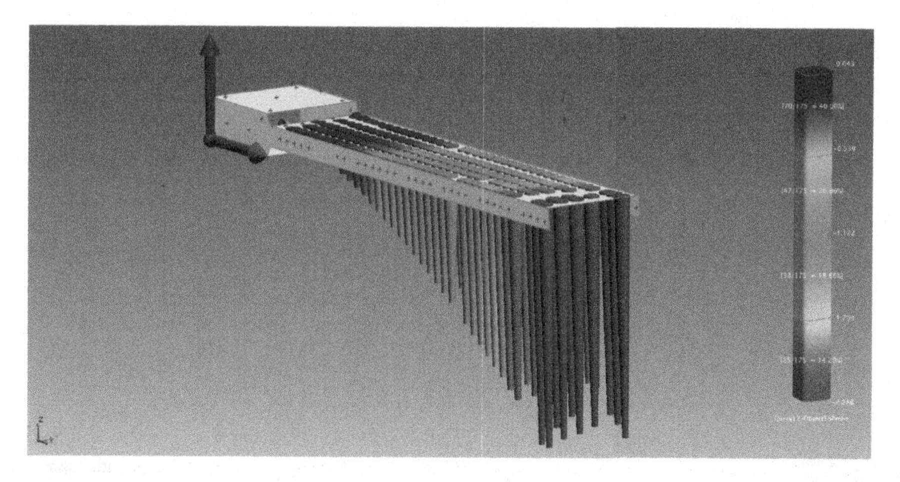

Fig. 12.24 Comparison between experimental points acquired for the deformed configuration and the baseline CAD

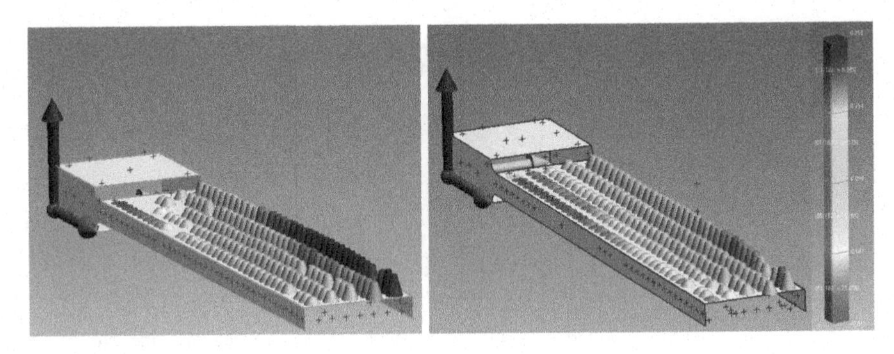

Fig. 12.25 Discrepancy obtained by removing the environment effect (top) and by RBF compensation (bottom)

the FEA estimation of the environment (Fig. 12.25 bottom). The maximum registered deviation with respect to the CAD is equal to 0.210 mm considering the data set acquired on the undeformed configuration, and results to be 0.178 mm considering the data set acquired on the deformed configuration and compensated with FEA environment estimation. Both results are within the targeted accuracy of the manufactured part. The effect of the compensation can be better understood by comparing the compensated experimental cloud with the one acquired on the undeformed geometry.

An histogram (Fig. 12.26) representing the differences between the vertical positions acquired on the rigid model and on the deformed one after RBF compensation, shows that the two datasets are very close each other. In order to establish how the detail of the FEA dataset used for the compensation can affect the

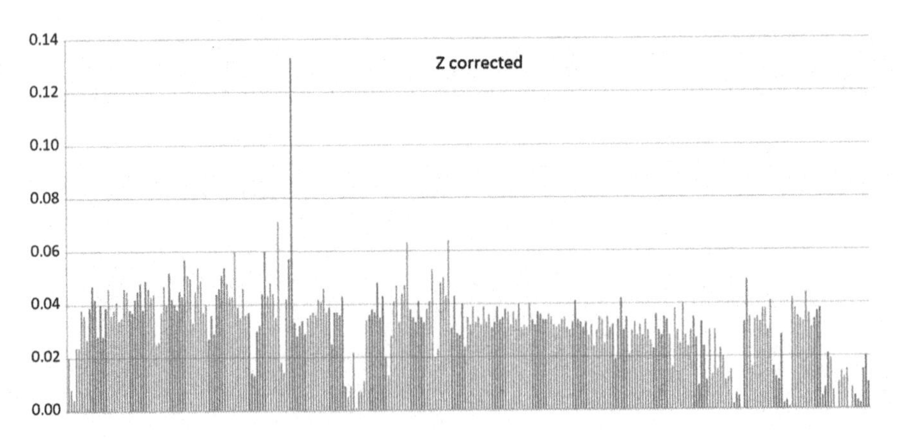

Fig. 12.26 Histogram with the difference of vertical positions of the same markers at undeformed configuration and at the deformed one after the RBF compensation

process, the correction has been computed using all the available dataset (coarse, medium, fine). The compensation amount is not affected and even the coarse mesh is suitable for the workflow.

It is worth to notice that the imposed environment is very simple (bending and torsion) and that for more complex scenarios the calibration of density should be addressed as well. For implementation purposes this is a key point because a fast RBF implementation should be considered for complex parts that are usually modelled with FEM meshes that have up to millions of nodes.

12.7 RBF for the Interpolation of Hemodynamics Flow Pattern

Although phase contrast magnetic resonance imaging (PC-MRI) allows non-invasive quantitative map of blood flowing in the cardiovascular system, the potentiality of such imaging technique is just partially exploited due to the limits of some features related to spatial and temporal accuracy such as pixel size, slice-thickness as well as number of time frames per cardiac cycle (Ponzini et al. 2012).

A possible approach to overcome the limitations dealing with visualization, consists of the usage of interpolation strategies (Vennell and Beatson 2009). The effectiveness of RBF to interpolate complex blood flow fields (see Fig. 12.27) is showcased in the application following described. Such an application addresses the

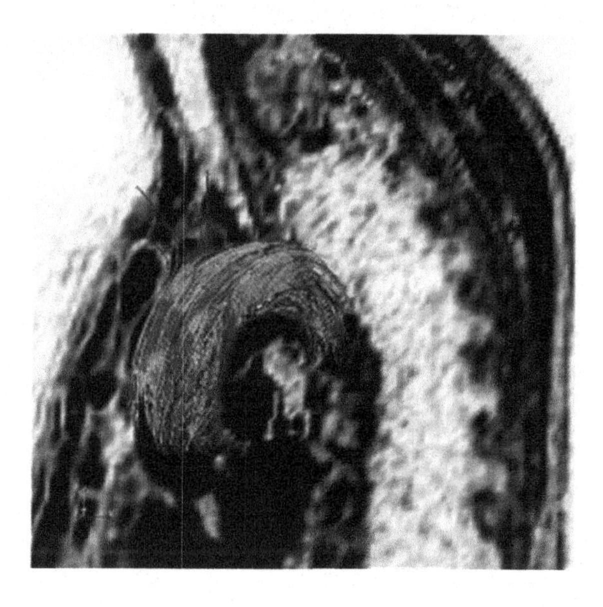

Fig. 12.27 Complex flow pattern measured in vivo

analysis of the percentage error in terms of flow velocity between the interpolated data with respect to the validated synthetic MRI-like data generated through an image-based computational model of the aortic hemodynamics (Morbiducci et al. 2011). In particular, selected sets of velocity data, taken from a patient specific image-based CFD model, are mapped onto equally spaced grids of imposed pixel size (matrices), slice thickness and number of time frames according to their placement in space and time. For spatial and temporal sampling, weighting Gaussian function and an arithmetic mean filter are respectively adopted.

The images built with the highest spatial and temporal accuracy, which are not feasible through in vivo tests, are considered to be the GS dataset. Given that, the GS represents the reference dataset to test, on the one hand, the accuracy of the RBF image interpolation and, on the other hand, to evaluate the quality of the coarser matrices (Ni, i = 1, 2, 3). The spatial accuracy of the GS matrix is equal to $0.5 \times 0.5 \times 0.5$ mm^3 while the temporal accuracy is 0.01 s, whilst the coarser images Ni have a spatial discretisation respectively of $1.0 \times 1.0 \times 1.0$ mm^3 (GS $\times$ 2), $2.0 \times 2.0 \times 2.0$ mm^3 (GS $\times$ 4) and $4.0 \times 4.0 \times 4.0$ mm^3 (GS $\times$ 8), and a temporal definition that range and from 0.02 to 0.07 s respectively.

To interpolate flow field data a direct RBF fit adopting the radial function $\varphi(r) = r$ is implemented by means of a Mathcad worksheet. The specific choice of RBF is done because this is the smoothest interpolant and it can be accelerated using fast RBF algorithms so that the method can be effectively extended to handle real life applications.

The source points to compute the RBF field are defined by selecting a set of points inside the domain boundaries through a sub-sampling method based on a prescribed radius set considering the typical in vivo measurements (4 mm), and by excluding the zero velocity points falling outside the domain boundaries through a delimiting domain. The generated source points are shown in Fig. 12.28 (complete slice on the left and enlarged detail on the right), where the red dots are the points selected as input for the RBF interpolation (about 2400), while the black dots represent the points of the original dataset in which the flow field is evaluated by the interpolation process (about 58,000).

Since the synthetic dataset points are characterised by a spatial resolution higher than the RBF dataset, they can be then used as a reference to evaluate the error of the interpolated data of the absolute value of the flow velocity and, accordingly, to understand whether the RBF interpolation, at typical in vivo spacing, can be reliably employed for the reconstruction of the actual flow field.

Referring to spatial interpolation, Fig. 12.29 shows the results of the mapping procedure at peak systole for a single "D" slice centred on the geometry model whose reference image pixel size is taken equal to 0.5×0.5 mm^2. The overall features of the velocity component are in good agreement as expected. Using the CFD-based data generator the analyst is then able to perform a very detailed point wise analysis and to quantify the percentage error in each point of the slice.

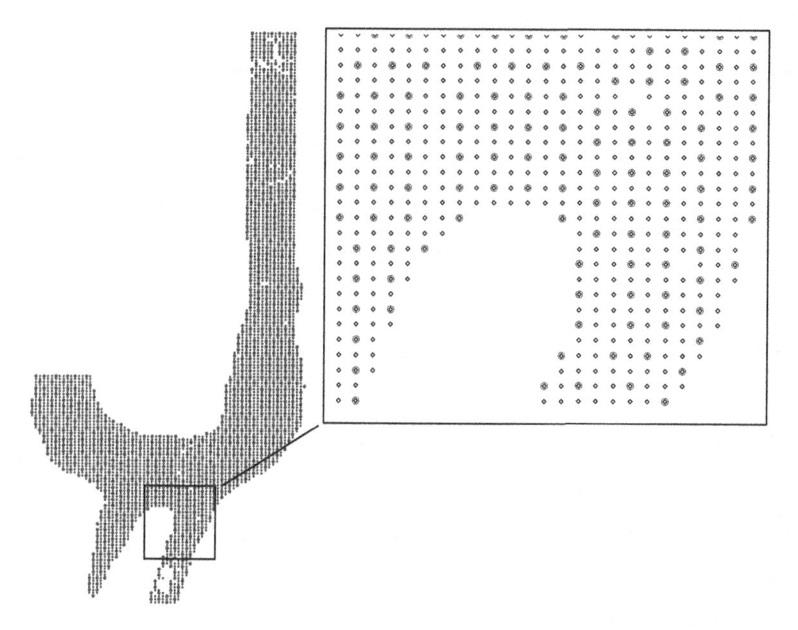

Fig. 12.28 Generated points of GS dataset (black dots) and RBF source points (red points)

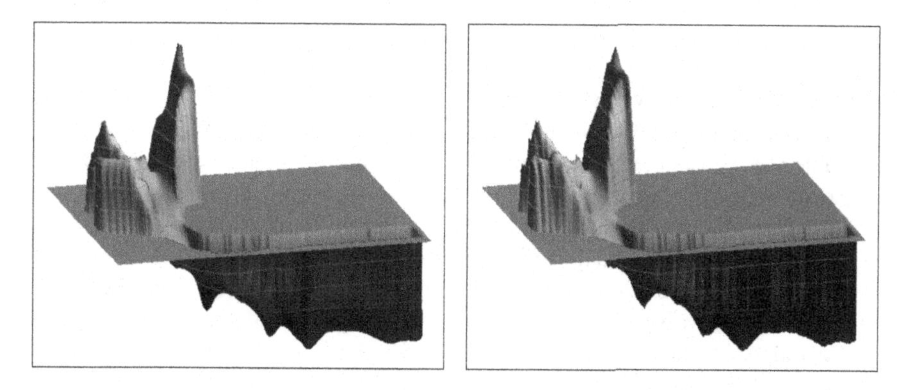

Fig. 12.29 Qualitative comparison of reference data (left) with RBF interpolated data (right)

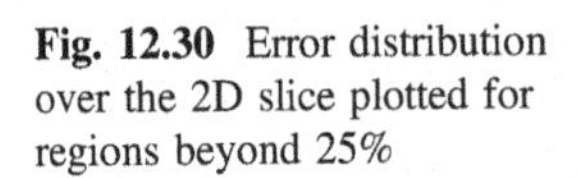

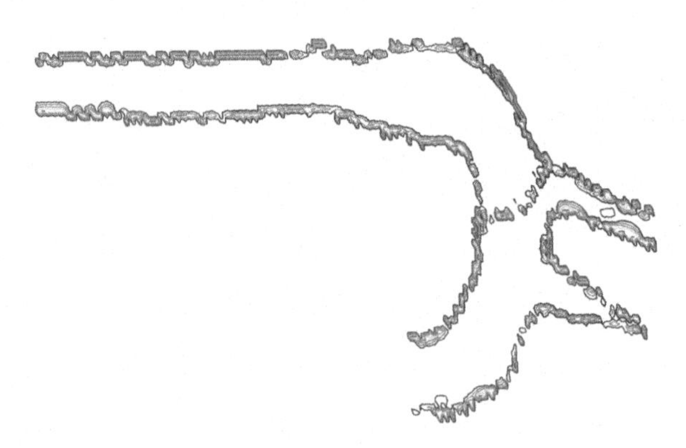

Fig. 12.30 Error distribution over the 2D slice plotted for regions beyond 25%

A large part of the points' population has a very low percentage error value. In detail, the distribution of the percentage error is the following one:

- 82% below 5%;
- 5% in the range 5–10%;
- 8% in the range 10–20%;
- 4% in the range 20–50%;
- 1% over 50%.

Figure 12.30 depicts the error percentage over the 2D slice. As visible, the flow field at the bulk points is very well described by the interpolated data, while the error is rather relevant at wall vessel location. This kind of configuration needs to be kept in mind when in vivo computation are performed including vessel location quantities such as the wall shear stress (WSS) related ones.

As far as temporal interpolation is concerned, the data obtained through RBF are available at each sampled time frame. The flow information at an arbitrary position in the domain for a given intermediate time is recovered interpolating the data already interpolated with those computed at a certain number of time frames.

Further investigations could span the sensitivity of other image parameters such as pixel size, slice thickness and number of time frame per heart cycle. All these parameters are involved in MRI in vivo data sampling and must be kept under consideration in the evaluation of the performance of any image processing algorithm.

To quantify the error of the RBF interpolation in function of the specific RBF used for the fit, a parametric study concerning the peak systole instant adopting the N3 spatial resolution (4 mm) is carried out. To ease the comprehension of the difference in the accuracy connected to the spatial resolution, Fig. 12.31 respectively shows the flow field at the peak systole for the GS, N1, N2 and N3 levels of spatial discretization.

The accuracy of the RBF interpolation of the flow field based on N3 coarse dataset is here evaluated by performing the interpolation at various resolutions,

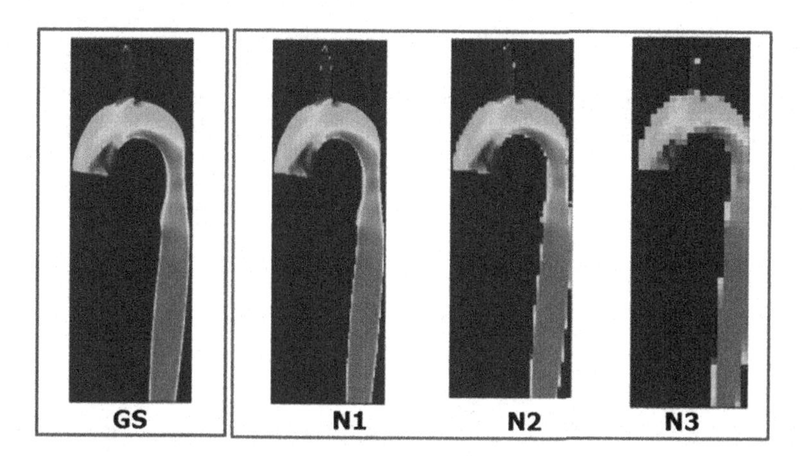

Fig. 12.31 Peak systole image with different levels of spatial discretization

namely N2, N1 and GS, and comparing how the interpolated N3 data match with available reference value.

In Table 12.9 the data of a parametric study concerning the error percentage of 33 interpolation configurations (IDs), obtained using 11 types of RBF, are summarised. In particular, 3 RBF with compact support (C0, C2, C4) having each 3 different radius $\left(4\sqrt{2}, 8\sqrt{2}, 12\sqrt{2}\right)$, 1 linear (LINEAR) and 1 cubic with total support (CUBIC) were employed. For all runs the RBF fit is performed on the coarse distribution N3 (4 mm) at the peak systole instant using just the velocity values different from zero. The DX parameter refers to 2D spatial distance and is equal to 0.5, 1 and 2 mm respectively.

The best result in terms of highest percentage of points with a percentage error less than 5% is gained using the cubic RBF (ID = 11, 22, 33). Good results are also guaranteed by the C2 function with the largest value of the radius (ID = 6, 17, 28).

Figure 12.32 shows the best and the worst interpolation case, namely those respectively with ID 11 and 29.

Referring to RBF with compact support, the density of the matrix is 2.96% 6.25% 10.1% for the three values of the radius considered in the study respectively, whilst it is 95.4% for the RBF with full support (along the principal diagonal zero values are present).

It is very important to remark that the meshless nature of the methods allows to improve the accuracy. As described, the predicted flow field exhibits a very low percentage error while the wall boundaries are poorly resolved. A substantial improvement of the interpolation could be achieved placing extra RBF points at wall location that impose to the flow the desired wall conditions. Furthermore, the analytic nature of the fit allows to evaluate spatial derivatives in closed form enriching substantially the information extracted from the experimental data flow.

Table 12.9 PC-MRI case: sensitivity study summary

ID	DX (mm)	Compact/Total support	Radius (mm)	%err < 5	5 < %err < 10	10 < % error < 20	20 < %err < 50	50 < %err < 100	%err > 100
1	0.5	C0	$4\sqrt{2}$	63.789	11.998	10.22	8.656	2.45	2.887
2	0.5	C0	$8\sqrt{2}$	73.598	9.403	7.165	5.727	1.588	2.52
3	0.5	C0	$12\sqrt{2}$	74.39	9.349	6.752	5.641	1.419	2.448
4	0.5	C2	$4\sqrt{2}$	65.506	11.772	9.222	8.221	2.396	2.883
5	0.5	C2	$8\sqrt{2}$	73.32	10.65	6.567	5.452	1.587	2.424
6	0.5	C2	$12\sqrt{2}$	75.224	9.811	5.868	5.385	1.415	2.297
7	0.5	C4	$4\sqrt{2}$	58.541	14.926	9.952	9.961	3.337	3.283
8	0.5	C4	$8\sqrt{2}$	72.675	10.637	6.729	5.773	1.724	2.461
9	0.5	C4	$12\sqrt{2}$	74.7	10.458	5.9	5.249	1.419	2.274
10	0.5	LINEAR	–	74.89	9.215	6.63	5.588	1.324	2.353
11	0.5	CUBIC	–	77.799	8.506	5.439	4.742	1.452	2.062
12	1	C0	$4\sqrt{2}$	62.92	12.328	10.55	9.098	2.599	2.504
13	1	C0	$8\sqrt{2}$	73.131	9.466	7.631	5.986	1.53	2.257
14	1	C0	$12\sqrt{2}$	74.158	9.086	7.422	5.853	1.32	2.161
15	1	C2	$4\sqrt{2}$	64.29	12.216	9.96	8.526	2.543	2.466
16	1	C2	$8\sqrt{2}$	72.959	10.48	7.095	5.814	1.549	2.104
17	1	C2	$12\sqrt{2}$	74.799	9.874	6.468	5.415	1.473	1.971
18	1	C4	$4\sqrt{2}$	57.312	15.351	10.189	10.763	3.403	2.982
19	1	C4	$8\sqrt{2}$	72.112	10.52	7.402	6.217	1.568	2.181
20	1	C4	$12\sqrt{2}$	74.321	10.448	6.391	5.473	1.454	1.914
21	1	LINEAR	–	74.354	9.282	7.081	5.933	1.301	2.048
22	1	CUBIC	–	77.484	8.635	5.686	5.093	1.359	1.742
23	2	C0	$4\sqrt{2}$	62.405	10.511	10.322	10.322	3.788	2.652

(continued)

Table 12.9 (continued)

ID	DX (mm)	Compact/Total support	Radius (mm)	%err < 5	5 < %err < 10	10 < % error < 20	20 < %err < 50	50 < %err < 100	%err > 100
24	2	C0	$8\sqrt{2}$	71.943	9.194	6.54	7.962	1.991	2.37
25	2	C0	$12\sqrt{2}$	72.417	9.289	6.635	7.962	1.517	2.18
26	2	C2	$4\sqrt{2}$	61.553	12.027	9.848	10.322	3.504	2.746
27	2	C2	$8\sqrt{2}$	71.917	9.867	6.452	7.875	1.898	1.992
28	2	C2	$12\sqrt{2}$	73.649	9.1	5.498	8.246	1.611	1.896
29	2	C4	$4\sqrt{2}$	53.03	16.667	9.47	12.689	4.451	3.693
30	2	C4	$8\sqrt{2}$	71.185	9.953	6.445	8.436	1.706	2.275
31	2	C4	$12\sqrt{2}$	73.081	9.668	5.687	8.152	1.517	1.896
32	2	LINEAR	–	72.581	9.108	5.787	8.729	1.423	2.372
33	2	CUBIC	–	76.705	7.292	5.303	7.386	1.515	1.799

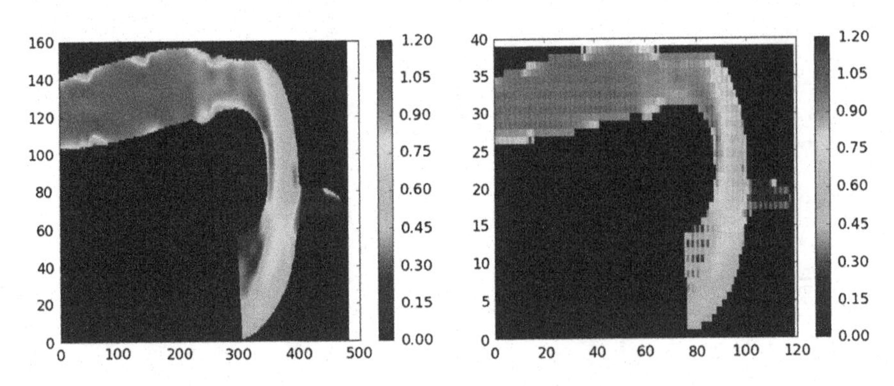

Fig. 12.32 ID 11 and 29: reference and interpolated velocity values

References

Amodio D, Broggiato GB, Salvini P (1995) Finite strain analysis by image processing: smoothing techniques. Strain 31(3):151–157

Arad N, Dyn N, Reisfeld D, Yeshurun Y (1994) Image warping by radial basis functions: application to facial expressions. CVGIP Graph Models Image Process 56:161–172

Besnard G, Hild F, Roux S (2006) Finite-element displacement fields analysis; from digital images: application to Portevin–Le Châtelier bands. Exp Mech 46:789–803

Biancolini M, Salvini P (2012) Radial basis functions for the image analysis of deformations. In: Computational modelling of objects represented in images: fundamentals, methods and applications III proceedings of the international symposium, Rome, pp 361–365. ISBN: 978-041562134-2

Biancolini ME, Brutti C, Chiappa A, Salvini P (2015) Post-processing strutturale mediante uso di radial basis functions. Associazione Italiana per L'Analisi delle Sollecitazioni – AIAS, 44° Convegno Nazionale, 2–5 Settembre, Università di Messina

Editor (1992) Biographical sketch of Rolland L. Hardy. Comput Math Appl 24(12):ix–x

Flusser J, Suk T, Zitová B (2009) Moments and moment invariants in pattern recognition. Wiley, USA

Halpern AB, Billington DP, Adriaenssens S (2013) The ribbed floor slab systems of Pier Luigi Nervi. In: Obrębski JB, Tarczewski R (eds) Proceedings of the International Association for Shell and Spatial Structures (IASS) symposium 2013, "BEYOND THE LIMITS OF MAN" 23–27 Sept, Wroclaw University of Technology, Poland

Hardy RL (1971) Multiquadric equations of topography and other irregular surfaces. J Geophys Res 76(8):1905–1915

Hardy RL (1990) Theory and applications of the multiquadric-biharmonic method, 20 years of discovery, 1968–1988. Comput Math Appl 19(8/9):163–208

Howland RCJ (1930) On the stress in the neighborhood of a circular hole in a strip under tension. Philos Trans R Soc Lond 229:49–86

Innovmetric (2017) Polyworks. https://www.innovmetric.com/

Jessop HT, Snell C, Allison IM (1959) The stress concentration factors in cylindrical tubes with transverse cylindrical holes. Aeronaut Q 10(4):326–344

Maggiolini E, Livieri P, Tovo R (2015) Implicit gradient and integral average effective stresses: relationships and numerical approximations. Fatigue Fract Eng Mater Struct 38:190–199

Morbiducci U, Ponzini R, Rizzo G, Iannaccone F, Gallo D, Redaelli A (2011) Synthetic dataset generation for the analysis and the evaluation of image-based hemodynamics of the human aorta. Med Biol Eng Comput 50(2):145–154

New River Kinematics (2017) Spatial analyzer. http://www.kinematics.com/spatialanalyzer/
Poncet L, Bellesia B, Oliva AB, Rebollo EB, Cornelis M, Medrano JC, Harrison R, Lo Bue A, Moreno A, Foussat A, Felipe A, Echeandia A, Barutti A, Caserza B, Barbero P, Stenca S, Da Re A, Ribeiro JS, Brocot C, Benaoun S (2015) EU ITER TF coil: dimensional metrology, a key player in the double pancake integration, fusion engineering and design, vol 98–99, pp 1135–1139. In: Proceedings of the 28th symposium on fusion technology (SOFT-28)
Ponzini R, Biancolini ME, Rizzo G, Morbiducci U (2012). Radial basis functions for the interpolation of hemodynamics flow pattern: a quantitative analysis. In: Computational modelling of objects represented in images III: fundamentals, methods and applications. Rome, Italy, 09/05/2012–09/07/2012. ISBN: 9780415621342
Schultz KJ (1941) On the state of stress in perforated plates. Ph.D., Technische Hochschule, Delft
Semeraro L (2017) Alignment and metrology challenges at ITER. In: "3rd PACMAN workshop", CERN
Sutton MA, Turner JL, Chao YJ, Bruch A, Chae TL (1991) Full field representation of discretely sampled surfaces deformation for displacements and strain analysis. Exp Mech 31(2):168–177
Sutton MA, Orteu JJ, Shreier HW (2009) Image correlation for shape, motion and deformation measurements. Springer, Berlin. ISBN 0387787461
Tarisciotti C, Biancolini ME (2013) Quando le eqauzioni dell'elsaticità progettano le strutture. Analisi e Calcolo. N 59
Vennell R, Beatson R (2009) A divergence-free spatial interpolator for large sparse velocity data sets. J Geophys Res 114:C10024. https://doi.org/10.1029/2008JC004973
Zienkiewicz OC, Taylor RL, Zhu JZ (2005) The finite element method: its basis and fundamentals, 6th edn. Butterworth-Heinemann, UK

Chapter 13
Data Mapping Using RBF

Abstract The present chapter details a generic mapping algorithm based on RBF. Mapping allows to interpolate information known at discrete locations across not-matching numerical grids of the same geometrical entities. After an overview of mapping procedures, examples of CFD computed pressure field mapping onto structural mesh adopted for FEA based CSM (see Chap. 11) is demonstrated starting with a simple reference geometry (a catenoid) and then with actual complex three-dimensional shapes of aircraft wings. The volume mapping is then addressed with a specific focus of the mapping of electro magnetic (EM) loads onto structural meshes. Examples in which magnetic loads are already available as force density or as magnetic field and density current (so that load density is computed on the FEA as a Lorentz Force) are presented. Practical examples involving the toroidal field (TF) coil of DEMO superconductor magnet are shown with details of the stress response of the structure subjected to the EM loads.

13.1 Data Mapping Background

The optimization of vector mapping across non matching meshes is a very important topic especially for multi-physics applications where the quantities obtained on a numerical grid, typically referred to as source, need to be transferred across a different model, commonly called target, as exemplified in Fig. 13.1. Some among the applications for which vector mapping is relevant are, for instance, the update of an high frequency (HF) model of an electronic system to account for the deformations obtained using a thermo-structural model, and the transfer of the magnetic loads or temperature distributions onto the structural FEM model of a component.

M. E. Biancolini, *Fast Radial Basis Functions for Engineering Applications*,
https://doi.org/10.1007/978-3-319-75011-8_13

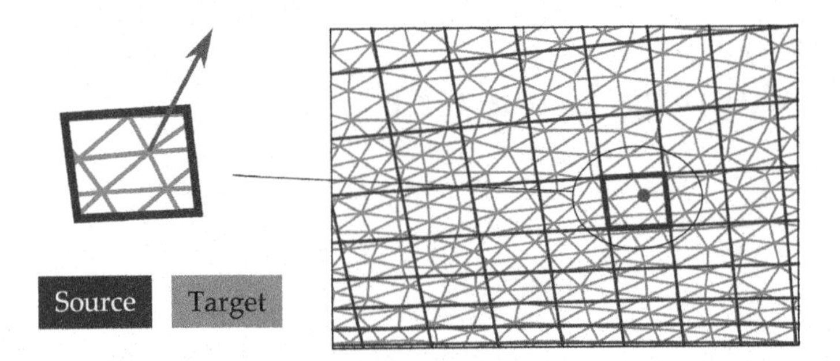

Fig. 13.1 Example of mapping solutions thorough non-matching meshes

In general, mapping methods need to fulfil the following listed requirements:

- accuracy: capability to guarantee no loss of load magnitude and direction during the transfer process;
- flexibility: capability to handle dissimilar meshes including the cases fine-to-coarse and coarse-to-fine;
- performance: capability to manage very large models in a reasonably short time.

The latter requirement acquires a particular significance when treating high-fidelity analyses dealing with FSI, for instance, because the generation and use of large-size meshes for CFD and FEM models (hundreds millions in industrial applications and even some billions in research studies) are mandatory needed.

13.2 Mapping of Pressure in Two-Way FSI Analyses

13.2.1 Introduction and Review of Mapping Schemes

In the case that data mapping is used for carrying out FSI analyses according to the two-way approach, a technique to transfer the aerodynamic loads from the wetted surfaces of the CFD mesh to the corresponding elements of the FEM surface mesh, needs to be used. Examples already exposed in Sect. 10.5 are focused on the global workflow and rely on the mapping tools included in the commercial CAE software adopted. In this section the focus is on the advanced mapping method developed in the framework of the RIBES project (Biancolini et al. 2017).

The aerodynamic loads typically consist of pressure and shear stresses, whilst the common boundaries of the computational domains have, in general, a non-matching discretization because of the different requirements of the generation of the CFD and FEM meshes. As such, the locations of the application of the forces computed by the CFD solver, that are extracted from the cells adjacent to the wall boundaries and available in the form of vectors positioned on a cloud of points, typically differ from the FEM grid points on which the loads have to be applied.

An interpolation between the two domains is then required with a consequential introduction of an error. The minimization of the uncertainness associated to this process depends on the accuracy of the mathematical approach chosen to perform such an interpolation.

A good review about load transfer schemes can be found in Jiao and Heath (2004) and Jaiman et al. (2006) where a great focus about load conservation and error estimation is given. The following families of load transfer methods are commonly used:

- **Point-wise interpolation and extrapolation**: the load is interpolated at the sources so that it becomes available as a point function and then used to compute load at the target. The method is flexible because several strategies can be selected to interpolate the source. The main drawback is that equilibrium is not guaranteed (Jiao and Heath 2004). Numerical experiments show that a wise optimization of the interpolator leads to very good results.
- **Point-to-element projection schemes**: a direct connection between the closest source element and the target point is defined. Value at target is then obtained using shape functions of the connected source element. The main complexity is related to neighbour searching. Although such a search may be optimised using space partitioning methods as octal tree decomposition, local and global equilibrium are not guaranteed because the connection between the source and target relies on local quantities and not on surface averaged ones. Nevertheless Samareh (2007) demonstrates that excellent results can be achieved on test cases relevant for the aeronautical field.
- **Area weighted averaging**: the value of a target element is defined as the weighted average of the values of the source elements in contact, where the weights are the areas of the intersections between the source and target elements. Conservation is guaranteed. The method can be classified as a special case of common refinement schemes (Jao and Heat 2004; Jaiman et al. 2006) they have proven to be conservative both globally and locally, producing results more accurate than point to element approach.
- **Mortar elements methods**: they are based on a weak formulation. A Galerkin approach is taken in order to minimize the difference of the integral of pressure fields over the source and target domains (Cebral and Löhner 1997). Since such integration is performed relying on numerical quadrature, several authors (Jiao and Heath 2004) acknowledge the need for a common grid (common refinement) to properly perform the said procedure. Nodes left out the common mesh can be treated directly by shape functions extrapolation or resorting to enforce consistency techniques (Wang et al. 2016).
- **Interface reconstruction methods**: the benefit of the high quality mapping may be achieved adopting the implicit surface definition based on RBF interpolation (Jin et al. 2003). This approach is usually adopted to reconstruct a geometry that comes for a laser scan in reverse engineering applications as the zero isosurface of a 3D scalar function interpolated using RBF. On-surface points and normal are used to generate off-surface points. The scalar function is defined so that is

zero at on-surface point and non-zero at off-surface points in the positive (negative) side with a typically value of 1 (−1).

13.2.2 Standard and High Fidelity RBF Mapping

RBF can be quickly adopted for the implementation of a point-wise interpolation method, here defined as "Standard RBF mapping". The pressure available on the CFD mesh is handled as a scalar function known at certain CFD mesh locations (usually the centre of face element where the information is computed). Such locations are used as RBF sources using as input the local computed pressure so that the RBF fit makes this scalar function available in space. Equation (2.1), here adapted in (13.1), expresses the pressure as a meshless function.

$$s_p(\boldsymbol{x}) = \sum_{i=1}^{N} \gamma_i^p \varphi(\|\boldsymbol{x} - \boldsymbol{x}_{s\,i}\|) + \beta_1^p + \beta_2^p x + \beta_3^p y + \beta_4^p z \tag{13.1}$$

The meshless field can be used to interpolate the pressure at FEM mesh nodal positions (13.2), or at FEM elements centroids. In the cases where the FEM mesh is composed by shell elements, the pressure load can be directly applied, whilst if the mesh consist of solid the relevant face has to be loaded instead. The topological complexity can be avoided adding a skin of shell elements on the wetted surface, i.e. the free faces of the solid mesh, so that the load transfer can be operated in any case on a surface mesh.

$$\boldsymbol{p}_{\boldsymbol{node\_fem}} = s_p\left(\boldsymbol{x}_{\boldsymbol{node\_fem}}\right) \tag{13.2}$$

Usually there are many options to map the load onto the FEM surface elements. When the mesh density are similar the FEM element can be loaded with a constant pressure; however this approach can be not accurate enough when FEM model mesh is coarser than the CFD one. In such situations the pressure spatial gradients accurately captured by the finer CFD mesh should be represented by applying a variable pressure distribution on the face of the FEM element. The FEM solver Nastran, for instance, accepts both input with the PLOAD2 and PLOAD4 cards; PLOAD2 receives a uniform pressure on the element and centroid valued is usually adopted, whereas PLOAD4 accepts different values of pressure at connected nodes (3 for CTRIA3, 3 nodes shells; 4 for CTRIA4 4 nodes shells).

As will be shown in the test cases detailed in the successive sections, the standard RBF mapping approach provides outputs in which a discrepancy between the values of the resultants forces due to interpolation is present. Such an error can be made negligible introducing an high fidelity approach successfully developed during RIBES, a Clean Sky research project (European Commission 2014; RIBES Project 2014). The main features of the high quality mapping method proposed in the RIBES project, combines several features of the pointwise, area weighted averaging and mortar elements methods (see Sect. 13.1). It is based on a pointwise representation of the source that allows to accurately interpolate both the source value (for instance pressure) and the source geometry itself as a point function. So, the original flow solution becomes meshless and it is defined onto a meshless representation of the surface that plays a role similar to the mortar elements surface and common refinement mesh. Such interpolator is then tuned with respect to the original source solution and mesh, so that both local and global equilibrium are met. A set of functions specifically introduced are used to represent the map applied on the target, whose coefficients are accordingly obtained enforcing the equilibrium between the target and the source models.

The method consists of the following main Steps:

1. An RBF interpolator suitable to accurately reproduce the load at source to be mapped (scalar or component of vector fields) is defined.
2. The RBF interpolator above described is applied to the nodes of the target mesh, supplying a pointwise map of pressure.
3. Superimposed clouds of source and target nodes are covered by a population of fuzzy sets (FSs) which include nodes with a given degree of membership (Zadeh 1965).
4. Load acting on each node of the target is calculated imposing the equilibrium between the source and target resultants of each FS through a set of correction coefficients. The calculation of such coefficients keeps into account the partition of nodes among the FSs.

The partition of unity (POU) method (Babuška and Melenk 1997) introduced in Sect. 3.5 could be used to arrange the source and target point sets into overlapping subdomains. The procedure consists of the decomposition of the original datasets into small RBF problems taking in mind that, for moderate amounts of data, a single RBF transfer could be considered as concerning the whole domain. The field, defined as a set of values at the corresponding centroids or nodes of the source mesh, is interpolated by RBF. In the case that force vectors fields are present, the three components are interpolated by three independent RBF solutions. In each subdomain, the interpolation problem is locally solved and the force field between the source subdomain and its target counterpart is exchanged. A set of local solutions is finally obtained that, in order to recover the continuity of the field have to be finally combined together through blending functions. As already exposed in Sect. 3.5, the smoothness of the global solution can be guaranteed adopting polynomial blending functions which are obtained from a set of smooth functions W_i by a normalization procedure, $w_i(x) = \frac{W_i(x)}{\sum_j W_j(x)}$, where the condition $\sum W_i = 1$ has to be satisfied.

The weighting functions W_i can be defined as the composition of a distance function d_i and a decay function V_i. The distance function has to satisfy the condition $d_i(x) = 1$ at the boundaries of a subdomain. The decay function is defined through the distance function, and its degree can defined by the user. Examples of decay functions with growing degree of continuity (C^0, C^1, C^2) are already given in

Eq. (3.6) here rewritten as (13.3) that can be applied to the spherical domain case according to Eq. (3.7) here rewritten as (13.4) where $\boldsymbol{r}(x)$ is the distance from the centre and R is the radius of the sphere.

$$\begin{aligned} \boldsymbol{V}^0(\boldsymbol{d}) &= 1 - \boldsymbol{d} \\ \boldsymbol{V}^1(\boldsymbol{d}) &= 2\boldsymbol{d}^3 - 3\boldsymbol{d}^2 + 1 \\ \boldsymbol{V}^2(\boldsymbol{d}) &= -6\boldsymbol{d}^5 + 15\boldsymbol{d}^4 - 10\boldsymbol{d}^3 + 1 \end{aligned} \tag{13.3}$$

$$\boldsymbol{d}(\boldsymbol{x}) = \frac{\boldsymbol{r}(\boldsymbol{x})}{\boldsymbol{R}} \tag{13.4}$$

The shape of a subdomain can be arbitrarily chosen. Spherical subdomains can correctly subdivide space only if they deeply overlap each other otherwise some points could be left outside the subsets. Boxes best fit in Cartesian space and the overlap depth can be arbitrarily imposed. A zero overlap can be also set at the cost of a loss of smoothness. Overlapping spheres demonstrated to provide the best results, especially if the mapping has to be conducted from surface to surface on complex shaped parts and are adopted in all the examples exposed in the next sections.

The forces on the target mesh nodes are obtained by multiplying the interpolated forces density field by the area (or volume for space fields) of the corresponding target cells. The error in the equilibrium between source and target field is related to this point and depends on the differences between the two discretizations. A procedure able to smoothly recover the forces and moments equilibrium is then required. Source and target sets of nodes are organized into FS, namely a distribution of overlapped subdomains. Three corrective coefficients (one for each component along X, Y and Z) that locally force the equivalence between the resultants of the source and target subdomains (not necessarily equivalent to the subdomain used in the interpolation) are introduced for each FS. The continuity and the smooth transition of the coefficients between the subdomains are obtained by overlapping them and by the adoption of membership functions which weight the force associated to each node. Such membership functions are chosen to have the same formulation of the blending functions in the POU decomposition.

Assuming two overlapped subdomains, indicated as i and j, the corrected resultants are obtained according to Eq. (13.5) where the weight functions $w(x)$ are the same previously described.

$$\begin{cases} \boldsymbol{R}_x = \sum \boldsymbol{F}_{x,i}\boldsymbol{c}_{x,i}\boldsymbol{w}_i(\boldsymbol{x}) + \sum \boldsymbol{F}_{x,j}\boldsymbol{c}_{x,j}\boldsymbol{w}_j(\boldsymbol{x}) \\ \boldsymbol{R}_y = \sum \boldsymbol{F}_{y,i}\boldsymbol{c}_{y,i}\boldsymbol{w}_i(\boldsymbol{x}) + \sum \boldsymbol{F}_{y,j}\boldsymbol{c}_{y,j}\boldsymbol{w}_j(\boldsymbol{x}) \\ \boldsymbol{R}_z = \sum \boldsymbol{F}_{z,i}\boldsymbol{c}_{z,i}\boldsymbol{w}_i(\boldsymbol{x}) + \sum \boldsymbol{F}_{z,j}\boldsymbol{c}_{z,j}\boldsymbol{w}_j(\boldsymbol{x}) \end{cases} \tag{13.5}$$

The corrective coefficient is constant within the same subdomain and the Eq. (13.5) can be rearranged in the form of (13.6):

$$\begin{cases} R_x = c_{x,i} \sum F_{x,i} w_i(x) + c_{x,j} \sum F_{x,j} w_j(x) \\ R_y = c_{y,i} \sum F_{y,i} w_i(x) + c_{y,j} \sum F_{y,j} w_j(x) \\ R_z = c_{z,i} \sum F_{z,i} w_i(x) + c_{z,j} \sum F_{z,j} w_j(x) \end{cases} \tag{13.6}$$

The corrective coefficient for the selected subdomains are then expressed as (13.7):

$$\begin{aligned} c_{x,i} &= \frac{\left[\sum F_{x,i} w_i(x)\right]_{source}}{\left[\sum F_{x,i} w_i(x)\right]_{target}} & c_{x,j} &= \frac{\left[\sum F_{x,j} w_j(x)\right]_{source}}{\left[\sum F_{x,j} w_j(x)\right]_{target}} \\ c_{y,i} &= \frac{\left[\sum F_{y,i} w_i(x)\right]_{source}}{\left[\sum F_{y,i} w_i(x)\right]_{target}} & c_{y,j} &= \frac{\left[\sum F_{y,j} w_j(x)\right]_{source}}{\left[\sum F_{y,j} w_j(x)\right]_{target}} \\ c_{z,i} &= \frac{\left[\sum F_{z,i} w_i(x)\right]_{source}}{\left[\sum F_{z,i} w_i(x)\right]_{target}} & c_{z,j} &= \frac{\left[\sum F_{z,j} w_j(x)\right]_{source}}{\left[\sum F_{z,j} w_j(x)\right]_{target}} \end{aligned} \tag{13.7}$$

The reported correction is continuous and provides the mathematical equilibrium of local forces. It is then expected a low order of error on the moments.

In the following sections the application of both the standard and high quality RBF-based pressure mapping is showcased for three tests cases characterised by not conformal meshes. The first deals with a shape defined by a mathematical formula, whilst the remaining ones concern two relevant real world aeronautical test cases, respectively called DLR-F6 and HIRENASD (see Sects. 9.4.2 and 10.6.2). In all these test cases, data processing is carried out by means of the Matlab software.

13.2.3 Catenoid

The catenoid (Wang et al. 2016) is a doubly curved geometry obtained by rotating a catenary curve around an axis. Such a geometrical item is parametrically identified by (13.8) equations where $u \in [0, \pi]$ and $v \in [-1.5,1.5]$.

$$\begin{cases} x = \cos(u)\cosh(v) \\ y = v \\ z = \sin(u)\cosh(v) \end{cases} \tag{13.8}$$

Half the catenoid's shape is taken as geometry on which apply pressure mapping. A coarse and a fine mesh of such a geometry, composed by triangular elements only, have been suitably generated to eventually get a very different resolution as shown in Fig. 13.2. In particular, the coarse mesh has 338 elements with 196 nodes, whilst the fine mesh has 2738 elements with 1444 nodes.

Both the coarse to fine and fine to coarse mapping are considered for a pressure field assigned on the source nodes using with the formula (13.9).

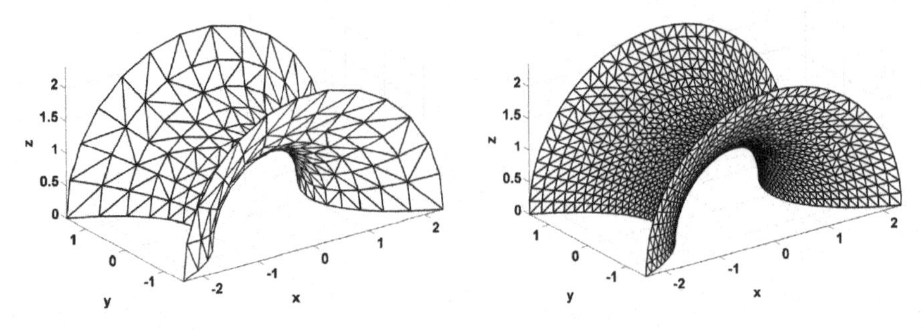

Fig. 13.2 Source (left) and target (right) meshes of the catenoid shape

$$p = \sin(3x + 3y) \tag{13.9}$$

The interpolation of pressure field on target through standard RBF is carried out adopting a simple spline as RBF kernel, namely by imposing $\varphi(\mathrm{r}) = \mathrm{r}$. The assigned source pressure field and the target pressure field for the case coarse to fine calculated through the standard RBF interpolation at each mesh node are depicted in the first row of Fig. 13.3 on the left and middle side respectively; a detail of the interpolation at yz plane cut is given on the right. The fine to coarse case results are in the second row of Fig. 13.3. In both situations (coarse to fine and fine to coarse) a very good mapping of the assigned function is observed; a special attention should be posed to the pressure map on the fine mesh, which is better resolved and looks to be very similar in the top row, i.e. after mapping, and in the bottom one, i.e. direct function input. The faithful reproduction of the pressure on the catenoid can be better understood examining the plot on the section that shows a very small difference between the pressure curves.

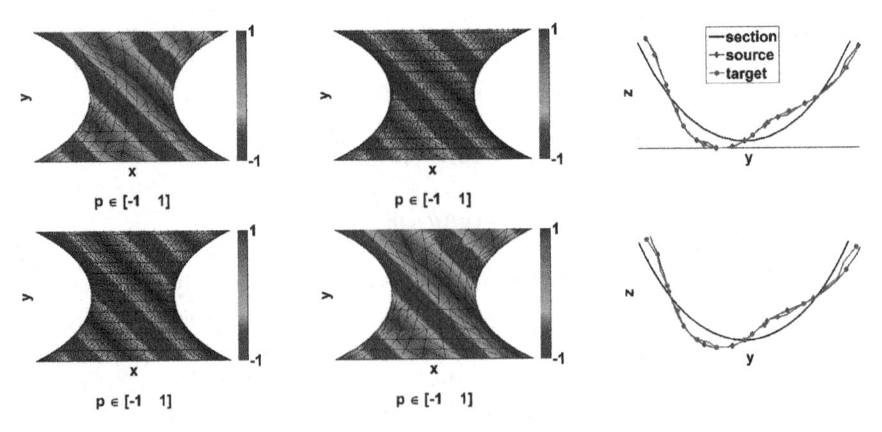

Fig. 13.3 Catenoid case: source (left) and target (middle) pressure field calculated through the standard RBF interpolation providing a comparison in the middle section (right). Top row is for coarse to fine, bottom row is for fine to coarse

The force and moment resultants with respect to target centre of gravity (CG) of the source mesh, and force and moment resultants of the pressure field interpolated on the target mesh are compared in Table 13.1 for the coarse to fine interpolation, and in Table 13.2 for the fine to coarse one. A very good balance of the load after mapping is registered for x and y directions (maximum difference is about 7%). The z direction resultants of the applied distribution should be zero on the continuum catenoid and are very sensitive to the mesh spacing. Despite the high percentage error, the absolute error is very small.

The target pressure results calculated through the FS correction, applied after the RBF interpolation, look so close to the ones of Fig. 13.3 that are not reported in full with the exception of the case coarse to fine for which a comparison of the pressure profile in the cross section is given in Fig. 13.4 where a comparison between the results computed by the standard RBF method and the FS corrected ones is shown. Discrepancies between target and source are mainly due to the mesh resolution and the difference between the two methods is very small and not easy to be appreciated.

Details of how accurate is the corrected method can be understood by inspecting Tables 13.3 and 13.4 which shows that the error on resultant loads is zero and that the error on resultant moments is less than 6% for x and y components. Even in this

Table 13.1 Catenoid case with standard RBF interpolation: comparison of the loads resultants for the coarse to fine case

	Fx	Fy	Fz	Mx	My	Mz
Source	2.1445	2.0219	0.0109	1.6435	−2.3335	0.0462
Target	2.1491	2.1300	0.0002	1.6411	−2.2581	−0.0030
Error (%)	0.214	5.347	−97.806	−0.147	−3.233	−106.543

Table 13.2 Catenoid case with standard RBF interpolation: comparison of the loads resultants for the fine to coarse case

	Fx	Fy	Fz	Mx	My	Mz
Source	2.1335	2.1802	0.0007	1.7633	−2.2353	−0.0029
Target	2.1412	2.0270	0.0109	1.6343	−2.3238	0.0408
Error (%)	0.359	−7.024	1475.296	−7.314	3.961	−1480.554

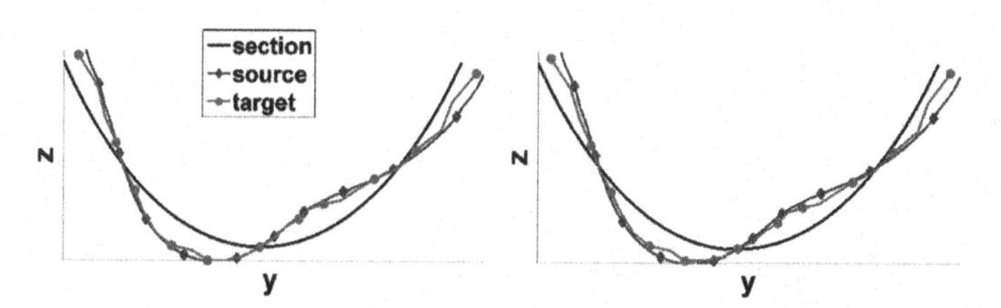

Fig. 13.4 Catenoid case: the pressure profile of the fine to coarse mapping calculated through the standard RBF interpolation (left) is compared with the FS correction of RBF interpolation (right)

Table 13.3 Catenoid case with the FS correction of RBF interpolation: comparison of the loads resultants for the coarse to fine case

	Fx	Fy	Fz	Mx	My	Mz
Source	2.1445	2.0219	0.0109	1.6435	−2.3335	0.0462
Target	2.1445	2.0219	0.0109	1.6474	−2.2944	0.0410
Error (%)	0	0	0	0.240	−1.677	−11.178

Table 13.4 Catenoid case with the FS correction of RBF interpolation: comparison of the loads resultants for the fine to coarse case

	Fx	Fy	Fz	Mx	My	Mz
Source	2.1335	2.1802	0.0007	1.7633	−2.2353	−0.0029
Target	2.1335	2.1802	0.0007	1.6638	−2.2231	0.00004
Error (%)	0	0	0	−5.639	−0.548	−101.417

case the numerical noise around zero for the z direction introduces a very high percentage error but that can be considered negligible if absolute values are observed.

13.2.4 *DLR-F6*

The second application deals with the aeronautical field and pertains the so-called DLR-F6 model (Laflin et al. 2005) which represents a wing-body-nacelle configuration of a subsonic transport type aircraft. The DLR-F6 wind tunnel model, depicted in Fig. 13.5, was employed in the NASA Drag Prediction Workshop (DPW) (NASA 2003) launched with the main purpose to assess the state-of-art of the computational methods as practical aerodynamic tools. In the DPW programme, the DLR-F6 was obtained adding the nacelle and pylon geometries to the previous model referred to as DLR-F4 (NASA 2001).

The CFD mesh has 226,923 cells and 123,520 nodes, whilst the FEM mesh has 20,262 elements and 10,175 nodes. To understand the difference of the resolution between the source (CFD) and target (FEM) mesh, a detail of them including the wing, the pylon and the nacelle is shown Fig. 13.6.

Figure 13.7 shows the pressure distribution on the DLR-F6 surfaces of the CFD model (left) and on the FEM surface mesh calculated with standard RBF interpolation.

The force and moment resultants calculated through the pressure field on the CFD model (source mesh) and those determined by means of the standard RBF mapping on the FEM mesh (target mesh) are compared in Table 13.5. It can be noticed that an error on drag (F_X) prediction is around 30%, value not acceptable for design purposes.

Fig. 13.5 DLR-F6 wind tunnel model

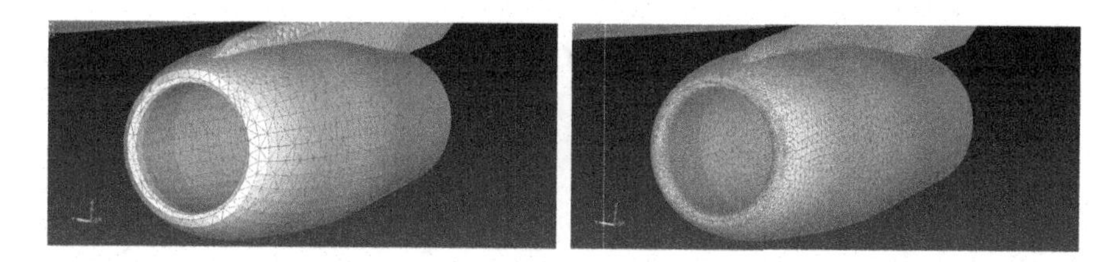

Fig. 13.6 Detail of the CFD and FEM meshes of the DLR-F6 model

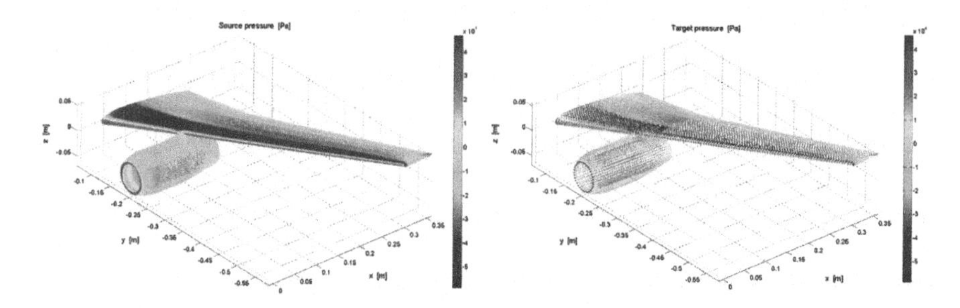

Fig. 13.7 DLR-F6 case with standard RBF interpolation: comparison of the pressure distribution

FS correction of the RBF interpolated data is applied using a fuzzy core at each target node and 100 source nodes inside each subset. Figure 13.8 depicts the pressure distribution on the DLR-F6 surfaces of the CFD model (left) and on the FEM surface mesh calculated with FS correction of the RBF interpolation, whilst the corresponding results in terms of resultants are summarized in Table 13.6.

Table 13.5 DLR-F6 case with standard RBF interpolation: comparison of the loads resultants

	Fx	Fy	Fz	Mx	My	Mz
Source	25.39	77.17	1331.5	−60.78	−52.96	3.340
Target	32.67	83.02	1308.4	−55.26	−51.19	2.586
Error (%)	28.7	7.6	−1.7	−9.1	−3.3	−22.6

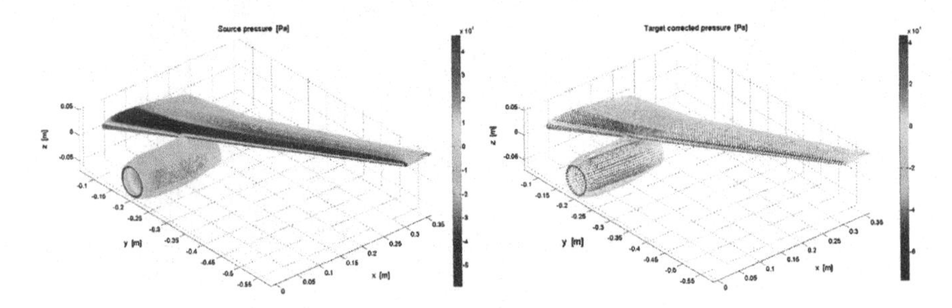

Fig. 13.8 DLR-F6 case with FS correction of the RBF interpolation: comparison of the pressure distribution

Table 13.6 DLR-F6 case with FS correction of the RBF interpolation: comparison of the loads resultants

	Fx	Fy	Fz	Mx	My	Mz
Source	25.39	77.17	1331.5	−60.78	−52.96	3.340
Target	25.39	77.17	1331.5	−60.72	−52.99	3.349
Error (%)	0	0	0	−0.1	0.1	0.3

A deeper analysis of the two proposed transfer steps can be performed by means of a comparison of the pressure profiles at three different spanwise cross sections of the wing (Fig. 13.9) calculated before and after the FS correction.

The first investigated section is near the wing-fuselage junction. The pressure distribution comparison is detailed in Fig. 13.10. The left plot provides a comparison of the RBF interpolated field (red lines) at the leading edge region versus the original one (blue lines), whilst the right plot compares the corrected pressure field (FS correction applied to the RBF field, as green lines) with the original pressure profile (blue lines).

It can be noticed that the FS correction produces a very soft modification of the interpolated pressure field which leads to a slightly increased of pressure magnitude across the leading edge area of the suction side.

Figures 13.11 and 13.12 report the same comparison for the two other sections: the nacelle pylon station (section 2) and a section near the tip (section 3) respectively.

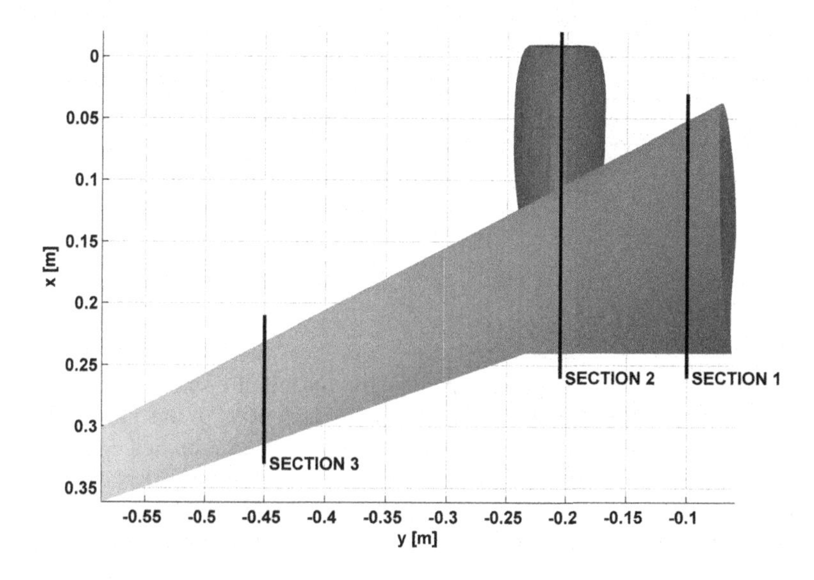

Fig. 13.9 DLR-F6 case: wing monitoring sections

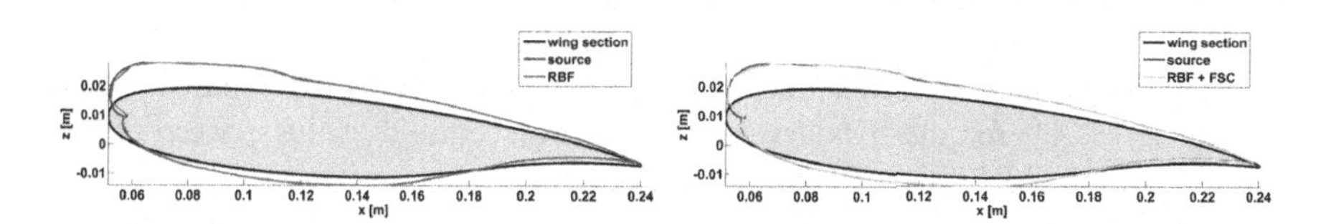

Fig. 13.10 DLR-F6 case: pressure distribution over the monitoring section 1 (Color figure online)

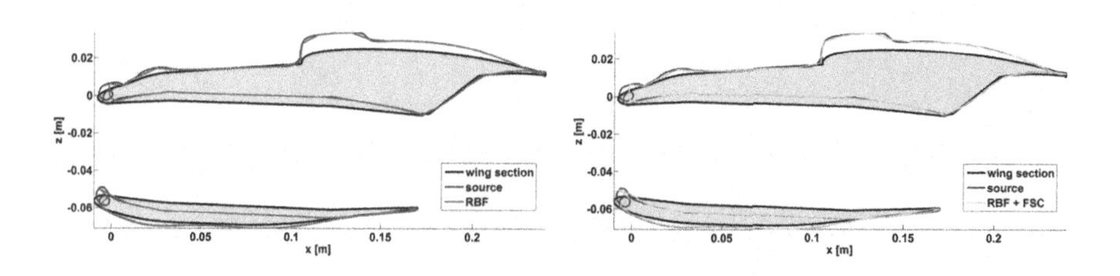

Fig. 13.11 DLR-F6 case: pressure distribution over the monitoring section 2

Analogous conclusions can be drawn for sections 2 and 3: FS correction locally tweaks the interpolated pressure field and its action becomes relevant, from a global point of view, when force and moment resultants have to be computed.

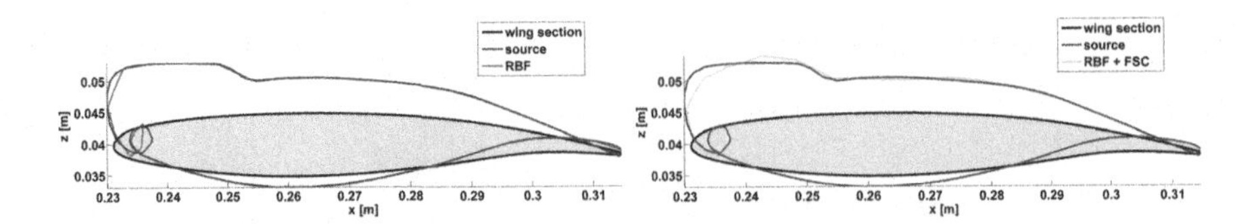

Fig. 13.12 DLR-F6 case: pressure distribution over the monitoring section 3

13.2.5 HIRENASD

The geometry of the third test case is particularly meaningful since it was conceived and designed specifically with the aim to asses and develop the most advanced numerical tools in the field of aeroelasticity, and was deeply studied and validated with numerical and wind tunnel tests in occasion of the Aeroelastic Prediction Workshop (AePW) (NASA 2012b). Using the high-fidelity CFD and FEM cases of HIRENASD (NASA 2012a) wind tunnel model made available by the AeWP committee, the pressure field was mapped from the wing surface mesh of the CFD case to the wing structural mesh of the FEM case. The ANSYS Fluent and NASTRAN solvers were used to generate the CFD and FEM model respectively.

To create the RBF solution representing the pressure map and to guarantee the high quality, the pressure values were exported both at nodal and centroids positions and combined for each surface element. A total of 27,270 RBF source points, of which 13,709 for mesh nodes and 13,561 for centroids values, were finally created. In Fig. 13.13 such RBF centres are previewed.

Exploiting the meshless nature of the RBF solution, nodal loads on the NASTRAN FEM grid were interpolated directly using an external algorithm written in C programming language. To properly map loads from source to target elements, congruent surfaces were defined both for the RBF solution calculation and the nodal interpolation taking advantage of the Nastran PID number. To this end,

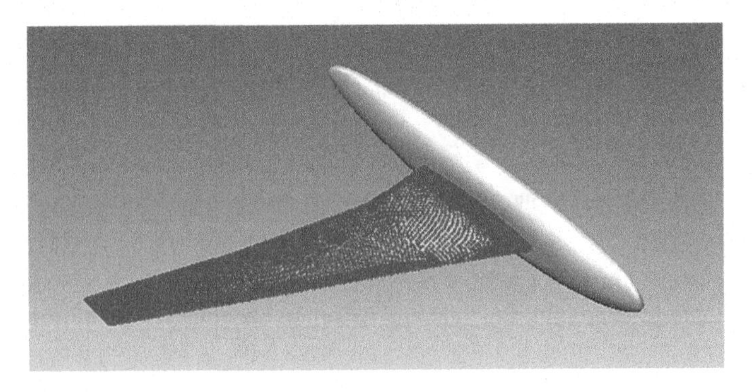

Fig. 13.13 Source point definition for the pressure load mapping on HIRENASD

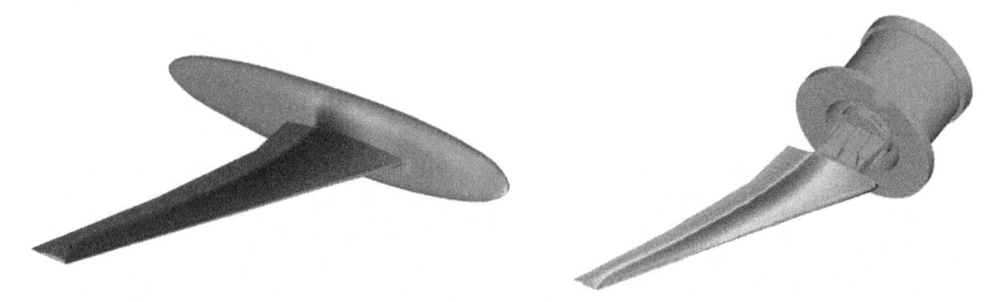

Fig. 13.14 Pressure distribution transfer from the CFD model to the FEM model (top view)

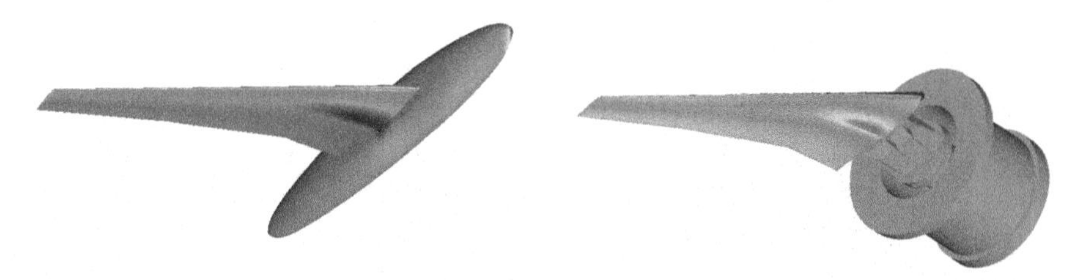

Fig. 13.15 Pressure distribution transfer from the CFD model to the FEM model (bottom view)

surface elements were defined in NASTRAN format using CTRIA3 and CQUAD4 bulk data entries.

Interpolated results were finally written in Nastran format using PLOAD2 or PLOAD4 entries that assign, respectively, a constant pressure at the centroid or a variable pressure prescribed at the nodes.

The load transfer results in terms of pressure contours are showed in Figs. 13.14 and 13.15, where the CFD model generated through ANSYS Fluent and the FEM model prepared by means of Femap pre-processor are compared.

Table 13.7 reports the resultant forces and moments, computed with respect to the target mesh centre of gravity, as well as the errors introduced by the interpolation process (without correction) between the two non-conformal domains.

The FS correction has been applied using a FS core at each target node and including 15 source nodes in each subset. Figure 13.16 reports the original map of pressure for the source domain (left) and its interpolated and corrected counterpart for the target domain (right).

Table 13.7 Errors of interpolation between the two non-conformal domains

	Fx	Fy	Fz	Mx	My	Mz
Source	271.67	150.65	5520.38	410.12	−214.43	56.89
Target	240.11	178.88	5578.96	364.18	−195.71	71.85
Error (%)	−11.62	18.73	1.06	−11.20	−8.73	26.31

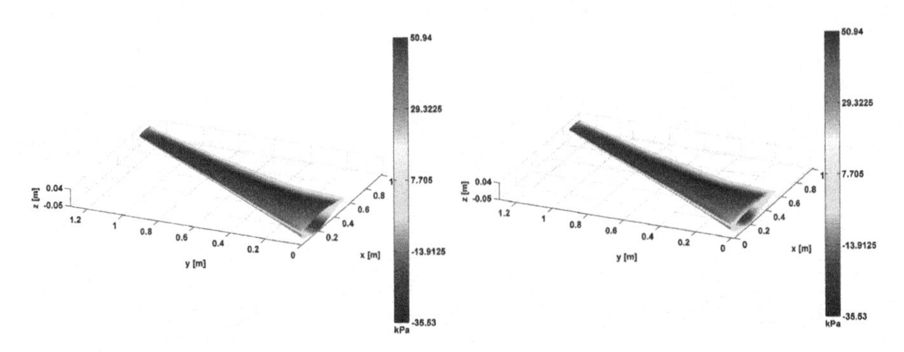

Fig. 13.16 Pressure on source (left) and target (right) points after FS correction

Table 13.8 Errors of corrected interpolation

	Fx	Fy	Fz	Mx	My	Mz
Source	271.67	150.65	5520.38	410.12	−214.43	56.89
Target	270.36	151.61	5518.17	409.74	−214.97	55.95
Error (%)	−0.48	0.63	−0.04	−0.09	0.25	−1.64

Table 13.8 reports the errors on resultant forces and moments obtained with the correction procedure above described. The errors, in absolute terms, are below 2.0% along all directions.

13.3 Mapping of Electro-magnetic Loads in FEM Analyses

In several engineering applications, the magnitude of the loads caused by electro-magnetic (EM) fields can be of a degree such to induce harmful effects to the components of the system to be analysed. Among others, these unwanted consequences may include mechanical damages caused by overheating, material melting as well as the sudden detachment of pieces of material.

These peculiar characteristics have to be tackled when dealing with the design of the magnets and conductors of fusion reactors. In this context, since structural mechanics and electromagnetics play together a crucial role, very stringent multi-physics specifications in a complex loading conditions framework have to be unavoidably taken into account. Given the need for considering the coupling of the just cited physics, if a stress analysis concerning the loading due to the EM field is required, the RBF based mapping methods can be used to effectively accomplish the transfer of available EM loads to the structural FEM model.

This latter is the specific simulation scenario deepened in the following sections which detail the development of a meshless RBF mapping algorithm and its successful application to enable the stress analysis of the DEMOnstration fusion reactor (DEMO) (Biancolini et al. 2015). In particular, the implemented method allows the analyst to compute and map the Lorentz loads by converting the scattered data of the magnetic field **B** and current density **i,** known at a given set of points, in point functions available everywhere. In such a manner, a magnetic force density can be determined and used to apply loads on a target distribution (nodes of the FEM mesh that can have an arbitrary shape) characterised by a size and spacing different from the source points distribution.

The case of mesh deformation, where the three components of the displacement field are interpolated as three different RBF according to Eq. (2.9), can be adapted in Eq. (13.10) which express the three components of **B**.

$$\begin{cases} B_x(\boldsymbol{x}) = \sum_{i=1}^{N} \gamma_i^x \varphi(\|\boldsymbol{x} - \boldsymbol{x}_{s\,i}\|) + \beta_1^x + \beta_2^x x + \beta_3^x y + \beta_4^x z \\ B_y(\boldsymbol{x}) = \sum_{i=1}^{N} \gamma_i^y \varphi(\|\boldsymbol{x} - \boldsymbol{x}_{s\,i}\|) + \beta_1^y + \beta_2^y x + \beta_3^y y + \beta_4^y z \\ B_z(\boldsymbol{x}) = \sum_{i=1}^{N} \gamma_i^z \varphi(\|\boldsymbol{x} - \boldsymbol{x}_{s\,i}\|) + \beta_1^z + \beta_2^z x + \beta_3^z y + \beta_4^z z \end{cases} \quad (13.10)$$

The same workflow has been recently adopted for the study of the Divertor Tokamak Test facility (DTT) proposal (ENEA 2015); details of the coupled magnetic structural loads can be found in the paper by Di Zenobio et al (2017).

13.3.1 Introduction of the DEMO Project Context

To give a clear and significant answer to aggressive programmes launched by several countries and aiming at fusion electricity production, in 2012 the European Commission requested European Fusion Development Agreement (EFDA) to prepare a technical roadmap to fusion as a sustainable and secure energy source by 2050. The working team that was set up produced a roadmap report (Romanelli et al. 2012), articulated in eight missions, to face the identified technical challenges. The first stage of the roadmap, called ITER, foresees that the proposed candidate solutions have to be firstly developed at small and medium size and, secondly, to demonstrate to be effective. Following ITER, DEMO is expected to produce net electricity for the grid starting in the early 2040s. The 2013 arrangement of DEMO reactor is shown in Fig. 13.17 (Bruzzone et al. 2014).

In the framework of ITER, the result of the so-called PROCESS system code (Kovari et al. 2011) has been taken as the basis for the preliminary mechanical studies of the DEMO toroidal field (TF) coil system (Kovari et al. 2011). At each instant of the plasma scenario, the mechanical load condition of the TF large D-shaped magnet depends on the current that is flowing in the TF magnet, in the poloidal field (PF) and in the central solenoid (CS) magnets (Knaster et al. 2010) as well as on the plasma current itself. As a matter of fact, the design of the DEMO

Fig. 13.17 General arrangement of DEMO reactor

magnets and conductors is a crucial issue for the overall engineering design of such a large fusion machine. The magnetic field load acting on the DEMO TF coil system superconductor has to be evaluated to feed the FEM model in view of determining the stresses in the mechanical structure. To gain the sought increase of flexibility in using FEM codes, a meshless approach taking advantage from the RBF ability to interpolate everywhere a generic scalar function known at a given set of points (source points), has been implemented and is hereinafter described.

13.3.2 Overview of the DEMO TF Coil System Design Scenario

The analysed DEMO TF coil system magnet has the winding pack (WP) layout that was proposed by ENEA. Such a WP is constituted by wind and react cable-in-conduit rectangular-shaped conductors (W & R CICC) with a reduced void fraction to minimise the size and cost of the whole superconductor. The 232 conductors are arranged in 9 graded double layers, namely having two rows of conductors each with a different dimension along the symmetry axis of the WP, wound with a Nb_3Sn/NbTi hybridization. The dimension of the reference section of the single conductor and the WP layout are respectively shown on left and right side of Fig. 13.18.

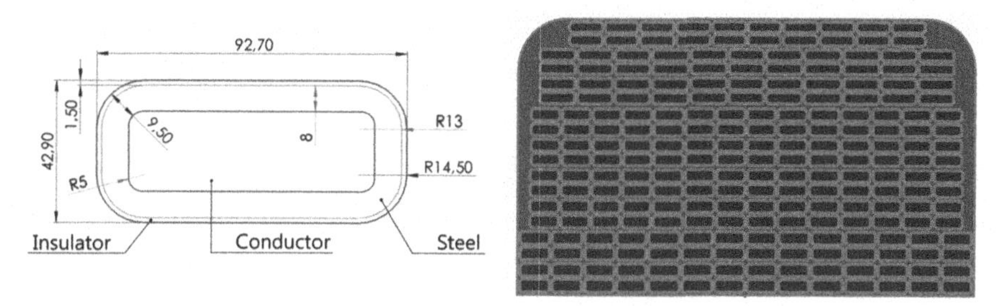

Fig. 13.18 Conductors geometry (left) and ENEA solution for WP (right)

Relating to the structural model, to consistently evaluate the accuracy and robustness of the proposed RBF mapping approach, a complete structural model of the system was used. In detail, two FEM models identifying the same superconductor and scenario were employed. Such models, referred to as coarse and fine FEM model, also include inter-coil structures and differ only for the order of the mesh elements. In particular, the coarse model was used for carrying out the preliminary FEM analyses and has a mesh composed of linear elements, whereas the fine model, utilised in the final stress assessment, has a mesh composed of parabolic elements. The FEM models mesh are depicted in Fig. 13.19.

Regarding the FEM analyses set-up, the cyclic symmetry condition was set with a wedge contact model positioned at the vertical cyclic symmetry axis of the reactor (axis Z), a contact model was imposed between the casing and the smeared model of the WP (conductor with material having homogeneous characteristics) and a

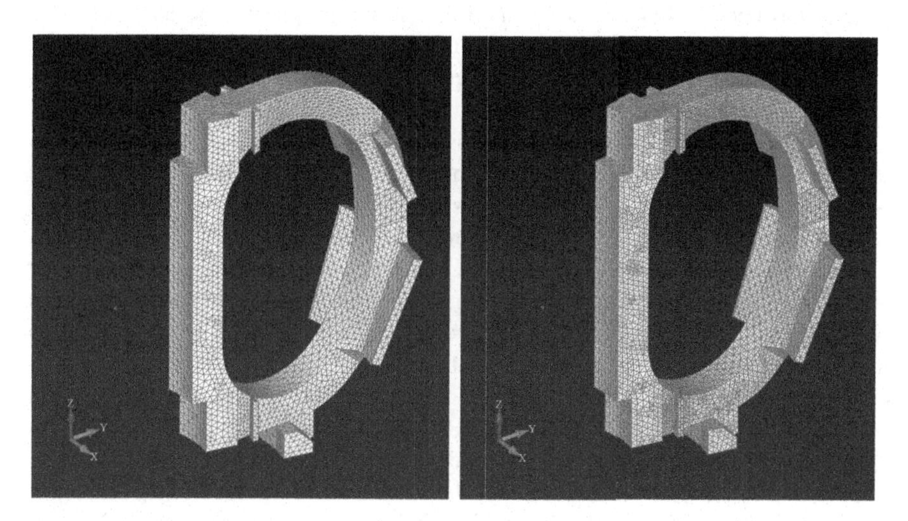

Fig. 13.19 Fine (left) and coarse (right) version of the complete FEM model of the DEMO TF coil system magnet

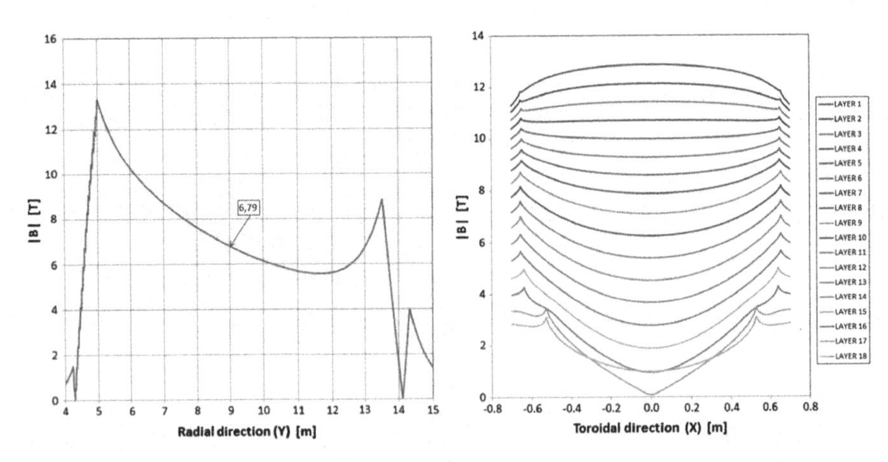

Fig. 13.20 Radial (left) and toroidal (right) profiles of the absolute value of the EM field B

force of 30 MN was assigned to simulate the effect of the pre-compression. A more detailed description of the FEM modelling can be found in (Muzzi et al. 2014).

The EM field **B** and **i** computation has been performed by means of the TOSCA solver, a magnetostatic 3D static EM field computation code whose numerical solver is based on the analytical integration of the Biot-Savart law over the conductor cross-section where the current density is assumed to be uniform. Fields calculation has been iterated using 18 current carrying elements positioned in the part of the magnet close to the centre of the structure (inner leg) and characterised by a thickness equal to the actual strand bundle. The contribution of the plasma current has been included in the numerical solution by placing the plasma axis in the radial direction Y at a distance of 9 m from the axis Z of the reactor. Furthermore, the peak field for every current carrying element has been calculated.

The magnetic field **B** absolute value along the radial direction Y laying on the equatorial plane of the D-shaped magnet is shown in Fig. 13.20 (left). The profile is reported through the whole reactor volume, to verify that the vacuum field at B(Y) is at the requested value, say 6.79 T, as for the process output (Bruzzone et al. 2014) with the purpose to prove the reliability of the model. The same EM parameter along the toroidal direction (X) on the magnet equatorial plane and for each layer of the inner leg is shown Fig. 13.20 (right). The EM model has been implemented including CS and PF coils contribution at the end-of-flat-top (EOF) instant of the scenario (Turtù et al. 2013).

To automate the generation of the cloud of source points needed to calculate EM parameters and accordingly apply the RBF interpolation, a procedure driven by a Mathcad worksheet has been implemented. In specific, such a worksheet takes in input the internal path of the WP and the cross section of the D-shaped magnet previously extracted through the Femap software by processing the CAD model data of the casing (Biancolini et al. 2015). Once these data are loaded, the procedure first calculates the mathematical expression of the curvilinear abscissa of the

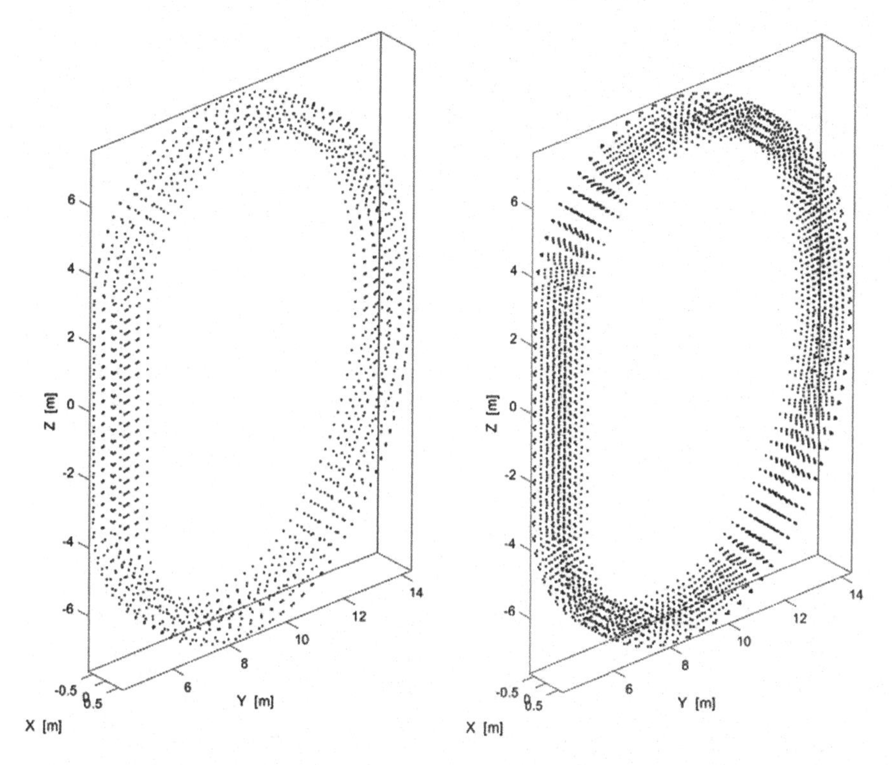

Fig. 13.21 Coarse (left) and fine (right) set of the clouds of points

DEMO TF coil central line using a piecewise function in which each segment is a circular arc. Successively, when a point distribution on the cross section is assigned, the wanted 3D cloud of points is finally generated. In addition, thanks to curvilinear abscissa definition the tangent vector, required to define the direction of the current, is calculated and exported for each point of the cloud.

To enrich the assessment of the study, two clouds of points with a different resolution have been generated: a coarse set with 1512 points created imposing 14 points over the cross section, and a finer set with 3325 points generated imposing 35 points over the cross section. The coarse and fine set cloud of points that were finally created are shown in Fig. 13.21 on the left and right side respectively.

13.3.3 DEMO TF Coil System FEM Analyses Through the Mapping Procedure Based on RBF

As earlier introduced, the application of data mapping is enabled by a workflow in which the software Mathcad is used for processing the foreseen inputs and outputs.

That workflow allows to couple the EM and FEM model adopting a process consisting of the following steps:

- Step1: The cloud of points where the exact values of **B** and **i** need to be evaluated and then interpolated, is generated through the Mathcad worksheet. It is important to underline that although this stage relies on the geometrical model of the magnet, it can be conducted using a general purpose meshing software as well.
- Step2: The exact values of **B** and **i** at the points of the cloud, available after the Step1 completion, are calculated using the TOSCA software.
- Step3: The RBF cubic function ($\varphi(r) = r^3$) is used to generate a fit in which input data are the points' position and the **B** and **i** vectors at each point of the cloud. RBF allows to interpolate **B** and **i** fields and to provide such parameters as point functions. In such a way, the local Lorentz force density $\boldsymbol{dF}(\boldsymbol{x})$ can be computed using Eq. (13.11) where **x** is the position of a point in the space.

$$\boldsymbol{dF}(\boldsymbol{x}) = \boldsymbol{i}(\boldsymbol{x}) \times \boldsymbol{B}(\boldsymbol{x}) \tag{13.11}$$

- Step4: As the field is available as a density of force and FEM mesh nodes are used as receivers, the volume of each target point is required to be known to finalise loads assignment. Nodal volumes of the target FEM mesh, supposing the conductor to behave as a smeared material, are computed on a node by node basis (i.e. ID x y z volume).
- Step5: Nodal loads at FEM nodes are evaluated using the nodal volumes and the interpolated force density through the following relationship (13.12) where Vol_{inod} and $\boldsymbol{x}_{inod}$ are respectively the volume and the position of the node $inod^{th}$. Nodal loads are then exported using the FEM solver format. In the specific case, the format is given in the respect of the FORCE card for the NX Nastran solver in which node ID and FX, FY and FZ components are specified.

$$\boldsymbol{F}_{inod} = Vol_{inod}\boldsymbol{dF}(\boldsymbol{x}_{inod}) \tag{13.12}$$

13.3.4 DEMO TF Coil System Analyses Results

The main results of the analyses of the FEM TF coil system are described in the following sections. They concern the local verification of interpolated magnetic field as well as other outputs of interest for design such as the evaluation of resultant loads acting along and on the TF coil, and the stress analysis of the complete system.

13.3.5 Local Validation of Interpolated Magnetic Field

The ability of the interpolated field to locally reproduce the expected profile of the actual EM field has been checked by comparing the absolute values of the magnetic field **B** obtained through the mapping algorithm with the coarse and fine datasets, with the ones computed by means of TOSCA using the complete 3D model. In particular, the compared values are expressed as function of the radial coordinate Y that lays on the coil equatorial plane and passes through the WP centreline. The comparison, depicted in Fig. 13.22, shows that the interpolation is very accurate where the RBF points are placed, i.e. inside the inner (4.2 m < Y < 5 m) and outer (13.5 m < Y < 14.3 m) legs, whereas a substantial discrepancy is observed over the remaining ranges.

A detail of the just reported comparison over the inner leg region is presented Fig. 13.23. The interpolation of both datasets is able to smoothly represent a filtered value of the EM parameter, although the local variation due to WP interlayer gaps is not captured.

Figure 13.24 shows the comparison between the value of B evaluated through the TOSCA software at the centre of the cable bundle, namely at the middle of each

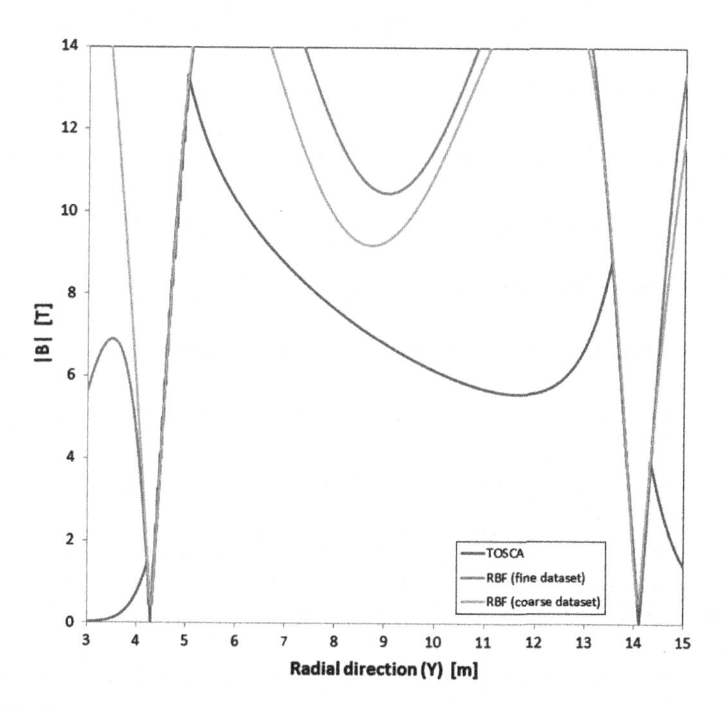

Fig. 13.22 Comparison of the module of B at the coil equatorial plane as function of the radial coordinate

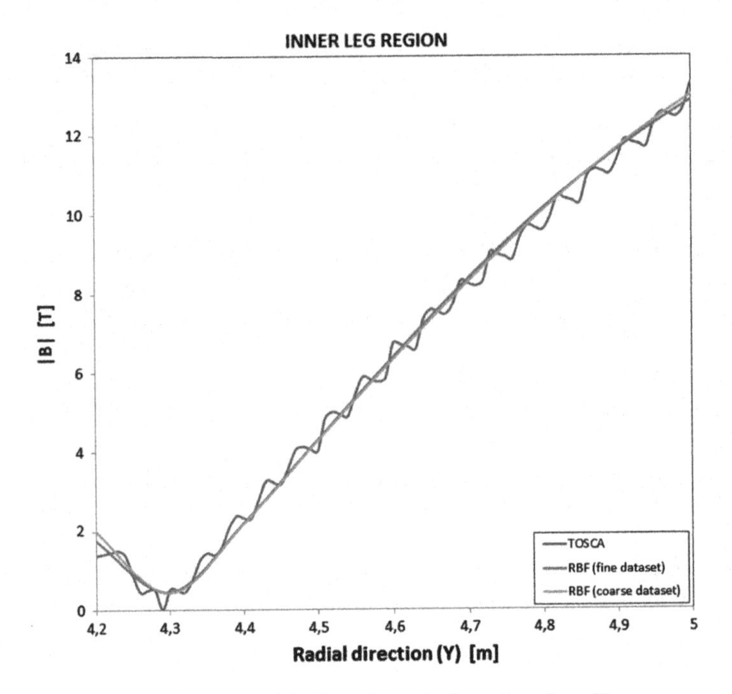

Fig. 13.23 Comparison of the B module throughout the inner leg along Y

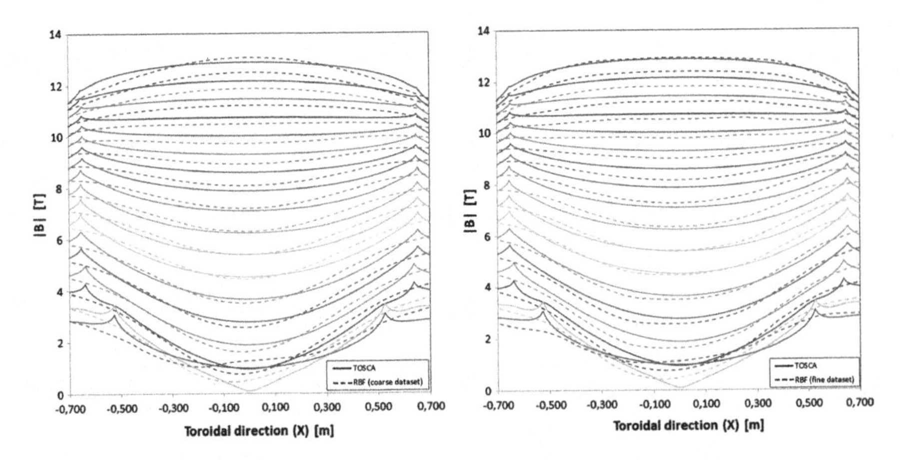

Fig. 13.24 Comparison of B values along the toroidal coordinate for each WP layer for the coarse (left) and fine (right) datasets

WP layer belonging to the coil equatorial plane in the inner leg, and expressed as function of the toroidal coordinate X and the interpolated values obtained with the coarse and fine datasets. Both datasets assure a good accuracy.

13.3.6 Resultant Loads Acting Along the TF Coil

An interesting feature of the proposed approach is that, due to the meshless nature of the method, the loads can be applied to a generic target (Step5). This capability can be thus used to evaluate load resultants at an arbitrary cross section of the D-shaped path of the magnet (see Fig. 13.25).

This feature has been used to evaluate the resultant forces along the superconductor. An example of the localization of a local 2D model in the full 3D model is presented in Fig. 13.26. Although this option is not pursued in this book, it is worth stating that such a feature can be adopted to efficiently transfer loads on an high fidelity 2D model of the cross section as addressed in other works (Biancolini et al. 2014).

Adopting a uniform distribution of 492 points over the cross section, as that shown in Fig. 13.25 on the left, the just mentioned method enables the computing of both the toroidal and radial force along the curvilinear abscissa of the magnet. The comparison between the evaluated profiles, respectively depicted in Fig. 13.27, shows, especially for the toroidal force, a slight dependence on the cloud points density.

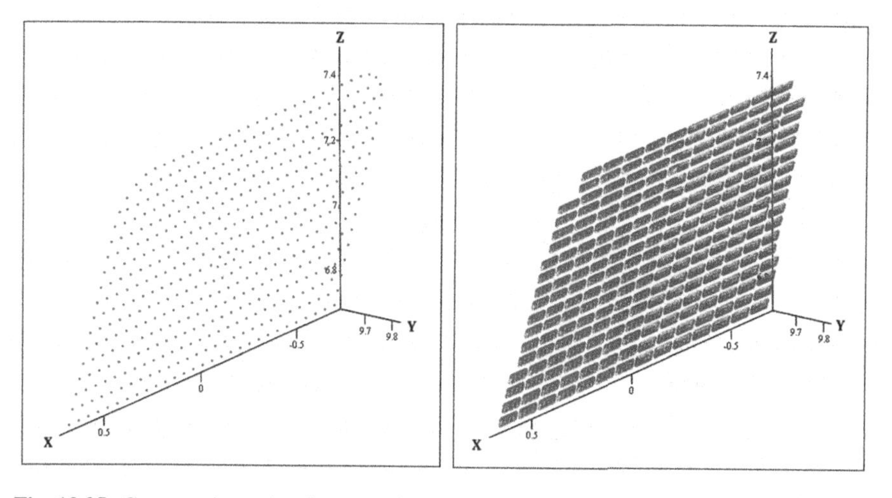

Fig. 13.25 Cross section points in a generic D position: smeared section (left) and full detailed section (right)

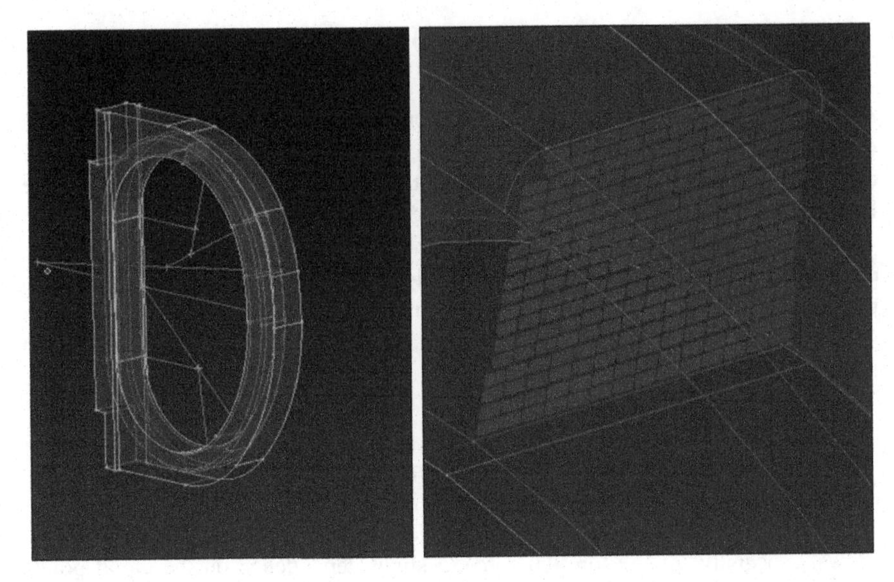

Fig. 13.26 Full detailed 2D cross section into 3D model (left) and zoom (right)

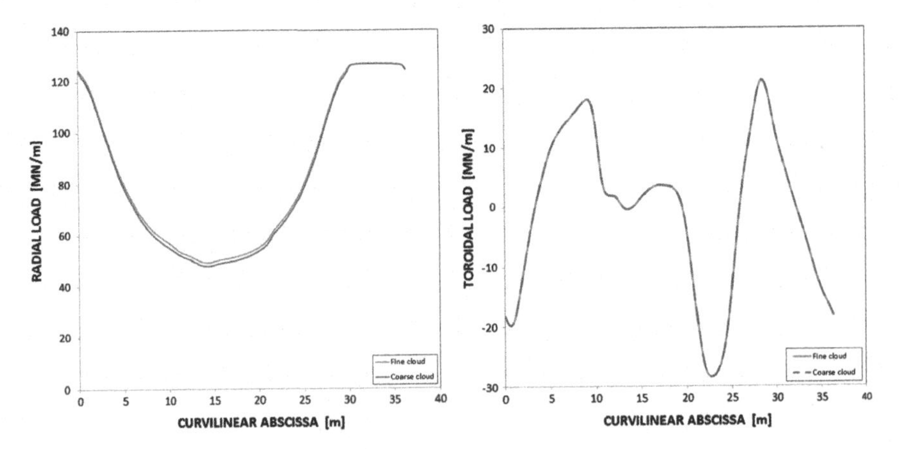

Fig. 13.27 Radial load (left) and toroidal load (right) at each cross section along the coil system WP

13.3.7 Resultant Loads Acting on the TF Coil System WP

The flexibility of the load transfer workflow can be also employed to properly transfer data on different targets. Given that, the EM force field has been computed using the coarse and fine dataset on two different meshes of the WP: a coarse mesh with 4365 nodes (Fig. 13.28 left) and a fine mesh with 38,638 nodes (Fig. 13.28 right).

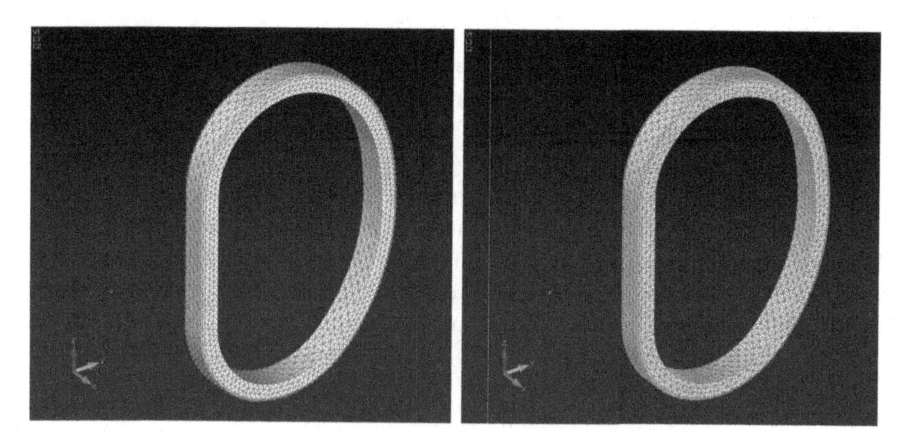

Fig. 13.28 Coarse (left) and fine mesh model (right) of the WP

Table 13.9 Verification of resultant force using different dataset and different target meshes

	F_X [MN]	F_Y [MN]	F_Z [MN]	Module [MN]
Coarse dataset/coarse WP mesh	−14.18	−894.00	−0.07	894.11
Coarse dataset/fine WP mesh	−14.89	−896.60	0.17	896.72
Fine dataset/coarse WP mesh	−16.11	−881.10	0.48	881.25
Fine dataset/Fine WP mesh	−16.33	−884.30	1.18	884.45

The robustness of data mapping can be evaluated by inspecting the resultant forces obtained on such targets for both datasets. Those outputs are summarised in Table 13. 9.

The first important finding is that the coarse dataset captures better than the fine one that the vertical resultant should vanish. The interpolation error is lower than the one achieved using the fine cloud. Resultants sensitivity with respect to target resolution is very low. Out of plane loads look to be underestimated using the coarse cloud of points, whilst radial load exhibits an opposite behaviour but with a minor discrepancy. For all investigated scenarios the mapping task is accomplished with success and the scattering of resultant loads is limited enough.

13.3.8 Stress Analysis of the Complete System

As observed in the previous section, data mapping introduces some undesired scattering on resultants values. In order to investigate the effects of such a scattering on stress assessment and on mechanical results a series of FEM analyses have been performed, by comparing the results obtained using the coarse and fine datasets, to generate input loads for a coarse and a fine FEM models.

Quantities monitored for this demonstration are the maximum stress and the maximum displacement on the conductor 3D model with smeared properties. The gained outputs for the coarse to coarse case are reported in Fig. 13.29 according to a colour maps visualization.

The mechanical analyses results for all the possible combinations are summarized in Table 13.10. As expected, the coarse FEM predicts a stress level lower than the fine FEM, but exhibits a similar stiffness of the fine one (a stiffer behaviour is expected using a coarse FEM mesh), with negligible differences. The density of the cloud used for mapping has a very small effect on both the stress and displacement demonstrating a good stability of the mapping approach with respect to results relevant for the stress assessment. Using for example the fine cloud instead of the coarse one, the stress increases less than 0.1% and the displacements difference is lower than 0.7%, regardless of the mesh refinement.

The outputs presented in the previous table demonstrate how the mapping tool is able to manage different kinds of mesh of the same geometry. Since the percentage difference is less than 3%, the maximum displacements match quite well. The maximum stress level is different, but this is due to the different FEM mesh convergence rather than the accuracy of mapped loads. Furthermore, the difference between the coarse and fine dataset on the same mesh model is less than 0.1%.

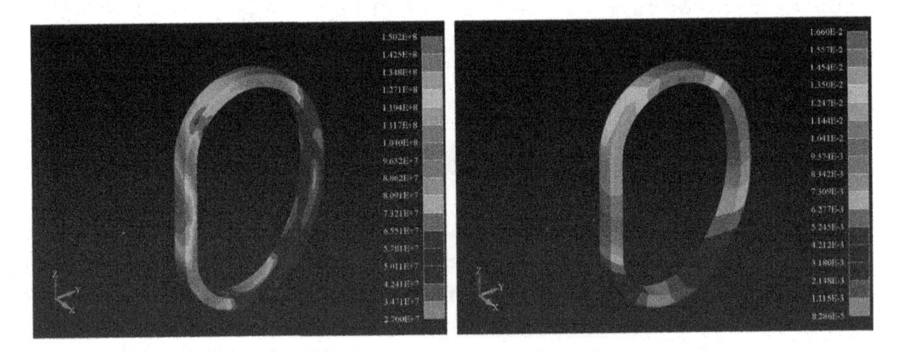

Fig. 13.29 Results for coarse dataset and coarse mesh: von Mises stress (left) and total translation (right)

Table 13.10 Von Mises stress and total translation computed using different dataset and different target meshes

	Von Mises stress [MPa]	Total translation [mm]
Coarse dataset/coarse mesh	150.20	16.60
Coarse dataset/fine mesh	210.30	16.96
Fine dataset/coarse mesh	150.30	16.69
Fine dataset/Fine mesh	210.40	16.85

References

Babuška I, Melenk JM (1997) The partition of unity method. Int J Numer Meth Eng 40(4):727–758. https://doi.org/10.1002/(SICI)1097-0207(19970228)40:4:727::AID-NME86:3.0.CO;2-N

Biancolini M, Viola I, Ramirez S (2014). Alla ricerca delle regolazioni ottimali di una vela mediante mesh morphing, A&C. ANALISI E CALCOLO, ISSN: 1128-3874

Biancolini ME, Brutti C, Giorgetti F, Muzzi L, Turtù S, Anemona A (2015) A new meshless approach to map electromagnetic loads for FEM analysis on DEMO TF coil system. Fusion Eng Des 100:226–238, https://doi.org/10.1016/j.fusengdes.2015.06.031

Biancolini ME, Cella U, Groth C, Chiappa A, Giorgetti F, Nicolosi F (2017) Progresses in fluid structure interaction numerical analysis tools within the EU CS RIBES Project. In: EUROGEN 2017 international conference, 13–15 September, Madrid, Spain

Bruzzone P, Sedlàk K, Stepanov B, Muzzi L, Turtù S, Anemona A, Harma J (2014) Design of large size, force flow superconductors for DEMO TF coils. IEEE Trans Appl Supercond 24 (2014):4201504

Cebral JR, Löhner R (1997) Conservative load projection and tracking for fluid-structure problems. AIAA J 35(4):687–692. https://doi.org/10.2514/2.158

Di Zenobio A, Albanese R, Anemona A, Biancolini ME, Bonifetto R, Brutti C, Corato V, Crisanti F, della Corte A, De Marzi G, Fiamozzi Zignani C, Giorgetti F, Messina G, Muzzi L, Savoldi L, Tomassetti G, Turtù S, Villone F, Zappatore A (2017) DTT device: conceptual design of the superconducting magnet system. Fusion Eng Des, ISSN 0920-3796, https://doi.org/10.1016/j.fusengdes.2017.03.102

ENEA (2015) http://fsn-fusphy.frascati.enea.it/DTT/downloads/Report/DTT_ProjectProposal_July2015.pdf

European Commission (2014) RIBES Project, http://cordis.europa.eu/project/rcn/192637_en.html

Jaiman RK, Jiao X, Geubelle PH, Loth E (2006) Conservative load transfer along curved fluid—solid interface with non-matching meshes. J Comput Phys 218:372–397

Jiao X, Heath MT (2004) Common-refinement-based data transfer between non-matching meshes in multiphysics simulations. Int J Numer Meth Eng 61:2402–2427. https://doi.org/10.1002/nme.1147

Jin X, Sun H, Peng Q (2003) Subdivision interpolating implicit surfaces. Comput Graph 27:763–772. https://doi.org/10.1016/S0097-8493(03)00149-3

Knaster J, Baker W, Bettinali L, Jong C, Mallick K, Nardi C, Rajainmaki H, Rossi P, Semeraro L (2010) Design issues of the pre-compression rings of ITER. Adv Cryog Eng 56(2010):145–154

Kovari M, Kemp R, Knight P, Ward D (2011) The PROCESS fusion reactor systems code. EURATOM/CCFE Fusion Association, Culham Science Centre, UK. Available online at: http://www.ccfe.ac.uk/powerplants.aspx

Laflin KR, Klausmeyer SM, Zickuhr T, Vassberg JC, Wahls RA, Morrison JH, Brodersen OP, Rakowitz ME, Tinoco EN, Godard JL (2005) Data summary from second AIAA computational fluid dynamics drag prediction workshop. J Aircraft 42(5):1165–1178

Muzzi L, Anemona A, della Corte A, Di Zenobio A, Turtù S, Bruzzone P, Sedlak K, Stepanov B, Brutti C, Biancolini ME, Reccia L, Harman J (2014) Assessment studies and manufacturing trials for the conductors of DEMO TF coils. IEEE Trans Appl Supercond 25(3):1–5

NASA (2001) Proceedings of the 1st AIAA CFD drag prediction workshop, https://aiaa-dpw.larc.nasa.gov/Workshop1/workshop1.html website accessed on 29 Sept 2017

NASA (2003) 2nd AIAA CFD drag prediction workshop, https://aiaa-dpw.larc.nasa.gov/Workshop2/workshop2.html. website accessed on 29 Sept 2017

NASA (2012a) HiReNASD model, https://c3.ndc.nasa.gov/dashlink/static/media/other/HIRENASD_base_legacy.htm

NASA (2012b) 1st AIAA aeroelastic prediction workshop, https://c3.ndc.nasa.gov/dashlink/static/media/other/AEPW_legacy.htm

RIBES Project (2014) http://ribes-project.eu/. Accessed on 23 Nov 2017

Romanelli F, Barabaschi P, Borba D, Federici G, Horton L, Neu R, Stork D, Zohm H (2012) A roadmap to the realisation of fusion energy, EFDA, ISBN 978-3-00-040720-8

Samareh JA (2007) Discrete data transfer technique for fluid-structure interaction, NASA Langley Research Center

Turtù S, Anemona A, Biancolini ME, Brutti C (2013) Electromagnetic and mechanical analysis of candidate prototype LTS conductors, winding pack and TF coil casing. *EFDA idm*, EFDA_D_2LNP3B v.2.0

Wang T, Wüchner R, Sicklinger S, Bletzinger KU (2016) Assessment and improvement of mapping algorithms for non-matching meshes and geometries in computational FSI. Comput Mech 57(5):793–816. https://doi.org/10.1007/s00466-016-1262-6

Zadeh LA (1965) Fuzzy sets. Inf Control 8(3):338–353. https://doi.org/10.1016/S0019-9958(65)90241-X

Printed in the United States
By Bookmasters